I0055790

Hydroponics and Protected Cultivation

A Practical Guide

Hydroponics and Protected Cultivation

A Practical Guide

Lynette Morgan

CABI

CABI is a trading name of CAB International

CABI
Nosworthy Way
Wallingford
Oxfordshire OX10 8DE
UK

CABI
WeWork
One Lincoln Street
24th Floor
Boston, MA 02111
USA

Tel: +44 (0)1491 832111
Fax: +44 (0)1491 833508
E-mail: info@cabi.org
Website: www.cabi.org

Tel: +1 (617)682-9015
E-mail: cabi-nao@cabi.org

A catalogue record for this book is available from the British Library, London, UK.

Library of Congress Cataloging-in-Publication Data

Names: Morgan, Lynette, 1970- author.
Title: Hydroponics and protected cultivation : a practical guide / Lynette Morgan.
Description: Boston, MA : CAB International, [2021] | Includes bibliographical references and index. | Summary: "A comprehensive, practical text which covers a diverse range of hydroponic and protected cropping techniques, systems, greenhouse types and environments. It also covers related subjects such as the use of indoor plant factories, vertical systems, organic hydroponics, aquaponics and other systems"-- Provided by publisher.
Identifiers: LCCN 2020046993 (print) | LCCN 2020046994 (ebook) | ISBN 9781789244830 (hardback) | ISBN 9781789244847 (ebook) | ISBN 9781789244854 (epub)
Subjects: LCSH: Hydroponics. | Greenhouse management. | Cropping systems.
Classification: LCC SB126.5 .M674 2021 (print) | LCC SB126.5 (ebook) | DDC 631.5/85--dc23
LC record available at https://lccn.loc.gov/2020046993
LC ebook record available at https://lccn.loc.gov/2020046994

References to Internet websites (URLs) were accurate at the time of writing.

ISBN-13: 978 1 78924 483 0 (hardback)
        978 1 78924 484 7 (ePDF)
        978 1 78924 485 4 (ePub)

Commissioning Editor: Rebecca Stubbs
Editorial Assistant: Emma McCann
Production Editor: Shankari Wilford

Typeset by SPi, Pondicherry, India

Contents

_____

Acknowledgements   **xiv**

1 Background and History of Hydroponics and Protected Cultivation   **1**
   1.1   Protected Cropping   1
   1.2   The Future of Protected Cropping 3
   1.3   Background and History of Hydroponic Production 4
   1.4   Hydroponic Systems   6
   1.5   Substrate-Based Hydroponic Systems 7
   1.6   Organic Hydroponics   8
   1.7   Summary   9

2 Greenhouses and Protected Cropping Structures   **11**
   2.1   Introduction   11
   2.2   Glasshouses and Plastic Greenhouses 12
   2.3   Closed and Semi-Closed Greenhouse Structures 13
   2.4   Passive Solar Greenhouses   14
   2.5   Sustainable Greenhouse Design   15
   2.6   Cladding Materials   16
   2.7   Screen Houses, Net Houses, Shade Houses, Rain Covers and Other Structures 18
   2.8   Screen and Shade Nets   18
   2.9   Low Tunnels and High Tunnels   19
   2.10 Hot Beds and Cold Frames   20
   2.11 Floating Mulches, Row Covers, Cloche Covers, Direct Covers and Frost Cloth 23
   2.12 Greenhouse Site Planning   23
   2.13 Windbreaks   24
   2.14 Outdoor Hydroponic Systems   25
   2.15 Controlled-Environment Agriculture   27

3 The Greenhouse Environment and Energy Use   **30**
   3.1   Introduction – Environmental Modification in Protected Cropping 30
   3.2   Heating   30

3.3	Cooling	32
3.4	Lighting	32
3.5	Shading	34
3.6	Ventilation and Air Movement	35
3.7	Humidity	36
3.8	Carbon Dioxide Enrichment	38
3.9	Greenhouse Automation	39
3.10	Energy Use and Conservation in Protected Cropping	39
3.11	Energy Sources for Protected Cropping	41
	3.11.1 Renewable energy sources	41
	3.11.2 Geothermal energy	41
	3.11.3 Solar energy	42
	3.11.4 Passive solar energy	43
	3.11.5 Wind-generated energy	44
	3.11.6 Biomass and biofuels	44
4	**Greenhouse Operation and Management**	**47**
4.1	Introduction	47
4.2	Greenhouse Sanitation and Hygiene	47
	4.2.1 Hygiene and sanitation for food safety	47
	4.2.2 Food safety and compliance programmes	49
	4.2.3 Hygiene and sanitation for crop protection	49
4.3	Source Water Quality and Treatment	51
4.4	Greenhouse Biosecurity	52
	4.4.1 People movement, human activity and biosecurity measures	53
	4.4.2 Staff, visitors and hygiene	54
4.5	Waste Management and Disposal	54
	4.5.1 Disposal of greenhouse wastewater	55
	4.5.2 Treatment of wastewater	55
	4.5.3 Disposal of and reduction in organic waste	57
	4.5.4 Disposal of plastics	58
	4.5.5 Disposal of pesticides and agrochemical containers	58
4.6	Occupational Health and Safety	58
5	**Hydroponic Systems – Solution Culture**	**61**
5.1	Introduction – Solution Culture Systems	61
5.2	NFT – Nutrient Film Technique	61
	5.2.1 NFT crops	61
	5.2.2 Types of NFT systems	63
	5.2.3 Nutrient solution management in NFT	65
5.3	Deep Water Culture/Deep Flow Technique – Float, Raft or Pond Systems	66
	5.3.1 Management of DWC and DFT systems	69
5.4	Aeroponics	71
5.5	Vertical Systems	72
5.6	Aquaponics	72
5.7	Organic Solution Culture	74
5.8	Hydroponic Fodder Systems	74
5.9	Nutrient Chilling Systems	75
5.10	Automation for Solution Culture Systems	75
6	**Substrate-based Hydroponic Systems**	**77**
6.1	Introduction	77
6.2	Properties of Hydroponic Substrates	77

6.3	Open and Closed Soilless Systems	77
6.4	Common Hydroponic Substrates	79
	6.4.1 Stone wool (mineral wool, rockwool or glass wool)	80
	6.4.2 Coconut fibre (coir, palm peat, coco peat, coco)	80
	6.4.3 Peak, bark and sawdust	81
	6.4.4 Perlite	82
	6.4.5 Pumice and scoria	82
	6.4.6 Vermiculite	83
	6.4.7 Expanded clay	83
	6.4.8 Rice hulls	83
	6.4.9 Sand and gravels	83
	6.4.10 New substrates	84
6.5	Substrates and Water-Holding Capacity	84
6.6	Substrates and Oversaturation	85
6.7	Matching Substrates to Crop Species	85
6.8	Physical Properties of Soilless Substrates	86
	6.8.1 Bulk density	86
	6.8.2 Particle size distribution	87
	6.8.3 Total porosity	87
	6.8.4 Air-filled porosity	87
	6.8.5 Water-holding capacity or container capacity	88
6.9	Chemical Properties of Hydroponic Substrates	88
	6.9.1 pH	88
	6.9.2 Cation exchange capacity	89
	6.9.3 Specific ion contents, salinity and electrical conductivity	89
	6.9.4 Testing methods	89
6.10	Nutrient Delivery in Substrate Systems	91
	6.10.1 Drip irrigation	91
	6.10.2 Drip-irrigated systems – design and layout	92
	6.10.3 Ebb and flow (flood and drain) nutrient delivery systems	93
	6.10.4 Capillary watering systems	93
	6.10.5 Gravity-fed irrigation	93
	6.10.6 Nutrient dosing and injectors	94
6.11	Irrigation and Moisture Control in Substrates	94
	6.11.1 Substrate moisture, growth balance and deficit irrigation	96
6.12	Microbial Populations in Substrates	96
7	**Organic Soilless Greenhouse Systems**	**100**
7.1	Introduction – Organic Greenhouse Production	100
7.2	Organic Hydroponic Systems	101
7.3	Organic Hydroponic Nutrients	102
7.4	Microbial Mineralization of Organic Nutrients for Hydroponics	102
7.5	Anaerobic and Aerobic Processing of Organic Materials	103
7.6	Vermicast and Vermicomposting	104
7.7	Using Vermiculture Liquids in Hydroponics	105
7.8	Composting for Organic Nutrient Processing and Substrate Preparation	106
7.9	Organic Materials for Vermicast, Composting and Biodigester Systems	107
	7.9.1 Organic fertilizer/nutrient sources	107
	7.9.2 Animal sources of organic fertilizers	107
	7.9.3 Plant-based inputs	108
7.10	Aquaponics	108
7.11	Organic Hydroponic Production Systems	108
7.12	Biofilms in Organic Hydroponic Systems	111

7.13   Nutrient Amendments   111
7.14   Organic Certification in the USA   112
7.15   Organic Pest and Disease Control   112
7.16   Hybrid Systems   113
7.17   Issues Commonly Encountered with Organic Hydroponic Systems   113
7.18   Conclusions   114

8   Propagation and Transplant Production   **118**
8.1   Introduction   118
8.2   Propagation from Seed   118
       8.2.1   Hybrid seed versus open-pollinated seed   118
       8.2.2   Seed treatments – pelleting, coating and priming   119
       8.2.3   Seed storage   120
       8.2.4   Production of transplants from seed   121
8.3   Seedling Delivery Systems   121
8.4   Seeding Methods   125
8.5   Germination Problems   125
8.6   Transplant Production Systems   126
       8.6.1   Transplant production environment   127
       8.6.2   Seedling nutrition   128
8.7   Use of Plant Factories for Seedling Transplant Production   129
8.8   Organic Transplant Production   129
8.9   Transplant Establishment   130
8.10   Grafting   130
8.11   Vegetative Propagation   131
8.12   Tissue Culture   132
       8.12.1   Tissue culture techniques and methods   133

9   Plant Nutrition and Nutrient Formulation   **136**
9.1   Water Quality and Sources for Hydroponic Production   136
       9.1.1   Well water   136
       9.1.2   Surface water   137
       9.1.3   Rainwater   137
       9.1.4   City or municipal water supplies   138
       9.1.5   Reclaimed water sources   139
9.2   Water Testing   139
9.3   Water Analysis Reports   140
       9.3.1   pH and alkalinity   140
       9.3.2   Electrical conductivity   140
       9.3.3   Mineral elements in water supplies   140
9.4   Water Quality and Plant Growth   141
9.5   Water Treatment Options   142
9.6   Water Usage and Supply Requirements   142
9.7   Plant Nutrition in Hydroponic Systems   143
9.8   Essential Elements – Functions in Plants and Deficiency Symptoms   143
       9.8.1   Nitrogen   143
       9.8.2   Potassium   144
       9.8.3   Phosphorus   145
       9.8.4   Calcium   145
       9.8.5   Magnesium   145
       9.8.6   Sulfur   146

	9.8.7	Iron	146
	9.8.8	Manganese	146
	9.8.9	Boron	146
	9.8.10	Zinc	147
	9.8.11	Copper	147
	9.8.12	Chloride	147
	9.8.13	Molybdenum	147
9.9	Beneficial Elements	147
9.10	Nutrient Formulation	149
	9.10.1	The process of nutrient formulation	150
9.11	Hydroponic Nutrient Formulation – Nitrogen Sources	151
9.12	Common Hydroponic Fertilizers	152
	9.12.1	Calcium nitrate	152
	9.12.2	Ammonium nitrate	153
	9.12.3	Ammonium phosphate	153
	9.12.4	Urea	153
	9.12.5	Potassium nitrate	153
	9.12.6	Potassium sulfate	153
	9.12.7	Monopotassium phosphate	154
	9.12.8	Calcium superphosphate	154
	9.12.9	Magnesium sulfate	154
	9.12.10	Magnesium nitrate	154
	9.12.11	Iron chelates	154
	9.12.12	Manganese sulfate, manganese chelates	155
	9.12.13	Copper sulfate, copper chelates	155
	9.12.14	Zinc sulfate, zinc chelates	155
	9.12.15	Boric acid, borax	155
	9.12.16	Sodium molybdate, ammonium molybdate	155
	9.12.17	Nitric and phosphoric acids	155
9.13	Fertilizer Composition and Grades	155
9.14	Chelation of Trace Elements	156
9.15	Foliar Fertilizers	156
9.16	Electrical Conductivity	156
9.17	pH	158
9.18	Automation and Testing Equipment	159
9.19	Conditions Which Affect Nutrient Uptake Rates	159
	9.19.1	Temperature and humidity	159
	9.19.2	Time of day	160
	9.19.3	Light levels	160
	9.19.4	Root health and size	160
	9.19.5	Aeration and oxygenation	160
9.20	Plant Tissue Analysis	161
9.21	Fertilizer and Environmental Concerns	161
9.22	Water and Nutrient Solution Treatment Methods	161
	9.22.1	Ultraviolet disinfection	162
	9.22.2	Ozone	163
	9.22.3	Filtration	163
	9.22.4	Slow sand filtration	164
	9.22.5	Chlorine	165
	9.22.6	Hydrogen peroxide	165
	9.22.7	Heat	165
9.23	Surfactants	166

10 Plant Health, Plant Protection and Abiotic Factors   **170**
  10.1   Introduction   170
  10.2   Major Greenhouse Pests   170
         10.2.1   Whitefly   170
         10.2.2   Aphids   172
         10.2.3   Thrips   173
         10.2.4   Mites   174
         10.2.5   Caterpillars and leaf miner larvae   175
         10.2.6   Fungus gnats   176
         10.2.7   Nematodes   176
  10.3   Pest Control Options – Integrated Pest Management   177
  10.4   Selected Diseases of Hydroponic Crops   178
         10.4.1   *Botrytis*   178
         10.4.2   Mildew diseases   179
         10.4.3   *Pythium* root rot   180
         10.4.4   Wilt diseases   182
         10.4.5   Common bacterial diseases   183
         10.4.6   Virus diseases   183
  10.5   Abiotic Factors and Physiological Disorders   184
         10.5.1   Temperature damage   184
         10.5.2   Light   185
         10.5.3   Root-zone abiotic factors   186
         10.5.4   Irrigation water quality and salinity   186
         10.5.5   Chemical injury (phytotoxicity)   187
         10.5.6   Ethylene   188
  10.6   Cultural Practices Causing Abiotic Disorders   188
  10.7   Identification of Abiotic Disorders   188
  10.8   Crop-Specific Physiological Disorders   189
         10.8.1   Blossom end rot   189
         10.8.2   Tipburn   190
         10.8.3   Bolting   191
         10.8.4   Fruit shape and splitting/cracking disorders   191

11 Hydroponic Production of Selected Crops   **196**
  11.1   Introduction   196
  11.2   Hydroponic Tomato Production   196
         11.2.1   Hydroponic systems for tomato production   198
         11.2.2   Tomato propagation   199
         11.2.3   Tomato environmental conditions   199
         11.2.4   Tomato crop training systems   199
         11.2.5   Tomato crop steering   201
         11.2.6   Tomato pollination and fruit development   202
         11.2.7   Tomato crop nutrition   203
         11.2.8   Tomato pests and diseases   203
         11.2.9   Tomato yields   204
  11.3   Hydroponic Capsicum Production   204
         11.3.1   Capsicum propagation   205
         11.3.2   Capsicum systems of production   205
         11.3.3   Capsicum pollination, fruit set and fruit development   206
         11.3.4   Capsicum training   206
         11.3.5   Capsicum crop nutrition   207
         11.3.6   Capsicum pests, diseases and physiological disorders   207
         11.3.7   Capsicum harvesting and yields   208

11.4   Hydroponic Cucumber Production   208
        11.4.1   Cucumber propagation and production   208
        11.4.2   Cucumber environmental conditions   209
        11.4.3   Cucumber training and support systems   210
        11.4.4   Cucumber crop nutrition   210
        11.4.5   Cucumber harvesting and yields   210
        11.4.6   Cucumber pests, diseases and physiological disorders   211
11.5   Lettuce and Other Salad Greens   212
        11.5.1   Lettuce propagation and production   213
        11.5.2   Lettuce environmental conditions   214
        11.5.3   Lettuce crop nutrition   214
        11.5.4   Lettuce pests, diseases and physiological disorders   216
        11.5.5   Lettuce harvesting and handling   217
11.6   Production of Hydroponic Micro Greens   217
        11.6.1   Harvesting micro greens   219
11.7   Hydroponic Strawberry Production   219
        11.7.1   Strawberry propagation   220
        11.7.2   Strawberry production systems   220
        11.7.3   Strawberry plant density, pruning, pollination and
                 fruit growth   221
        11.7.4   Strawberry production environment   222
        11.7.5   Strawberry crop nutrition   222
        11.7.6   Strawberry pests, diseases and disorders   223
        11.7.7   Strawberry harvest and postharvest handling   223
11.8   Hydroponic Rose Production   224
        11.8.1   Rose production systems and planting material   224
        11.8.2   Rose plant density, pruning and plant management   225
        11.8.3   Rose growing environment   225
        11.8.4   Rose crop nutrition   225
        11.8.5   Rose pests, diseases and disorders   226
        11.8.6   Rose harvesting   226

12   Plant Factories – Closed Plant Production Systems   229
12.1   History and Background   229
12.2   Advantages of Plant Factories   230
12.3   Criticisms of Plant Factories   232
12.4   Costs and Returns   233
12.5   Domestic and Other Small-Scale Plant Factories   233
12.6   Crops Produced Including Pharmaceuticals   234
12.7   Vertical or Multilevel Systems, Including Moveable Systems   235
12.8   Crop Nutrition in Plant Factories   236
12.9   Plant Factory Environments   238
12.10   Lighting   239
12.11   Environmental Control and Plant Quality in Plant Factories   241
12.12   Automation and Robotization   241
12.13   New Innovations   242

13   Greenhouse Produce Quality and Assessment   246
13.1   Background – Produce Quality and Testing   246
13.2   Components of Crop Quality   247
13.3   Quality Improvement   247

13.4	Cultural Practices to Improve Greenhouse Produce Quality	248
	13.4.1 Nutrient solution electrical conductivity levels, salinity and deficit irrigation	248
	13.4.2 Calcium and potassium and compositional quality	249
13.5	Environmental Conditions and Produce Quality	250
	13.5.1 Light levels and produce quality	250
	13.5.2 Temperature and produce quality	252
	13.5.3 Nutrient solution chilling	253
13.6	Genetics and Produce Quality	253
13.7	Quality Testing and Grading Methods	254
	13.7.1 Colour analysis	254
	13.7.2 Total soluble solids (Brix) testing	254
	13.7.3 Dry weight percentage	256
	13.7.4 Acidity and pH	256
	13.7.5 Flavour quality – aroma and taste	257
	13.7.6 Sensory evaluation of compositional quality	257
	13.7.7 Volatiles testing – aroma	259
13.8	Nutritional Quality	260
13.9	Biologically Active Compounds	260
13.10	Texture and Firmness Quality Assessment	261
13.11	Microbial Quality and Food Safety	262
13.12	Mycotoxins and Contaminants	262
13.13	Heavy Metals and Chemical Contamination	263
13.14	Naturally Occurring Toxins	263
13.15	Nitrate in Leafy Greens	263
14	**Harvest and Postharvest Factors**	**268**
14.1	Harvesting	268
	14.1.1 Harvest maturity	268
	14.1.2 Hand harvesting	268
	14.1.3 Robotic harvesting of greenhouse crops	270
14.2	Postharvest Handling, Grading and Storage	272
	14.2.1 Pack houses	272
	14.2.2 Washing, cleaning and sanitation	273
	14.2.3 Size and shape grading	273
	14.2.4 Manual grading	273
	14.2.5 Colour sorting and grading	274
	14.2.6 Automated colour and grading systems	275
	14.2.7 Grading other produce – cut flowers	275
14.3	Fresh-Cut Salad Processing	276
14.4	Shelf-Life Evaluation	278
14.5	Packaging	278
14.6	Postharvest Cooling	279
14.7	Postharvest Handling Damage	280
14.8	GAP – Good Agricultural Practices in Postharvest Handling	281
14.9	Postharvest Storage	281
	14.9.1 Postharvest physiology during storage	281
	14.9.2 Storage systems	282
	14.9.3 Refrigeration and cool storage	282
	14.9.4 Controlled and modified atmosphere storage	283
	14.9.5 Modified atmosphere packaging	284

14.10   Postharvest Disorders   284
          14.10.1   Temperature injury   284
          14.10.2   Ethylene injury   285
          14.10.3   Other postharvest storage disorders   285
          14.10.4   Storage decay   285
14.11   Food Safety and Hygiene   286
14.12   Ready-to-Eat, Minimally Processed Produce   286
14.13   Certification and Food Safety Systems   287
          14.13.1   Documentation and recall programmes   288
14.14   Postharvest Developments   288

Index   **291**

Acknowledgements

I would to thank Urban Crop Solutions (Belgium) and the Eden Project (UK) for supplying images for use within this publication. I would also like to thank Simon Lennard (Suntec NZ Ltd) for assistance with images and diagrams, and for the content guidance provided throughout the writing process.

1 Background and History of Hydroponics and Protected Cultivation

1.1 Protected Cropping

Protected cultivation of horticultural crops involves the use of structures, barriers, films, mulches, screens, glass and other materials to provide a modified and more favourable environment for optimal plant growth. The main objectives of this environmental modification are multiple and include protection from damaging natural elements such as wind, rain, hail, snow, frost, cold/high temperatures, excessive light, insects and predators, as well as providing conditions which increase yields and quality. Further advantages of modern protected cultivation structures now incorporate the efficient use of scarce water, fertilizer, energy and land resources with greater productivity per unit area, allowing production in regions otherwise unfavourable for cropping and for out-of-season supply of fresh local produce worldwide. More recent innovations in the 21st century have included the continued development of the 'closed environment' greenhouse allowing growers complete control of all environmental factors and high-value crops grown on a large scale inside warehouses or indoor areas using only artificial light, intensive climate control and hydroponic growing methods.

While modern, high-technology protected cropping such as greenhouses incorporate a vast array of computer-controlled equipment and processes for precise environmental modification, the earliest forms of such structures were basic and mostly aimed at protecting sensitive crops from cold damage. Early Roman gardeners grew cucumbers under frames glazed with oiled cloth or sheets of mica, plants were transported outside into the sun while contained in wheeled carts and taken back inside at night to prevent cold damage. This method was reportedly used to grow cucumber fruit for the Roman emperor Tiberius Caesar (AD 14–37) (Pliny the Elder, 77 CE). By the 1300s–1500s, rudimentary greenhouse-type structures were being built in Italy and France to house exotic crops and grow flowers with minimal environmental modification, along with 'glass bells' to house individual plants. By the 1600s, the first fully heated glasshouses were being used in Europe, the most well-known example being 'orangeries': solid-walled structures, using glass on the southern side to trap sunlight, with stoves to provide additional heat. Greenhouses using hot water for heating, improved glass panelling and construction techniques were also developed in Europe in the late 1600s, allowing a rapid expansion in forcing crops during the 1700s–1800s (Fig. 1.1).

In China, Japan and Korea, glasshouses were built as a low structure with glass only on the roof and southern wall, the northern and side walls were constructed of either concrete or adobe embanked with bales of rice straw for insulation (Wittwer and Castilla, 1995). It was in the late 19th century that large, expansive and often elaborate glasshouses and conservatories were built to house extensive collections of rare and exotic plants. The conservatory at Kew Gardens (Fig. 1.2), the Crystal Palace in London and the New York Crystal Palace are examples of Victorian glasshouses. By the early 1900s glasshouse production was starting to expand worldwide, there were an estimated 1000 glasshouses in the USA and by 1929, there was 550 ha of vegetables raised under glass (Wittwer and Castilla, 1995). By the 1960s, the Netherlands had emerged as the world leader in production under glass with an estimated 5000–6000 ha; by 2005 glasshouse area had increased to occupy over 10,500 ha or 0.25% of the total land area in the Netherlands (Costa and Heuvelink, 2005).

© L. Morgan 2021. *Hydroponics and Protected Cultivation* (L. Morgan)
DOI: 10.1079/9781789244830.0001

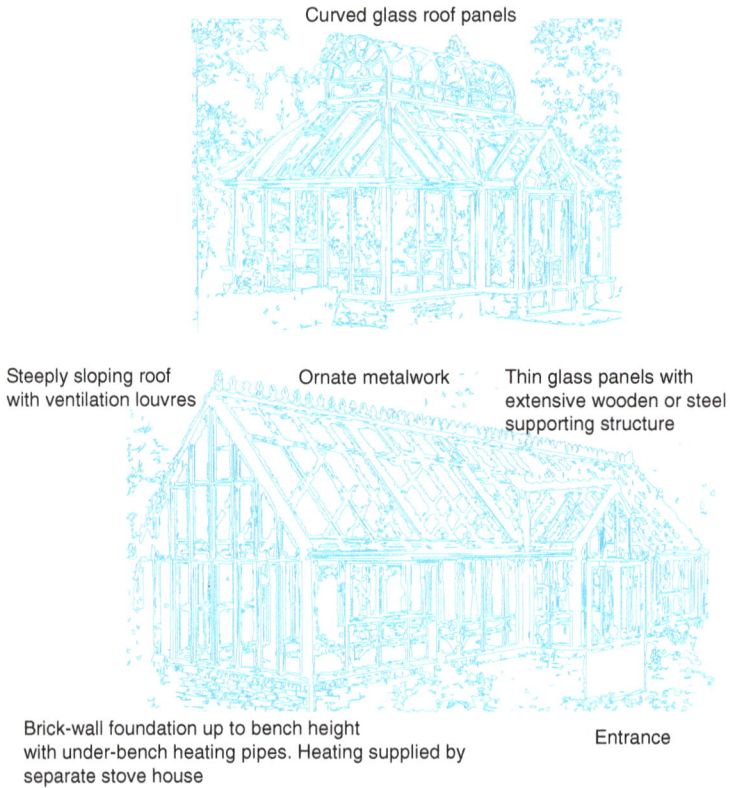

Curved glass roof panels

Steeply sloping roof
with ventilation louvres

Ornate metalwork

Thin glass panels with
extensive wooden or steel
supporting structure

Brick-wall foundation up to bench height
with under-bench heating pipes. Heating supplied by
separate stove house

Entrance

Fig. 1.1. Examples of Victorian glasshouse construction.

Fig. 1.2. The conservatory at Kew Gardens, London.

By the 1950s and 1960s greenhouse technology was changing rapidly with the increased availability of plastic cladding films, the development of drip irrigation in Israel and the gradual uptake of soilless cultivation (hydroponic methods). Tunnel or hoop houses began to make an appearance as low-cost alternatives to traditional glass-clad structures which resulted in many more small farmers having access to protected cultivation methods. By the 1970s polyethylene films were developed with improved ultraviolet (UV) inhibitors and a longer life-span, with gutter-connected greenhouses coming into increased use by the 1980s and 1990s. In the 20th century, significant increase in greenhouse production of a wide range of high-value crops was occurring in Asia and Mediterranean countries, largely fuelled by the development of plastic for non-heated greenhouse construction which expanded into large areas of Almeria in Spain, Italy and China. By 2010, the estimated protected cultivation area worldwide was 1,905,000 ha of greenhouses and 1,672,000 ha of low tunnels and floating covers; this huge increase in area under cultivation in recent decades was largely due to expansion in China (Castilla, 2013).

By 2019, the estimated global protected agricultural area was 5,530,000 ha, with 496,800 ha utilized for greenhouse vegetable production worldwide (HortiDaily, 2019). In recent times the largest areas in greenhouse vegetable production are Europe (173,561 ha), South America (12,502 ha), North America (7288 ha) and Asia (224,974 ha) (FreshPlaza, 2017). The type of greenhouse cladding is highly dependent on climate and region: 61% of greenhouses in Northern Europe are glass clad, in the Americas 20% and in Asia only 2% utilize glass as the main greenhouse covering (Parrella and Lewis, 2017).

1.2   The Future of Protected Cropping

The current trend of expansion of protected cropping structures into regions not previously utilizing this technology is likely to continue as consumers demand regular and consistent supplies of fresh, high-quality and often out-of-season produce. The limitations on land, water and energy, and restrictions regarding food production and safety, environmental concerns and conservation mean that protected cropping structures that maximize use of limited resources and produce increasingly higher yields per unit of area will become more common in many regions. Greenhouse technology, particularly with regard to energy conservation, efficient running via automated computer control

systems, robotics and improved management, is continually developing and will see more efficient structures and greater yielding crops as a result. One of the most rapidly advancing technologies is in greenhouse design and modern cladding materials. New claddings, films and panels are continually being developed which not only increase energy efficiency, but are also targeted for specific purposes and have an extended lifespan before needing replacement. The latest are those cladding films which selectively exclude certain wavelengths of light; this may be in order to retain heat, reduce the occurrence of certain crop pests and diseases, or to facilitate improved crop growth and productivity. Plastics used in protected cultivation – which have been an increasing concern regarding disposal, particularly of temporary row covers and mulches – are being developed which will biodegrade once discarded or are able to be recycled, thus lowering the impact on the environment.

Greenhouse horticulture is dominated by energy usage; whether it be a labour-based, low-technology system or large-scale conventional production, energy is required to grow crops. There is also a general long-term trend towards using more energy to provide food, although there are some exceptions and caveats (Wood et al., 2006). Energy input into horticultural operations has come under increasing scrutiny in recent years as the heavy reliance on fossil fuels and ever-rising costs of energy sources have put pressure on growers to become more energy efficient. Energy-use reductions and improvements in energy efficiency have become more important due to a shortage of energy reserves, concerns over environmental issues and carbon dioxide (CO_2) emissions, and the continued reliance on non-renewable resources such as fossil fuels. Much of the current research into energy utilization in horticulture is focusing not only on improved energy efficiency, but also on alternatives to non-renewable energy sources. These include the use of solar, wind, geothermal, biomass energy and hydro generated electricity which can all play a role in providing renewable energy sources for horticultural production.

Energy use within horticultural systems is complex and with the growing awareness of finite and ever more costly fossil fuel resources, the importance of energy use has become a food security and environmental concern. Energy input and efficiency analysis comparison of different crops and production systems, use of renewable energy resources and less reliance on energy-intensive fertilizers and other materials are all under review worldwide in an attempt to improve energy optimization in the horticultural industry.

Along with food production, innovative uses of protected structures and modern greenhouse design have seen in recent times the construction of large 'biospheres' for educational, conservation, recreational and tourism purposes. These include the Eden Project in the UK (Fig. 1.3) and the award-winning glass biome conservatories in the Gardens by the Bay complex in Singapore which create vast indoor controlled environments replicating many of the climatic conditions on Earth, growing plants native to those regions all on the same site. In urban areas, particularly those in climates with extremes of heat and cold, large, climate-controlled planted parks and food production facilities sited inside architecturally designed protected cropping structures are likely to increase in popularity, making use of advances in greenhouse structural and cladding technologies (Fig. 1.4). In order to conserve energy and running costs, completely enclosed indoor growing environments sited in warehouses and other industrial urban buildings for the production of local fresh food such as salad vegetables and herbs is a growing

trend, particularly in hot, dry climates which make greenhouse cropping difficult and expensive. Since protected cropping can produce many times the yield of outdoor field production, particularly where vertical systems are used for suitable small crops, high-rise greenhouses with a limited land footprint in areas of land scarcity are also likely to be a growing trend as technology develops further.

1.3 Background and History of Hydroponic Production

Soilless cultivation of a wide range of crops involves the practice of growing plants in containers, beds, trays, chambers or channels of a soilless medium which may be either liquid or solid. Soilless culture systems encompass a wide range of horticultural production methods from potted nursery crops in solid substrates, drip-irrigated greenhouse vegetable crops to water culture methods, the latter of which are more correctly termed 'hydroponics'. In modern times, hydroponics has become the term used to cover many forms of soilless production, both where a solid medium is used to support the plant and where solution culture only is employed.

Growing plants in containers of soilless medium is an ancient practice which has been utilized throughout the age of agriculture by many civilizations. Almost 4000 years ago the Egyptians documented the use of containers to grow and transfer mature trees from their native countries of origin to the king's palace when local soils were not suitable for particular plants (Naville, 1913).

Fig. 1.3. The biome structures of the Eden Project, UK. (Photo courtesy of Ben Foster/Eden Project.)

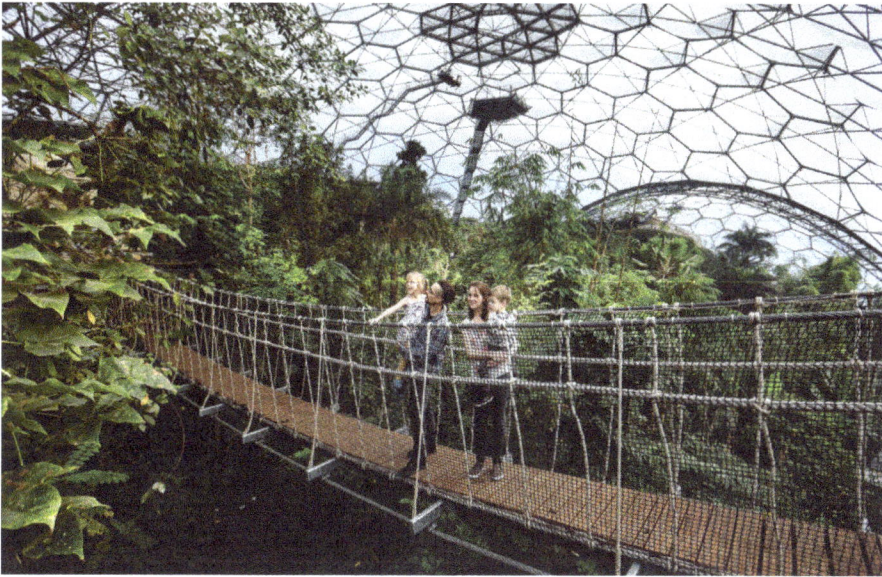

Fig. 1.4. Inside the rainforest biome of the Eden Project, UK. (Photo courtesy of Steve Tanner/Eden Project.)

Early examples of 'hydro-culture' include the floating gardens of the Aztecs of Mexico and those of the Chinese and the hanging gardens of Babylon (Resh, 1987). Despite the development of early soilless systems, it was not until the middle of the 19th century that scientists began to experiment with and understand how to create nutrient solutions of a known chemical composition which supported plant growth. Earlier attempts in the 17th and 18th centuries had determined that 'earth not water was the matter that constitutes vegetables' as was published in 1699 in John Woodward's account of 'Some thoughts and experiments concerning vegetation' (Sholto-Douglas, 1985). By 1842, nine essential nutrient elements for plant growth had been listed and later discoveries by German botanists Julius von Sachs and Wilhelm Knop led to the development of standardized nutrient solutions created by dissolving inorganic salts into water. These methods were expanded further by a number of scientists and rapidly became a standard research and teaching technique to study plant growth and nutrient uptake. From 1865 to 1920 a number of nutrient formulae were developed and tested for soilless culture of plants; some of these, such as that devised by D.R. Hoagland (1920), are still in use in modern agriculture today.

Aside from use as a research tool, true commercial application of solution culture began in 1928 with the work of William Frederick Gericke of the University of California at Berkeley. Gericke devised a practical system of solution culture and was the first to term this system 'hydroponics' from the Greek words *hydro* meaning water and *ponos* meaning working. By 1938 hydroponics was emerging as a commercial method of crop production with growers in the USA installing soilless culture beds. However, these early attempts suffered from a lack of technical information and limited availability of the correct materials and many did not succeed. Over the next few decades, it was the work of a number of researchers in the USA, England and France that documented the success and technical information required to make hydroponic systems a commercial reality on a worldwide basis. By the outbreak of the Second World War in 1939 there was renewed interest in soilless culture as a means of providing beleaguered countries with extra supplies of home-grown produce (Sholto-Douglas, 1985). In 1944, the US Air Force was utilizing hydroponic installations to grow vegetables at isolated bases such as at Ascension Island. By the 1960s, Dr Allen Cooper of England was developing the soilless system of NFT (nutrient film technique), a solution culture system in widespread use today for the production of a number of crops. More recent developments include the use of soilless culture systems in urban agriculture to grow an ever-increasing range of food, ornamental and medicinal crops and as part of the controlled ecological life-support systems (CELSS) research programme of the US National Aeronautics and Space Administration (NASA).

1.4 Hydroponic Systems

Hydroponic culture as a method of plant production has seen widespread acceptance in the horticultural industry for a number of reasons. First, it improves yields and quality of crops and allows avoidance of many soil-borne diseases which were common in greenhouse soil monoculture. soilless culture also allows more efficient use of water and fertilizer resources when managed well, particularly when used in closed or recirculating systems. Oxygen, nutrients and moisture levels can be more easily controlled and optimized in many soilless substrates, allowing easier crop management. The disadvantages of hydroponic production include the higher capital and running costs as compared with many conventional soil cultivation systems and the increased requirement for grower skill to correctly manage and monitor the technology in use. Disposal of spent nutrient solution is also a disadvantage as this can pose an environmental risk; however, increased use of closed systems, nutrient recycling and improved nutritional control is helping to overcome this concern.

Soilless culture using mineral nutrient solutions may be classified into two main categories: those that use a solid medium or substrate to support the plant's entire root system and those which use only a liquid medium or solution culture (sometimes termed 'hydroculture'). Further system classifications are based on the method of nutrient solution delivery such as drip irrigated, ebb and flow, capillary fed, continuous flow or aeroponic misting. Hydroponic systems may also be 'open' or 'closed' depending on whether the nutrient solution is discharged to waste after passing through the substrate and root system (open system), or collected and recirculated through the crop on a regular basis (closed system).

The majority of hydroponic systems, both substrate-based and solution culture, are operated inside greenhouse or crop protection structures as this minimizes the effects of rainfall, wind and other climatic factors on crop growth. Use of greenhouse technology for soilless culture allows a higher degree of climate control with heating and/or cooling provided to maximize crop growth and yields. Additional technology such as CO_2 enrichment and artificial lighting of the growing environment is also commonly used in many climates to boost production in hydroponic systems. In suitable climates, soilless culture production systems may be established outdoors with little or no overhead protection (Fig. 1.5). In the tropics, shade house or insect mesh structures help prevent excessive heat build-up while providing a suitable climate for crop growth. Outdoor hydroponic benches covered with plastic cloche frames are utilized for small framed crops such as lettuce and herbs in some temperate areas, while indoor hydroponic systems set up in large warehouses using artificial lighting and environmental control are established in a range of climates, often where excessive heat or cold or lack of land is an issue for greenhouse production. High-technology controlled environment agriculture (CEA) is a combination of horticultural and engineering techniques that optimize hydroponic crop production, crop quality and production efficiency (Falah et al., 2013). Under CEA plants are grown in hydroponic systems where lighting, nutrient supply temperature and humidity are strictly controlled at optimum levels via a computer.

Fig. 1.5. An outdoor NFT production system.

Worldwide hydroponic cropping operations can vary from large, corporate producers running many hectares of greenhouse systems particularly for crops such as tomatoes, cucumber, capsicum and lettuce, to smaller-scale growers growing fresh product for local markets only. Hydroponic produce may be exported, shipped across continents, distributed into supermarket or chain-store marketing systems or sold directly by the grower to consumers, local restaurants, catering companies, food processors or (a more popular option for small-scale growers) farmers' markets.

While the majority of commercial hydroponic systems are set up and run to produce large volumes of fruit, vegetables, flowers, foliage and herbs for fresh consumption, there is an increasing trend in the development of 'urban hydroponics': small-scale hydroponic systems set up in urban environments where soil and space are limited to provide smallholding and hobby ornamental and vegetable cultivation, intensive high-value crop production, as well as in- and outdoor beautification and for greening up walls and

roofs in residential areas. In many instances, urban hydroponics may be used to reduce air pollution (Schnitzler, 2013). Urban spaces such as vacant lots, flat roofs or terraces enable people to grow and consume what they plant, but also to sell or trade produce for income, allowing higher yields to be obtained from otherwise unproductive areas and soil-based systems. Hydroponic systems are also used for therapy, rehabilitation and educational purposes for those with mobility issues or learning disorders and in schools to teach the basics of biology and horticultural production in limited spaces.

1.5　Substrate-Based Hydroponic Systems

The initial shift towards soilless, substrate-based cultivation systems was largely driven by the proliferation of soilborne pathogens in intensively cultivated greenhouse soils (Raviv and Lieth, 2008). This trend was further driven by the fact that soilless substrates allowed a greater degree of control over a range of plant growth factors such as root moisture levels, oxygenation, improved drainage and ability to precisely control nutrition. Higher yields, crop consistency and greater product quality were more easily achievable in soilless systems and the technology rapidly gained acceptance as a commercial greenhouse production method through the 1970s. The main purpose of the substrate in soilless systems is to provide plant support, allowing roots to grow throughout the medium, absorbing water and nutrients from the applied nutrient solution. Worldwide, a vast array of soilless media has been tested, developed, blended and manufactured for use under hydroponic production. The type of soilless substrate selected often depended on what materials were available locally as shipping bulky media long distances is costly. However, many substrates such as rockwool, perlite and coconut fibre gained acceptance rapidly and are now shipped worldwide to high-technology greenhouse and hydroponic growers in many different countries.

Soilless growing mixes have long been used as a growing medium by horticulturalists, mostly for the production of seedlings and young plants requiring additional nurturing before being planted out into the field. Early growing media were largely composed of well-composted organic matter or leaf litter; however, other natural materials such as sand and animal manures were often incorporated to improve drainage, nutritional status and the physical structure of compost-based mixes. The development of a commercial container substrate industry was largely based around peat mining with this material still in widespread use today. The value of peat for gardening and plant production was reported as early as the 18th century (Wooldridge, 1719; Perfect, 1759) and peat was the primary organic component of the first standardized growing medium for plants in containers (Lawrence and Newell, 1939). By the 1950s the standardized 'UC growing mixes' based on peat and sand combinations were developed at the University of California (Baker, 1957). Further research in the 1970s developed peat as both a component of a wide range of container mixes and as a growing medium in its own right for a range of fruit, vegetable and flower crops. By the 1990s the heavy reliance on high-quality peat for both hydroponic substrates and as a component of potting and container mixes saw a rapidly increasing demand for mined peat with raising costs associated with the use of this material. Over the last few decades, concerns over the availability of peat in the future have seen the development of a number of new container and substrate media based on renewable resources and waste products from other industries. The sustainability of peat mining has been questioned as it is harvested from peatlands and thus has resulted in the rapid depletion of wetlands, creating the loss of a non-renewable resource (Fascella, 2015). Recent research on the development of peat replacements for potting and container mixes, as well as a soilless hydroponic medium, has resulted in an increased interest in waste recycling and the use of different organic materials as economically viable, low-cost growing substrates.

1.6 Organic Hydroponics

In recent years, the possibility of 'organic hydroponics' or organic soilless production has become a topic of much debate. In some countries, such as the USA, organic soilless systems, even those using NFT or solution culture, are certifiable as organic despite not making use of soil. However, in much of the rest of the world soilless systems are currently not certifiable as organic due to the absence of soil which is considered to be the 'cornerstone' of organic production. Where organic soilless systems are considered to be allowable, these typically incorporate the use of natural growing substrates such as coconut fibre which may be amended with perlite, compost or vermicasts. These substrates are irrigated with liquid organic fertilizers which do not contain non-organic fertilizer salts such as calcium nitrate, potassium nitrate and the like. Many of the organic nutrient solutions are based on seaweed, fish or manure concentrates, allowable mineral fertilizers, processed vermicasts, or plant extracts and other natural materials. Using organic fertilizers to provide a complete and balanced nutrient solution for soilless production is a difficult and technically challenging process and these systems are far more prone to problems with nutrient deficiencies, particularly in high nutrient-demanding crops such as tomatoes. Aquaponics, which uses organic waste generated during fish farming processes to provide nutrients for crop growth via bacterial mineralization, is sometimes considered a form of organic hydroponics when no additional fertilizer amendments are added.

1.7 Summary

Along with new types of protected cropping structures, materials and technology, the range and diversity of hydroponic crops grown are also expanding. While the greenhouse mainstays of nursery plants, tomatoes, capsicum, cucumber, salad vegetables and herbs will continue to expand in volume, newer, speciality and niche market crops are growing in popularity. These include new cut flower species, potted plants and ornamental crops, and a growing trend in the commercial production of medicinal herbs using high-technology methods such as aeroponics. Exotic culinary herbs such as wasabi, dwarf fruiting trees and spices such as ginger and vanilla are now grown commercially in protected cropping structures, while many home gardeners continue to take up hydroponics and protected cropping as both a hobby and a means of growing produce. Protected cropping and hydroponic methods will further their expansion into hostile climates which never previously allowed the production of food.

References

Baker, K.F. (1957) The UC system of producing healthy container grown plants: through the use of clean soil, clean stock, and sanitation. *California Agricultural Experimental Extension Service, Manual 23*. University of California, Agricultural Experimental Station, Extension Service, Berkeley, California.

Castilla, N. (2013) *Greenhouse Technology and Management*, 2nd edn. CAB International, Wallingford, UK.

Costa, J.M. and Heuvelink, E. (2005) Introduction: the tomato crop and industry. In: Heuvelink, E. (ed.) *Tomatoes*. CAB International, Wallingford, UK, pp. 1–20.

Falah, M.A.F., Khurigyati, N., Nurulfatia, R. and Dewi, K. (2013) Controlled environment with artificial lighting for hydroponics production systems. *Journal of Agricultural Technology* 9(4), 769–777.

Fascella, G. (2015) Growing substrates alternatives to peat for ornamental plants. In: Asaduzzaman M. (ed.) *Soilless Culture – Use of Substrates for the Production of Quality Horticultural Crops*. InTech Open. Available at: https://cdn.intechopen.com/pdfs-wm/47996.pdf (accessed 1 September 2020).

FreshPlaza (2017) Nearly half of the world's greenhouse vegetable area located in Asia. *FreshPlaza*, 2 May 2017. Available at: https://www.freshplaza.com/article/2174767/nearly-half-of-the-world-s-greenhouse-vegetable-area-located-in-asia/ (accessed 1 September 2020).

Hoagland, D.R. (1920) Optimum nutrient solutions for plants. *Science* 52(1354), 562–564.

HortiDaily (2019) World greenhouse vegetable statistics updated for 2019. *HortiDaily*, 2 January 2019. Available at: https://www.hortidaily.com/article/9057219/world-greenhouse-vegetable-statistics-updated-for-2019/ (accessed 1 September 2020).

Lawrence, W.J.C. and Newell, J. (1939) *Seed and Potting Composts*. Allen and Unwin, London.

Naville, E.H. (1913) The Temple of Deir el-Bahari (Parts I and III). In: *Memoir of the Egypt Exploration Fund*, Vol. 16. Egypt Exploration Fund, London, pp. 12–17.

Parrella, M.P. and Lewis, E. (2017) Biological control in greenhouse and nursery production: present status and future directions. *American Entomologist* 63(4), 237–250.

Perfect, T. (1759) *The Practice of Gardening*. Baldwin, London.

Pliny the Elder (77 CE) Vegetables of a cartilaginous nature – cucumbers. Pepones. In: *The Natural History, Book XIX: The Nature and Cultivation of Flax, and An Account of Various Garden Plants*; trans Bostock, J. and Riley, H.T. (1855) as *The Natural History. Pliny the Elder*. Taylor and Francis, London.

Raviv, M. and Lieth, J.H. (2008) *Soilless Culture: Theory and Practice*. Elsevier, London.

Resh, H.M. (1987) *Hydroponic Food Production*, 3rd edn. Woodbridge Press Publishing Company, Santa Barbara, California.

Schnitzler, W.H. (2013) Urban hydroponics – facts and vision. In: *Proceedings from SEAVEG 2012: Regional Symposium on High Value Vegetables in Southeast Asia: Production, Supply and Demand, Chiang Mai, Thailand, 24–26 January 2012*. World Vegetable Center, Tainan, Taiwan, pp. 285–298.

Sholto-Douglas, J. (1985) *Advanced Guide to Hydroponics*. Pelham Books Ltd, London.

Wittwer, S.H. and Castilla, N. (1995) Protected cultivation of horticultural crops worldwide. *HortTechnology* 5(1), 6–23.

Wood, R., Lenzen, M., Dey, C. and Lundie, S. (2006) A comparative study of some environmental impacts of conventional and organic farming in Australia. *Agricultural Systems* 89, 324–348.

Wooldridge, J. (1719) *Systema Horticulturæ: or the Art of Gardening*. Freeman, London.

2 Greenhouses and Protected Cropping Structures

2.1 Introduction

Protected cropping structure design is based on the local climate, crop and capital investment required and has features that will modify the environment suitable for high yielding production. This includes heat retention or loss, humidity control, supplementary light or shading, manual or computerized control of greenhouse variables and control over irrigation. Greenhouses for crop production through a cold winter period often differ considerably from structures used to produce crops in year-round hot, humid climates. Protected cropping structures can range from simple rain covers with open sides to fully enclosed, possibly twin skin, automated greenhouses and sophisticated glasshouses which are the basis of high-technology production.

A large proportion of greenhouse crops worldwide is still grown in relatively low-technology structures which represent a limited capital investment. These may be basic tunnel houses or wood- and plastic-roofed structures with little or no heating, forced ventilation or other forms of environmental control. These types of protected cropping structures are largely used in climates where heating is not required or only used for production during warmer times of the year when temperatures are suitable for crop growth (Fig. 2.1). The most extensive greenhouse industry worldwide exists in Asia where China has almost 55% of the world's plastic clad greenhouse area and over 75% of the world's small plastic tunnels (Costa et al., 2004). The Mediterranean region has the second largest area of greenhouses in the world, most being plastic-clad structures with minimal environmental control. The main factors which influenced the expansion of plastic-clad structures in the Mediterranean region were a mild climate with a high amount of solar radiation year-round, improved transportation to markets across Europe, high heating costs of greenhouses in Northern European countries and the low cost of materials for the construction of greenhouses (Grafiadellis, 1999). The most common type of greenhouse in Spain is the Parral type consisting of a vertical structure of rigid wood or steel pillars on which a double grid of wire is positioned to hold the plastic film, allowing for stability in high winds (Teitel et al., 2012). These low-cost structures may be multi-span but can have issues with water condensation on the inside of the cladding, a lack of sufficient roof ventilation and reduced light transmission due to the low slope of the roof (Teitel et al., 2012).

While the Mediterranean region is largely dominated by plastic-clad structures, high-technology glasshouses are more concentrated in the lower light climate of the Netherlands which has more than a quarter of the total area under glass worldwide (Peet and Welles, 2005). The Dutch greenhouse industry is associated with the Venlo greenhouse, which is largely used for vegetable, cut flower and potted plant production with either narrow- or wide-span designs (Teitel et al., 2012). Older-style Venlo greenhouse designs often had gutter heights of 2.5–3 m; however, more modern structures have seen an increase in gutter height to 5–7 m allowing for a greater degree of environmental control (Teitel et al., 2012). Venlo greenhouses are typically clad in glass, allowing for a high degree of light transmission, although some may be clad in other materials such as rigid plastic. High-technology structures are typically set up where the climate requires a greater level of environmental control for maximum yields and where

© L. Morgan 2021. *Hydroponics and Protected Cultivation* (L. Morgan)
DOI: 10.1079/9781789244830.0002

Fig. 2.1. Tomato greenhouse sited in a tropical climate.

markets have been developed which will pay premium prices for quality fresh produce.

2.2 Glasshouses and Plastic Greenhouses

Many climates experience lower air temperatures during the winter months, but with summer temperatures which are still higher than optimal and varying levels of radiation. In these conditions, greenhouses must be able to be economically heated in winter and cooled in summer. When cropping is concentrated only during the warmer months of the year, basic, unheated structures may be used for many crops. In these environments pad- and fan-cooled, plastic greenhouses with top vents and winter heating are used for many common crops such as tomatoes, capsicum and cucumber. Greenhouse designs for temperate and mid-latitude climates are designed to modify the environment for both seasonal and year-round variations in temperature. Efficient heating of air inside the greenhouse with insulation

and maintaining this heated air becomes the main consideration. Traditionally, heated pipes carrying hot water from boilers were the main method of heating and this is still effective for many crops. Heating may also use a system of plastic ducts at floor level which deliver warm air to the base of the plants. Greenhouses in temperate-zone climates usually incorporate fully clad side walls, roof and often side vents allowing large ventilation areas, computer control of environmental equipment such as heaters, shade or thermal screens, fogging and vents. Using modern plastic films and building technology has seen the development of twin layers of plastic which are inflated, offering improved insulation and a greater degree of environmental control.

Greenhouses sited in cold temperate and high latitude climates have large variations in day length and temperature. Day temperature may be below freezing for a large part of the year with very short day lengths, while coastal regions have short, mild summers with extended day lengths. Protected cropping structures for this type of climate require solid walls with well-built, comparatively steep solid roofs to carry snow loading which may otherwise collapse plastic-film-clad structures. Greenhouse structures in cold climates may have double-insulated walls and retractable thermal screens to assist with heat retention at night. Use of supplementary, artificial lighting is more common in these climates as low light intensity, snowfall and short days severely restrict incoming natural light levels for much of the year.

Greenhouse designs are varied; however, large industrial-scale installations often use gutter-connected greenhouses which allow for relatively easy expansion of the greenhouse as required. Gutter-connected greenhouses are composed of a number of bays running side by side along the length of the structure with the production area inside completely open and available for large-scale cropping, some of these structures are more than a hectare in size. The roof consists of a number or arches which are connected at the gutters where the bays meet (Fig. 2.2). Modern greenhouses have a high gutter height to allow for the production of long-term crops such as tomato and capsicum which may reach as tall as 3–4 m and to allow a greater degree of environmental control (Fig. 2.3). Gutter heights in modern greenhouses are often 5–6 m and claddings range from glass panels, polycarbonate sheets to single or double polyethylene film skins.

2.3 Closed and Semi-Closed Greenhouse Structures

A closed greenhouse structure completely eliminates ventilation and air exchange with the outside environment, thus must provide sufficient heating, cooling, humidity removal and CO_2 for maximum crop growth. The main objectives of a closed greenhouse system are to maintain high levels of CO_2 for crop photosynthesis, reduce pesticide requirements by preventing pests from entering the structure and save on energy costs for heating. While closed greenhouses appear to have significant advantages in cooler climates, the high requirement for cooling control during warm summer conditions

Fig. 2.2. Greenhouse design which consists of a number of connected arches.

Fig. 2.3. Modern greenhouses have a high gutter height to facilitate the growth of tall crops and give a good degree of environmental control.

and control of humidity are reported as being not economical to carry out (Zaragoza *et al.*, 2007; van't Ooster *et al.*, 2008). Closed greenhouse systems often harvest and store latent heat for use at night or when required for temperature control. Semi-closed greenhouses do not have 100% closure and maintain some air exchange with the outside environment largely for humidity removal; however, this percentage is still relatively low. Dennehl *et al.* (2014) reported that semi-closed greenhouses produced crop photosynthesis and yield increases of 20% or higher, while conferring advantages such as an improved degree of control of the greenhouse environment, reduced water requirements and reduced entry of pests and disease.

2.4   Passive Solar Greenhouses

Passive solar energy is widely used to heat greenhouse structures and requires no specialized photovoltaic (PV) cells. Passive solar heating is simply the collection of heat energy inside a greenhouse due to sunlight and this process can be effective for extending the growing season. Eighty-five per cent of the total greenhouse energy requirement is typically used for heating in cold climates, which is a function of the high heat loss from most greenhouse cladding materials (Harjunowibowo *et al.*, 2016). Passive heating technology uses thermal or heat banks which absorb the heat of sunlight during the day and re-radiate this into the growing environment at night to keep temperatures higher than those outside. Another example of a passive solar greenhouse still widely in use in China consists of a thick wall and partial roof on the north side of the structure that acts as a heat sink, absorbing solar energy during the day and releasing this at night, where a thermal blanket is used over the plastic cladding to retain the heat (Tong *et al.*, 2009).

A more advanced form of heat storage is the use of phase-change materials (PCMs) which require less space and have a higher heat capacity than traditional heat-bank materials such as water or concrete. Phase change refers to materials that change between a liquid and solid when absorbing or releasing heat and include a range of substances such as oils, paraffin and salt hydrates. PCMs are most suitable for passive temperature control in regions where there are large variations between outside day and night temperatures and can be used for both heating and cooling when required. Najjar and Hasan (2008) found that the use of a PCM inside a greenhouse to absorb excess heat decreased temperature by 3–5°C.

Thermal screens which are pulled across inside the top of the greenhouse to retain daytime heat well into the night are also part of using solar heating efficiently alongside energy-efficient claddings and heat-retentive measures such as inflatable twin-skin greenhouse designs. Heat-bank materials and thermal screens may not be sufficient to provide optimal temperature control in some climates; however, they supplement the energy cost of other types of heating systems. More advanced methods of passive solar energy collection are being developed which not only store and release heat, but also can contribute the cooling required inside a greenhouse structure (Dannehl et al., 2014).

2.5 Sustainable Greenhouse Design

The development of sustainable greenhouse structures and systems has become of increasing importance over recent years, driven by consumer awareness and concern over food production methods and the requirement for growers to remain economically competitive. Greenhouses provide the environmental control required to increase yields and quality of produce; however, this requires the use of considerable amounts of energy and generates large quantities of wastes to be disposed of (Vox et al., 2010). The installed energy power load of a greenhouse structure depends on local climate conditions and has been estimated at 50–150 W/m^2 in southern regions of Europe, 200–280 W/m^2 in northern and central

regions of Europe and up to 400 W/m^2 (heating, lighting and cooling) for complete microclimate conditioning of a greenhouse structure (Campiotti et al., 2012). There are an estimated 200,000 ha of greenhouses within Europe alone, of which about 30% are permanent structures and use fossil fuels for environmental control (Campiotti et al., 2012). With concerns over the reliance on non-renewable fossil fuels and CO_2 emissions, sustainability directives have become more focused on new greenhouse designs which incorporate new technology in cladding materials and use solar, geothermal and solid biomass energy sources.

While reducing energy use and CO_2 emissions are the main sustainability issues within the greenhouse industry, other factors include the safe disposal of waste plastics and claddings after use alongside overall waste reduction, reduction in water and fertilizer usage, prevention of drainage into groundwater and soil preservation, significant reduction in agrochemical use, effective management of the greenhouse environment by maximizing the use of solar radiation, air temperature, humidity and CO_2, and the use of renewable energy sources (Vox et al., 2010). Energy conservation in protected cropping can be achieved with use of the correct structure design for the local climate, incorporation of improved glazing materials, more efficient heating and distribution systems and new technologies in climate control, thermal insulation and overall greenhouse management. Locating greenhouses in regions with higher light, and in sheltered areas or installing windbreaks, can also assist with prevention of wind-induced heat loss and thus energy requirements for heating.

Thermal screens help retain heat by acting as a barrier between plants and the roof and also reducing the volume of air to be heated at night. Thermal screens are also used in summer to shade the crop from intense sunlight and thus reduce the energy required for cooling. Use of computerized environmental control inside greenhouse structures can also give improved energy efficiency by integrating heaters, fans, ventilation and humidity, controlling the activation of thermal and shade screens, and linking environmental inputs to irrigation timing.

2.6 Cladding Materials

Climate and cladding material determine the most suitable structure for greenhouse design, with many low-cost greenhouses utilizing a single span with plastic cladding. From the 1950s and 1960s onwards, the development of widespread, low-cost plastic technology allowed lighter and more diverse greenhouse frame structures to be used in greenhouse design where previously heavy glass panels required considerable support and framework. Low-cost materials such as wood and bamboo were able to be formed into plastic-clad, often unheated, greenhouse structures which rapidly increased in number in milder climates. Plastic films are often modified to allow maximum light transmission but conserve heat loss to the outside environment, and most incorporate some degree of UV resistance to breakdown or anti-condensation factors. Films with selectivity to certain wavelengths of radiation and co-extruded films made up of different layers of materials are becoming more common. Rigid plastic coverings are more expensive than film claddings, and include fibreglass, polycarbonate and polyvinyl chloride (PVC) panels. Glass claddings are used extensively where crops benefit from increased light transmission that this cladding provides and are common in structures such as the Dutch Venlo greenhouse design.

New technologies in greenhouse cladding films are continuing to be developed and many of these have distinct advantages for hydroponic cropping. The plastic materials most commonly used for greenhouse claddings are based on low-density polyethylene (LDPE), ethylene vinyl acetate (EVA) and plasticized PVC (Castilla, 2013), and these may be modified with additives for specific purposes. Plastic films permit photosynthetically active radiation (PAR) to penetrate into the greenhouse for crop photosynthesis; however, nearly half of the solar radiation energy is near-infrared radiation (780–2500 nm) which is a direct source of heat (Liu *et al.*, 2018). The transmission of near-infrared radiation raises the temperature inside the greenhouse, which may be beneficial in cooler winter climates, but is undesirable in tropical and subtropical regions. Claddings which reduce the transmission of near-infrared wavelengths have been trialled under warm-climate cropping conditions and it was found that temperatures can be reduced by 2–3°C compared with a greenhouse with a conventional polyethylene film (Liu *et al.*, 2018). It had been previously reported that issues with near-infrared blocking films included a reduction PAR levels entering the greenhouse for photosynthesis when infrared-blocking compounds in the film were increased (Teitel *et al.*, 2012). Near-infrared selective compounds can be incorporated into the cladding in a number of ways: as permanent additives to the cover material, as a temporary coating to the surface applied seasonally or as movable screens. Near-infrared radiation control has proven to be highly effective when diamond microparticles are used as a coating, giving a high transmittance with the shorter visible wavelengths and high reflectance in the near-infrared region (Aldaftari *et al.*, 2019). Temporary coatings and mobile screens offer the advantage of being applied during summer when heat reduction is required and removed to allow maximum heating under cooler conditions. Further improvements in the optimal qualities of greenhouse cladding films are expected to be made which allow further modification of the greenhouse environment and maximize crop responses to incoming radiation levels.

Other additives incorporated into greenhouse cladding films are those which alter the spectral quality of light entering a crop. These additives are designed to block or absorb some wavelengths that are not used by plants and transform them into wavelengths used in photosynthesis (Castilla, 2013). Some cladding films have been developed that can alter the ratio of red to far-red wavelengths which influences certain plant photomorphogenic processes. Cladding additive compounds which block UV radiation (280–400 nm) have been studied with respect to how this might affect the behaviour of certain insect pests by limiting their vision. It has been found that under UV-blocking greenhouse films and netting, lower numbers of whitefly (*Bemisia tabaci*), thrips (*Ceratothripoides clara-tris*) and aphids (*Aphis gossypii*) entered and were found in greenhouses compared with ones with higher UV intensity (Kumar and Poehling, 2006). Insect-vectored virus infection levels were also found to be significantly lower under UV-blocking claddings as compared with non-blocking greenhouse films (Kumar and Poehling, 2006). Other studies have found similar results, with the number of whiteflies trapped on sticky yellow plates under UV-absorbing film to be four to ten times lower and that of thrips ten times lower than the number trapped under regular films (Raviv and Antignum, 2004). While numbers of insect pests such as whitefly, aphids and thrips may be reduced under UV-blocking films, the effect on crop yields has shown no differences for crops such as tomatoes, capsicum and cucumber; however, it may reduce the purple/violet coloration of certain cut flower species (Messika *et al.*, 1998).

Apart from spectrum-selective cladding films, greenhouse plastics may be manufactured to diffuse light so that leaf burning under high radiation levels may be prevented in certain climates. Radiation is considered to be diffuse when it deviates by more than 2.5° from the direct incident radiation (Teitel *et al.*, 2012), with this referred to as 'turbidity'. Diffuse radiation increases the light uniformity at crop level and has been shown to increase yields in certain climates (Castilla and Hernandez, 2007). Diffuse light allows more radiation to be intercepted by the crop, particularly in the lower levels of the canopy, so that the overall assimilation rate is higher. Cladding films can provide turbidity up to 80%; however, lower percentages may be just as effective. Studies have shown that the composition of plastic greenhouse cladding films can not only affect yields but also quality of hydroponic produce. Petropoulos *et al.* (2018) found that cover materials significantly affected tomato fruit quality, particularly the sugar and organic acid contents as well as tocopherols and pigments, with higher sugar

contents found under single three-layer film and highest organic acids under seven-layer LDPE film and single three-layer film.

Greenhouse cladding films may also have additives incorporated into the polymers which modify the surface tension of the film; this may be to either repel dust on the outside of the structure or avoid issues with water droplets caused by condensation on the inside. The anti-drip properties are obtained by modification of the surface tension of the cladding film (Giacomelli and Roberts, 1993). As condensation forms, the water remains in a continuous sheet which flows towards the edge of the roof for collection, rather than forming droplets which fall on to the crop below increasing the risk of disease. This effect is achieved by addition of anti-fog additives such as surfactants into the plastic film material; however, this anti-fog effect reduces over time as the surfactants are extracted by the condensed water (Fernandez *et al.*, 2018).

While improved technology in greenhouse films continues to develop over time, plastic films still have a limited lifespan and require replacement, with most having a minimum useful life of 24 months (Giacomelli and Roberts, 1993) and many lasting for 3–4 years before replacement. Degradation of plastic greenhouse claddings is typically caused largely by exposure to UV radiation, despite the use of UV inhibitors incorporated into the film material. Films also lose transmittance over time through the action of dust, pollution, weathering and pesticide usage. The use of polyethylene films which are co-extruded with different layers can improve UV degradation and the mechanical properties of the film (Giacomelli and Roberts, 1993), giving a more versatile product and multiple beneficial properties.

Plastic films are the most widely used of the greenhouse materials due to their flexibility, lightweight nature and low costs; however, rigid structured plastics are strong and have a considerably longer life than most cladding films. Rigid sheets used for greenhouse cladding may be made from acrylic, fibreglass, polycarbonate and PVC. Polycarbonate and fibreglass sheets are typically either corrugated or reinforced and have high light transmission. Most also have a usable life of 10 years before light transmission declines. The main advantage of rigid plastic claddings is the fact these are lighter in weight than glass, with a high diffusion of light and high resistance to impacts and weather events such as hail. Rigid claddings also contain UV protection and allow transmission of short-wave infrared radiation which is an advantage in warmer climates and for summer cropping (Castilla, 2013).

Traditionally, before the development of plastics, all greenhouses were clad in glass panes which had a long lifespan and did not result any loss of light transmission over time provided cleaning was carried out. Glass maximizes light transmission, allowing all wavelengths (320–3500 nm) apart from a small portion of the UV to pass through, but is more expensive than plastic films and requires a stronger structure for supporting the cladding. Glass panels of various sizes may be used; however, large panels reduce structure shading and provide more light for the crop. For this reason, glasshouses are still popular in North-West Europe because of the economic advantage of higher light transmission (Peet and Welles, 2005).

Future developments for greenhouse cladding films are likely to include longer-lasting materials which retain high levels of light transmission over their usable lifespan, thus reducing volumes of plastic waste from the greenhouse industry. Other challenges include the development of improved thermal properties of claddings which retain more heat without the loss in light transmission that typically occurs. The use of more specific UV filters which limit infestation by pests without affecting pollinator activity is another likely development in the greenhouse industry (Fernandez et al., 2018).

2.7   Screen Houses, Net Houses, Shade Houses, Rain Covers and Other Structures

In climates with sufficient warmth for year-round cropping, screen/net or shade houses or open-sided rain cover structures are commonly used for plant production. For certain low-light species, shade houses may be utilized in a wider range of climates to modify the environment without completely controlling it. In dry, tropical desert environments where temperatures can be extremely high all year round with low humidity and minimal rainfall, the main environmental threat is wind carrying dust and sand. In these climates protected cropping structures may consist of poles set deeply into the ground, covered with high-tensile steel wires to from a basic framework over which a single layer of fine insect mesh is stretched and secured around the edges. This forms a shaded and insect-proof structure which lowers both excessive light and temperatures to more optimal levels with internal humidity increased via fogging and misting and air-movement fans promoting transpiration from the crop. In the humid tropics, which experience warm to hot air temperatures all year accompanied by high rainfall, light levels are often inconsistent and combined with short day lengths, and insect pressure is extremely high. Commonly utilized protected cropping structures under these conditions are often simple overhead rain covers with insect mesh sides. A saw-tooth greenhouse design allows good ventilation of hot air inside the greenhouse during clear days. Use of evaporative cooling under tropical climates is frequently not possible due to naturally high air humidity.

2.8   Screen and Shade Nets

Screen and shade houses may be covered in a diverse range of cladding materials depending on the degree of environmental modification required or other objectives such as insect screens for the exclusion of virus-transmitting insects and birds and a reduction in pesticide requirements. Shade screens and netting do not provide the degree of microclimate modifications achieved with plastic- and glass-clad greenhouses; however, they are an efficient tool for hydroponic cropping in many regions. Shading screens and nets modify the intensity of light entering the structure as well as air and root-zone temperature, humidity, CO_2 concentration, air turbulence and ventilation rate, and they also protect from hail and wind damage (Fig. 2.4). Compared with unprotected cultivation, screens and net houses reduce solar radiation by 15–39% and air velocity by 50–87%, increase relative humidity by 2–21%, and decrease air temperature by 2.3–2.5°C and evapotranspiration by 17.4–50% (Mahmood et al., 2018). Screens and net claddings fall into a number of different categories: the most common are composed of materials such as high-density polyethylene (HDPE), polypropylene and aluminized screens, as well as screens of different colours (black, white, blue, red and green), different levels of shading (10–50%) and mesh sizes.

More sophisticated screens and net claddings incorporate the use of spectral absorption for insect control, aluminized screens for reducing frost damage and photo-selective screens for modifications in fruit and vegetable attributes (Mahmood et al., 2018).

2.9 Low Tunnels and High Tunnels

Low and high tunnels are low-cost, plastic-covered, arched structures which differ in height depending on the crop being produced. Low tunnels, also termed 'row covers', are a form of temporary plasticulture structure, less than 1 m in height and are also referred to as 'cloches' (Fig. 2.5). These may be constructed with wire or galvanized metal frames and clips, or double-wire systems which sandwich the plastic film in place and allow it to be slid up at the sides for access to the plants and ventilation control. Older-style low tunnels may be constructed from timber frames with plastic attached to the top and sides.

High tunnels, also known as 'hoop houses', 'poly tunnels' or 'poly houses', are of sufficient height to walk in so that crops may be planted, trained and harvested from within. While high tunnels are a form of low-cost greenhouse, these are defined as being a 'free-standing or gutter connected structure, typically without any form of heating, cooling or

Fig. 2.4. Shade houses modify the intensity of light entering the structure as well as air and root-zone temperatures while providing protection from hail and wind damage.

Fig. 2.5. Low tunnels or cloches sited on NFT benches.

electricity, using natural or passive ventila-tion' (Fig. 2.6). The main objective of high tunnels is to provide a low-cost method of warming the air and protecting crops from wind and rain which extends the growing sea-son and improves produce quality for many crops, as they do not provide the degree of en-vironmental control of modern greenhouse structures. Taller crops such as tomato, capsi-cum and cucumber may be cultivated in high tunnels year-round, although a lack of venti-lation and temperature control can create issues mid-summer and mid-winter in many climates. Although high tunnels may have wind-up sides for natural air exchange and ventilation, warm air can rise and become trapped in the arched roof of the structure causing excessive heat build-up (Fig. 2.7). Cladding films for both high and low tunnels are similar, including EVA, PVC and standard greenhouse-grade polyethylene. Low tunnels in particular may be covered with perforated plastic film to facilitate air movement and

ventilation in frames which do not allow the plastic sides to be raised and lowered during the day. These types of high and low tunnel structures may also be covered with materials other than plastic film including shade cloth, insect mesh, the spun-bounded polyester or polypropylene used in floating row covers and frost cloth depending on the species being grown, the climate and the type of pro-tection required. Main crops grown world-wide in high tunnels are vegetables, including tomato, capsicum, cucumber, muskmelon, lettuce, squash and aubergine; however, other crops include small fruits such as straw-berries, tree fruits and cut flowers.

2.10   Hot Beds and Cold Frames

Cold frames are essentially miniature greenhouses used to extend the growing season, often to raise early transplants in spring. A traditional cold frame was usually

Fig. 2.6. High tunnels are a form of low-cost greenhouse using natural or passive ventilation.

constructed from brick or timber with a sloped, glass-roofed top, built low to the ground. Historically these were often part of the foundation brickwork along the southern wall of larger greenhouse, used to harden off seedlings before planting out. Cold frames are heated only by the sun and may be ventilated under warmer conditions by lifting of the hinged roof. While raising and hardening off seedlings are the main purposes of cold frames, they may also be used to grow cold-hardy winter vegetables, providing solar warming and protection from wind and heavy rain.

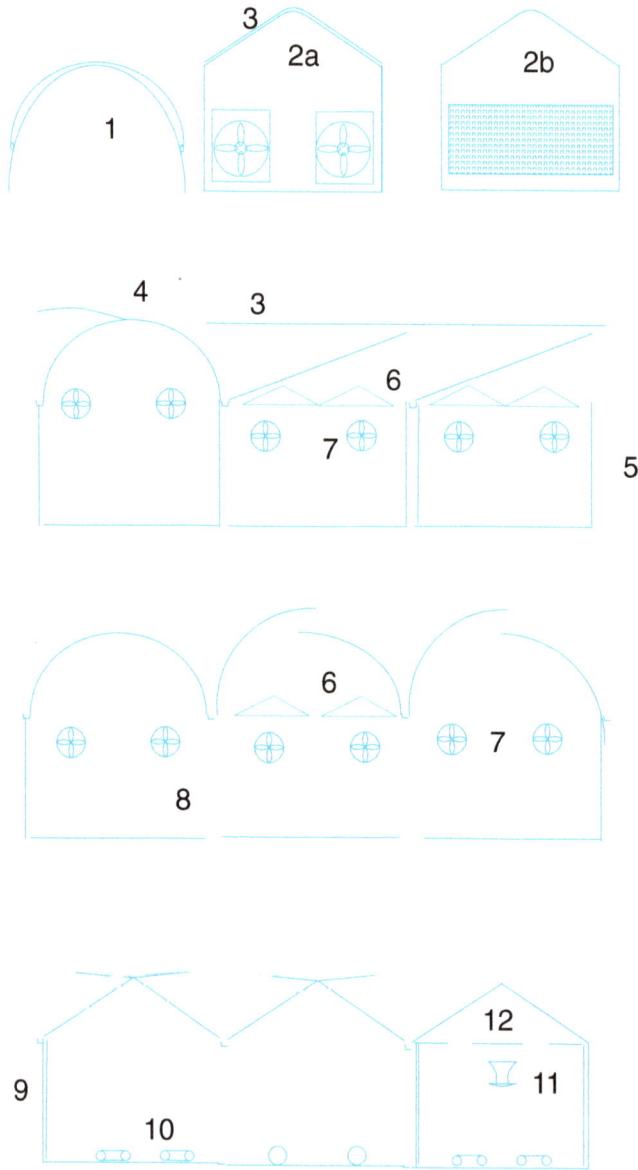

Fig. 2.7. Types of greenhouse structures: 1, tunnel house with cover partially retracted for ventilation; 2, gothic arch greenhouse with fan (a) and pad (b) evaporative cooling; 3, external shade cover; 4, high tunnel with adjustable roof ventilation; 5, sawtooth greenhouses with permanent roof ventilation; 6, evaporative misting; 7, horizontal air-movement fans; 8, gutter-connected high tunnels with and without permanent ventilation; 9, Venlo-style glass or plastic greenhouse with adjustable roof with or without side ventilation; 10, heating pipes (hot water) or ducted hot air; 11, vertical air-movement fan; 12, thermal screens across roof and sides.

Hot beds or hot boxes are similar in construction to cold frames; however, while cold frames are largely heated by the sun, hot beds use other forms of increasing the temperature inside the structure and are more suitable for warm-season plants such as tomatoes, melons, capsicums and aubergines. Heating of hot beds was traditionally carried out with pits or troughs of composting organic matter such as manure, placed directly under the base of the hot bed. The process of decomposition released sufficient heat to assist with germination and growth of seedlings. Electric hot beds use insulated heating cables to provide heat as required, while flue-heated structures used the burning of various fuels to create heat which was transferred into the hot bed. Steam and hot water heating pipes are also used in hot bed structures, particularly when associated with an adjacent heated greenhouse.

2.11   Floating Mulches, Row Covers, Cloche Covers, Direct Covers and Frost Cloth

Floating mulches, also known as 'direct covers' or 'row covers', are lightweight, inexpensive, easy-to-install covers placed directly over a crop which help trap heat from the sun, insulate against frost and act as a barrier to wind, insects and other predators. While a range of fabrics and plastics may be used in cover construction, the most common are perforated polyethylene approximately 1 mm thick or spun-bounded polyester or polypropylene which is manufactured into several different weights for varying usage. Perforated plastic covers have a pattern of holes in the material which assist ventilation and hence facilitate air movement through the cover and around the plants. The disadvantage of perforated covers is that heat may be lost rapidly via ventilation at night and small insects may gain entry to the crop below. Spun-bounded covers are manufactured from a thin mesh of synthetic fibres, usually white in colour, which retain heat from the sun and exclude insects while allowing

light and rain to penetrate. Cover weight can range from 10 to 70 g/m^2, with heavier covers used for frost protection (referred to as 'frost cloth') which may be permanent in winter or used only at night, and lighter weights for insect and bird exclusion as these have little heat-retentive properties. Floating row covers are manufactured in a range of widths and roll lengths depending on the crop they are to be applied to.

Due to their lightweight nature, row covers sit lightly over the surface of most crops and do not require any support structures, hoops or cloche frames to be attached to.

Apart from providing a protected and more favourable microclimate for early- and late-season crops and frost damage prevention in winter, floating row covers are commonly used for exclusion of flying insects which are major pests of certain plant species. It has been found that polypropylene floating covers almost completely excluded aphids and tarnished plant bugs from crisphead lettuce (*Lactuca sativa*), pests which cause significant plant damage (Rekika *et al.*, 2009). Other insect pests which are effectively controlled by lightweight floating row covers include brassica crop looper caterpillars and cabbage worms, flea beetles on aubergine, radish and other plants, sweet potato whitefly (*B. tabaci*), vegetable leaf miner (*Liriomyza sativae*) on cantaloupe (Orozco-Santos *et al.*, 1995) and many other flying insect pests which infest and damage crops, spread virus disease or lay eggs on foliage.

2.12   Greenhouse Site Planning

When planning a hydroponic greenhouse operation, there are many factors to consider aside from the type of structure, cladding, dimensions and level of technology. The site selected requires certain characteristics essential for the operation and running of a hydroponic greenhouse business. These include access to a suitable, high-quality water supply, site drainage and a level area with good access via roads and driveways. Sites which have some degree of shelter such as natural or artificial shelter belts are preferable where high winds are common as these can cause considerable damage to greenhouse claddings and structures. Shading from hills, neighbouring buildings and other structures needs to be considered so that maximum light levels can be obtained. Utilities such as electricity are required for most hydroponic operations and additional land may be necessary for water storage (tanks or reservoirs), waste disposal, pack-house facilities and storage buildings. Greenhouses should be sited with the correct orientation and with room for future expansion.

When designing a greenhouse operation, consideration needs to be given to the materials flow, internal movement of staff and produce, and efficient layout of all the facilities required. A commercial operation aims to optimize food safety and other risks by designing a system where materials and inputs move from the least hygienic areas (inwards goods bays, storage areas, and rubbish and waste receptacles) to the cleanest areas – storage of produce and outwards shipping bays where the final product is transported from the operation. This helps prevent cross-contamination between areas of the greenhouse operation and maintain hygiene in the harvested product. Inside the greenhouse, management and labour for hydroponic crop production are a major expense, thus any means to increase labour productivity and improve labour management are beneficial (Giacomelli, 2007). Larger greenhouses such as gutter-connected structures are generally more labour efficient than a number of smaller, stand-alone structures and centralized preparation areas for staff are also beneficial. Commercial greenhouses are divided into bays and typically designed with a wide central aisle from which rows of plants run. The central aisle provides passage for staff, harvested produce and other supplies into and out of the greenhouse. A head house is often incorporated into the greenhouse structure where facilities such as the pump room, nutrient testing and supplies are contained; this area may also serve as a grading and packing facility for harvested produce.

2.13 Windbreaks

Windbreaks may be installed surrounding protected cropping structures such as greenhouses, tunnels or shade houses to assist with prevention of damage by wind and to lower the heat loss from these. Windbreaks are particularly important for providing shelter for outdoor or cloche-covered hydroponic bench systems and provide some degree of microclimate modification. The improvement in microclimate that windbreaks provide includes less disease incidence, warmer temperatures, high humidity and more favourable conditions for the activity of pollinators such as bees and other insects. In arid and desert regions, windbreaks may be installed to assist with the prevention of wind erosion of soils, contamination of produce with wind-blown dust and particles as well as crop protection.

Windbreaks may be constructed from a range of materials, the most common being timber, metal or concrete structures and wire framework which support woven or knitted polyethylene or polypropylene windbreak material. On a smaller scale, windbreaks may be constructed from bamboo, brush stick screens, wooden stakes, canes or other natural materials. Live or natural windbreaks consist of a number of different tree species, selected, trimmed and maintained to provide the correct degree of wind prevention. Windbreak species include eucalyptus, cypress, casuarinas, poplar, oak, maple, acacia and willow; however, many other plant types may be used depending on local availability and climate. A combination of both artificial and natural windbreaks may be used for maximum efficiency in some circumstances. For windbreaks to be efficient and prevent damage the porosity or permeability of the material must be in the range of 40–50%, this slows the wind and reduces its intensity. A high-density or solid windbreak generates unwanted turbulence and eddies on the crop side of the structure, whereas allowing some permeability creates the correct degree of protection. The horizontal range of wind reduction by a windbreak, expressed in percentage, is proportional to its height and is practically independent of wind speed (Wittwer and Castilla, 1995).

2.14 Outdoor Hydroponic Systems

Most hydroponic production systems make use of some form of protected cropping structure whether that is a simple rain cover, shade house or a high-technology greenhouse; however, in milder climates systems may be set up outdoors (Fig. 2.8). In the early days of hydroponics, many systems, some quite large in scale, were set up directly outdoors and produced significant quantities of fresh fruit and vegetables. In the early 1940s the Gericke system of production was used for large-scale vegetable production on the rocky Pacific Islands with the objective of supplying US Navy troops during the Second World War (Rumpel and Kaniszewaki, 1998). Other extensive outdoor systems largely producing mostly small vegetables were set up in many other countries from the mid-1940s onwards. It was the development of plasticulture and high-technology greenhouse structures that largely saw commercial hydroponics being confined to protected cropping. More recently a growing interest in using soilless culture for the production of some fruits and vegetables in the open field has developed (Rumpel and Kaniszewski, 1998; Hochmuth et al., 2008; van Os et al., 2012).

The main disadvantage of outdoor production is that rainfall can not only dilute the nutrient solution but also regularly wet the foliage, leading to an increased risk of fungal and bacterial disease. Outdoor hydroponic production can take many forms, the most common being drip-irrigated substrate systems and NFT channels for small crops such as lettuce mounted on benches above the ground. Many smaller, hobby or backyard, urban or rooftop-type hydroponic and aquaponic systems may start out as outdoor production units and gradually add a degree of protected cultivation such as an open roof structure as costs allow. Despite being positioned outdoors, many systems use some form of environmental modification such as natural or artificial shelter belts to control

Fig. 2.8. An outdoor NFT lettuce production system.

winds and create a warmer microclimate, row covers placed directly on top of plants in frost-prone areas or plastic cloches over small plants such as lettuce and herbs.

Commercially, outdoor hydroponic production has not undergone the same degree of research as protected cropping systems; however, as a low-cost alternative, outdoor production has potential in a number of situations and for specific crops. The use of soilless culture methods in the open field has also been seen as a viable alternative where factors such as soil sterilization, limited water resources and issues with soil quality or contamination have arisen. Strawberry growers in North Florida, which has a mild winter climate, utilize outdoor soilless bag culture systems for out-of-season berry production. Soilless cropping outdoors provides an opportunity for many small producers to intensively produce strawberry outdoors without the investment required for extensive soil fumigation and mulching of soil production beds as well as extensive crop rotational programmes (Hochmuth et al., 2008). Since the withdrawal of the soil fumigant methyl bromide, control of soilborne diseases, weeds and nematodes has become an issue with field-based strawberry systems. Under this outdoor hydroponic system, strawberry plants receive some protection via polypropylene frost row covers as is typically used in field-based crops. Outdoor hydroponic strawberry crops, with use of nursery-raised plug plants instead of the traditional bare-rooted runner planting stock, establish faster and decrease the time to first harvest (Hochmuth et al., 1998). In a study by Hochmuth et al. (2008), it was found that an outdoor lay-flat bag culture system produced approximately twice the yields of the same cultivars grown in traditional field production systems.

A further benefit is that the same land can be used for hydroponic strawberry production year after year without the need for rotational programmes to control soil diseases and other issues. The growing medium can be replaced as required, preventing carryover of disease pathogens; however, Hochmuth et al. (2008) reported that it is possible to grow a second crop of a different species such as leafy vegetables in the same bag after the initial strawberry plants have been removed. The advantage of hydroponic cropping is that any outdoor vegetables grown via this method are free of any sand or grit compared with those grown directly in the field. While outdoor hydroponic cropping of high-value crops such as strawberries produced higher yields than soil-based systems, with the advantage of cost savings, challenges such as keeping irrigation systems clean and free from clogging and protecting polyethylene bags from premature degradation from intense UV exposure need to be considered during system design (Hochmuth et al., 2008). In another study with outdoor-grown hydroponic tomato and pepper crops, the design of the irrigation system was also seen as one the greatest challenges to outdoor cropping. This was largely due to the use of drip tape and emitters which needed to be positioned correctly, with a sufficient leaching rate of 15–20% and flushing of the irrigation system to remove solid particles and organic matter from the system (Hochmuth et al., 2002).

Outdoor hydroponic systems for other commercial crops include leafy vegetables and herbs, fruiting vegetables such as tomatoes, cucumbers, melons and capsicum. In the Netherlands restrictions on the emissions of nutrients and pesticides to surface and ground waters and the delivery of uniform high-quality products to retailers have restricted some open-field vegetable production systems (van Os et al., 2012). Outdoor soilless production systems where nutrient solutions are contained or collected (i.e. closed systems) are seen as one way of conforming to new environmental standards for the protection of water quality by preventing leaching of nutrients and reducing soil and water pollution. The increasing interest in open-field hydroponic production has largely been led by new regulations regarding the protection of water resources in areas of intensive agriculture (Rumpel and Kaniszewski, 1998). Germany has also introduced regulations which severely restrict the use of fertilizers, plant protection compounds and irrigation in protected zones and growers must develop new technologies to address these problems. Studies on the use of open-field hydroponic systems have consistently proven that growth, yields and quality of plants are generally higher than those from conventional soil-based culture (Rumpel and Kaniszewski, 1998). Van Os *et al.* (2012) examined a system for outdoor production of hydroponic leek which has strict market requirements for stem length and thickness and found that high-quality, clean product without soil contamination and with a strong reduction in labour needed for market preparation could be produced. Results of this trial found that a fast-growing crop with 4 crops/year and at least 80 plants/m² were achievable with a much shorter production time than growing in soil (van Os *et al.*, 2012). In a trial with outdoor-grown hydroponic tomato and pepper it was found that use of a perlite lay-flat bag system, irrigated with nutrient solution, produced higher yields than that from a standard, mulched, drip-irrigated production system (Hochmuth *et al.*, 2002). This system was also used to produce several crops continuously with the perlite bags, with winter squash and cucumber successfully grown in the same media following tomato and peppers; the yields of these second crops were also greater than those of the same vegetables grown in soil-based systems (Hochmuth *et al.*, 2002).

In milder climates, outdoor NFT and other solution culture systems are used for the commercial production of a number of crops, mostly lettuce, leafy greens, herbs and other vegetables. Many of these systems incorporate a standard NFT system with benches consisting of several adjacent channels positioned at a convenient working height. These open-air benches usually have the option of a small cloche placed directly over the channels clad in either plastic or shade cloth which provides protection from rain or excess light as required. The side of these cloches allows the cladding film or material to be slid up during the day for ventilation and to access the plants. Large-scale outdoor NFT operations typically have a central pump house containing a nutrient tank and dosing equipment from which the nutrient flows out and around the system, returning through a series of channels.

2.15 Controlled-Environment Agriculture

While the majority of protected cropping structures are clad in films and materials which transmit natural light for crop growth, in some regions protected cropping takes place indoors using only artificial lighting as the sole source of radiation for photosynthesis. An inhospitable outdoor climate can make use of greenhouses or other outdoor systems impractical, uneconomical or even impossible. A lack of natural light, very short day lengths, snow, ice and extremes of heat and/or cold can all make indoor cropping more economical in some regions than use of a greenhouse or tunnel structure. Indoor cropping systems may be set up in large, industrial-scale warehouses, in purpose-built facilities, educational and research laboratories, in urban areas where greenhouses are not practical, and even outside the Earth's atmosphere where NASA has grown fresh produce for astronauts onboard space missions (Stutte, 2006). Practically any building where there is sufficient ventilation, water, drainage and electricity to run the necessary equipment for crop production can potentially be used for plant cultivation. Commercial indoor protected cropping is most often carried out for high-value crops such as salad greens, herbs and potted plants which may be grown vertically or in 'tiers', many levels high, to utilize restricted indoor space more efficiently (see Chapter 12, this volume).

References

Aldaftari, H.A., Okajima, J., Komiya, A. and Maruyama, S. (2019) Radiative control through greenhouse covering materials using pigmented coatings. *Journal of Quantitative Spectroscopy and Radiative Transfer* 231, 29–36.

Campiotti, C., Viola, C., Alonzo, G., Bibbiani, C., Giagnacovo, G., *et al.* (2012) Sustainable greenhouse horticulture in Europe. *Journal of Sustainable Energy* 3(3).

Castilla, N. (2013) *Greenhouse Technology and Management*, 2nd edn. CAB International, Wallingford, UK.

Castilla, N. and Hernandez, J. (2007) Greenhouse technological packages for high quality crop production. *Acta Horticulturae* 671, 285–297.

Costa, J.M., Heuvelink, E. and Botden, N. (eds) (2004) *Greenhouse Horticulture in China: Situation and Prospects*. Ponsen & Looijen BV, Wageningen, The Netherlands.

Dannehl, D., Josuttis, M., Ulrichs, C. and Schmidt, U. (2014) The potential of a confined closed greenhouse in terms of sustainable production, crop growth, yield and valuable plant compounds of tomatoes. *Journal of Applied Botany and Food Quality* 87, 210–219.

Fernandez, J.A., Orsini, F., Baeza, E., Oztekin, G.B., Munoz, P., *et al.* (2018) Current trends in protected cultivation in Mediterranean climates. *European Journal of Horticultural Science* 83(5), 294–305.

Giacomelli, G.A. (2007) Greenhouse structures. *CEAC Paper #E-125933-04-01*. Controlled Environment Agricultural Center, College of Agriculture and Life Sciences, University of Arizona, Tucson, Arizona.

Giacomelli, G.A. and Roberts, W.J. (1993) Greenhouse covering systems. *HortTechnology* 3(1), 50–58. https://doi.org/10.21273/HORTTECH.3.1.50

Grafiadellis, M. (1999) Greenhouse structures in Mediterranean regions: problems and trends. In: Choukr-Allah, R. (ed.) *Protected Cultivation in the Mediterranean Region*. CIHEAM/IAV Hassan II, Paris, pp. 17–27.

Harjunowibowo, D., Cuce, E., Omer, S.A. and Riffat, S.B. (2016) Recent passive technologies of greenhouse systems: a review. Presented at *15th International Conference on Sustainable Energy Technologies*, Singapore, 19–22 July 2016. Available at: https://nottingham-repository.worktribe.com/output/800359 (accessed 1 September 2020).

Hochmuth, R.C., Leon, L.L., Crocker, T., Dinkins, D. and Hochmuth, G.J. (1998) Comparison of bareroot and plug strawberry plants in soilless culture in North Florida. *Florida Cooperative and Extension Service Report, SVREC 98-1*. University of Florida, Gainesville, Florida. Available at: https://svaec.ifas.ufl.edu/media/svaecifasufledu/docs/pdf/svreports/crops/strawberry/98-04.pdf (accessed 1 September 2020).

Hochmuth, G.J., Jasso-Chaverria, C., Horchmuth, R.C. and Stapleton, S.C. (2002) Field soilless culture as an alternative to soil methyl bromide for tomato and pepper. *Proceedings of the Florida State Horticultural Society* 115, 197–199.

Hochmuth, G.J., Leon, L.L., Dinkins, D. and Sweat, M. (2008) The development and demonstration of an outdoor hydroponic specialty crop production system for North Florida. *Florida Cooperative and Extension Service Report, SVREC 99-12*. University of Florida, Gainesville, Florida. Available at: https://svaec.ifas.ufl.edu/media/svaecifasufledu/docs/pdf/svreports/greenhousehydroponics/99-12.pdf (accessed 1 September 2020).

Kumar, P. and Poehling, H.N. (2006) UV blocking plastic films and nets influence vectors and virus transmission on greenhouse tomatoes in the humid tropics. *Environmental Entomology* 35(4), 1069–1082.

Liu, C.H., Chyung, A., Kan, C.J. and Lee, M.T. (2018) Improving greenhouse cladding by the additives of inorganic nano-particles. In: *Proceedings of 2018 IEEE International Conference on Applied System Invention (ICASI), Chiba, Japan, 13–17 April 2018*. Institute of Electrical and Electronics Engineers, Piscataway, New Jersey, pp. 673–677.

Mahmood, A., Hu, Y., Tanny, J. and Asante, E.A. (2018) Effects of shading and insect proof screens on crop microclimate and production: a review of recent advances. *Scientia Horticulturae* 241, 241–251.

Messika, Y.Y., Nishri, M., Gokkes, M., Lapidot, M. and Antignus, Y. (1998) UV absorbing films and aluminet screens – an efficient control means to block the spread of insect and pests in Lisianthus. *Dapey Meyda The Flower Growers Magazine* 13, 55–57.

Najjar, A. and Hasan, A. (2008) Modeling of greenhouse with PCM energy storage. *Energy Conversion and Management* 48(11), 3338–3342.

Orozco-Santos, M., Perez-Zamora, O. and Lopez-Arriaga, O. (1995) Floating row cover and transparent mulch to reduce insect populations, virus diseases and increase yield in cantaloupe. *Florida Entomologist* 78(3).

Peet, M.M. and Welles, G. (2005) Greenhouse tomato production. In: Heuvelink, E. (ed.) *Tomatoes*. CAB International, Wallingford, UK, pp. 257–304.

Petropoulos, S.A., Fernandes, A., Katsoulas, N., Barros, L. and Ferreira, I. (2018) The effect of covering material on yield, quality and chemical composition of greenhouse grown tomato fruit. *Journal of the Science of Food and Agriculture* 99(6), 3057–3068. https://doi.org/10.1002/jsfa.9519

Raviv, M. and Antignum, Y. (2004) UV radiation effects on pathogens and insect pests of greenhouse grown crops. *Photochemistry and Photobiology* 79(3), 219–226.

Rekika, D., Stewart, K.A., Boivin, G. and Jenni, S. (2009) Row covers reduce insect populations and damage and improve early season crisphead lettuce production. *International Journal of Vegetable Science* 15(1), 71–82.

Rumpel, J. and Kaniszewski, S. (1998) Outdoor soilless culture of vegetables: status and prospects. *Journal of Vegetable Crop Production* 4(1), 3–10.

Stutte, G.W. (2006) Process and product: recirculating hydroponics and bioactive compounds in a controlled environment. *HortScience* 41(3), 526–530.

Teitel, M., Montero, J.I. and Baeza, E.J. (2012) Greenhouse design: concepts and trends. *Acta Horticulturae* 952, 605–620.

Tong, G., Christopher, D.K. and Li, B. (2009) Numerical modelling of temperature variations in a Chinese solar greenhouse. *Computers and Electronics in Agriculture* 68(1), 129–139.

Van Os, E.A., van Weel, P.A., Bruins, M.A., Wilms, J.A.M., de Haan, J.J. and Verhoeven, J. (2012) System development for outdoor soilless production of leek (*Allium porrum*). *Acta Horticulturae* 974, 139–146.

Van't Ooster, A., van Henten, E.J., Gassen, E.G.O.N., Bot, G.P.A. and Dekker, E. (2008) Development of concepts for a zero-fossil-energy greenhouse. *Acta Horticulturae* 801, 725–732.

Vox, G., Teitel, M., Pardossi, A., Minuto, A., Tinivella, F. and Schettini, E. (2010) Sustainable greenhouse systems. In: Salazar, A. and Rios, I. (eds) *Sustainable Agriculture: Technology, Planning and Management*. Nova Science Publishers Inc., New York, pp. 1–79.

Wittwer, S.H. and Castilla, N. (1995) Protected cultivation of horticultural crops worldwide. *HortTechnology* 5(1), 6–23.

Zaragoza, G., Buchholz, M., Jochum, P. and Petez-Parra, J. (2007) Watergy project: towards a rational use of water in greenhouse agriculture and sustainable architecture. *Desalination* 211(1–3), 296–303.

3 The Greenhouse Environment and Energy Use

3.1 Introduction – Environmental Modification in Protected Cropping

Compared with general outdoor field cultivation, intensive, protected cropping in greenhouses and other structures such as indoor plant factories uses a higher proportion of energy inputs. This is largely attributed to the significant amounts of energy required for heating, cooling, artificial lighting and running of greenhouse climate control equipment and irrigation. For most greenhouse growers, energy costs are the third highest input for the production of most high-value intensive crops, with heating representing 70–80% of total energy consumption and electricity 10–15% (Sanford, 2011). In cold winter climates, energy costs of heated greenhouse production can be so high that it is often more energy efficient to import produce despite transportation costs. One study reported that open-field tomatoes shipped from Southern Europe to Sweden were more energy efficient than tomatoes produced in heated Swedish greenhouses (Carlsson-Kanyama et al., 2003). The specific energy input for the production of outdoor field tomatoes has been reported as being 18.02 MJ/m², while for greenhouse tomato production it was an average of 24.13 MJ/m² (Dimitrijevic et al., 2015). With environmental control of many greenhouses requiring large energy inputs, the efficiency of heating, cooling and other equipment becomes an important aspect of crop production which is continually improving with technological and material advances.

Greenhouse cladding and material selection play a major role in energy efficiency under protected cultivation. Common glazing materials for greenhouses include glass, polyethylene film, acrylic, polycarbonate and fibreglass, which may be used in single or multiple layers. Glass has the advantage of the highest light transmittance; however, the trade-off is a greater heat loss from this cladding and thus more energy is required to heat glasshouses than those clad in twin-skin polyethylene film. A double layer of poly film glazing loses heat at half the rate of single-paned glass but only has a lifespan of 3–4 years before replacement. Replacement of glazing materials every few years also represents an indirect energy cost due to the materials, manufacture and transport of such claddings. Further energy efficiencies in protected cultivation can be obtained with use of thermal screens and shade curtains. About 80% of greenhouse heating occurs at night, so increasing the resistance to night-time heat loss has a large impact on reducing energy required for heating (Sanford, 2011). Energy-efficient greenhouses often use thermal screens which are pulled across the greenhouse roof at night and may also be used to cover side walls. Thermal screens help retain heat by acting as a barrier between plants and the roof as well as by reducing the volume of air to be heated at night. Thermal screens are additionally used in summer to shade the crop from intense sunlight and thus reduce the energy required for cooling. Use of computerized environmental control inside greenhouse structures can also give improved energy efficiency by integrating heaters, fans, ventilation and humidity, controlling the activation of thermal and shade screens and linking environmental inputs to irrigation timing.

3.2 Heating

Providing sufficient heat for optimal crop growth, both in the root zone and the air surrounding the canopy, is one of the main advantages for protected cropping.

© L. Morgan 2021. *Hydroponics and Protected Cultivation* (L. Morgan) DOI: 10.1079/9781789244830.0003

While the majority of modern greenhouse heating is in the form of natural gas fuels, burning of other fuels such as oil, wood or coal is still in use. Electricity may be the main form of energy source in some locations, where this is a more cost-effective option than burning fuels or natural gas. Lesser utilized forms of energy in protected cropping include geothermal for heating in certain regions of the world, use of waste heat from adjacent industries such as power generation plants or other industries, use of solar energy and that generated by anaerobic digestion. A more common approach is using waste fuels from other industries, such as vine pruning, shell corn, rice hulls, coconut processing residue or wood or oil waste, to run boilers for heating greenhouses.

Boilers for piped hot water heating of greenhouses are widely used in soil-grown crops such as tomatoes as this not only heats the air surrounding the crop, but also the soil and root zone. Hot water heating is also used with hydroponic or soilless cultivation systems with heating pipes often doubling as rails for trolleys used by workers tending and harvesting the crop (Fig. 3.1). Since warm air tends to rise above the level of the crop and cool quickly, the method of delivery is important and a major component of the overall energy efficiency of the greenhouse. One of the most widely used systems of warm-air delivery inside protected cropping structures is the use of inflatable polyethylene ducts. Warm air fills and inflates these long cylindrical bags and is forced either up or down through large holes placed at regular intervals along their length. Warm air which is forced up from ground level through the canopy of the crop provides not only heat, but also maximum air flow around the base of the plants, assisting with humidity removal and disease control. All methods of air heating inside greenhouses require sufficient air movement and mixing to distribute heat evenly around the crop, this is assisted with the use of fans.

Fig. 3.1. Hot water heating pipes.

In soilless cultivation, particularly with solution culture hydroponics, the irrigation water or nutrient solution applied to the crop, or continually bathing the root zone, may be heated to increase nutrient uptake and growth rates. NFT, DFT (deep flow technique) and aeroponics, which do not incorporate any solid growing medium or substrate, make heating of the nutrient solution that surrounds the root the simplest method of root-zone temperature control. While heating the nutrient solution and warming the root zone does assist plant growth, it is not a complete substitute for air heating of the canopy above.

Solar energy, or heating from the sun inside a protected cropping structure, is the most common form of air heating for crop growth and may be trapped and stored for later use. Greenhouses in cooler climates are designed to capture as much solar air heating as possible and retain this with the use of heat-retentive cladding films; however, heat banks may also be used. In climates where days are warm but nights cold, construction materials which have the ability to store and release heat for additional night heating may be utilized as a thermal bank. These include heat-bank materials such as brick, ceramic or stone floors, or a bank wall or large volumes of water positioned to heat up during the day and re-radiate this heat at night.

3.3 Cooling

In tropical and subtropical climates, cooling becomes just as vital as heating in temperate-zone and cold cropping regions. Reducing the temperature inside a protected cropping structure is carried out by a number of methods, many of which are used in combination. Shading which reduces the amount of incoming solar radiation assists with temperature control, as does venting warmer, more humid air from inside the structure to the outside and drawing drier, cooler air in. Plants naturally cool themselves via transpiration, so facilitation of this process with sufficient air movement,

watering and humidity control assists keeping leaf temperatures within an optimal range. In climates where humidity allows, evaporative cooling of greenhouse structures is a common and effective method of temperature reduction and may be carried out with use of 'evaporative cooling walls' where air is drawn through a porous, wet wall or with fan and pad cooling systems (Fig. 3.2). Fogging, misting and wetting down of floors with a water spray are other methods of cooing the greenhouse environment through the use of evaporation. Misting systems for cooling and humidification can be used in both mechanically and naturally ventilated greenhouses, with the fine droplets created evaporating quickly into the air – a process that requires heat from the environment, thus providing the cooling effect. Evaporative cooing, however, relies on the air being sufficiently low in humidity to allow this method to work and the process increases the water vapour content of the air surrounding the crop which may lead to greater disease occurrence under certain conditions. In high-technology structures and with completely enclosed indoor cropping, air conditioner units may be used to cool the air and give precise climate control where outdoor conditions are characterized by extremely high temperatures and/or where evaporative cooling is not an option.

3.4 Lighting

Low natural winter light levels are a limiting factor for crop growth in many regions, and this issue is intensified in protected cropping structures where cladding films and materials and the associated framework block a percentage of the incoming light. This, combined with the high plant densities often utilized under protected cropping, means that light restricts growth for a portion of the year unless supplemented with artificial sources. Other climates may have issues with excessively high light levels, either year-round or during summer, which can place stress on certain crops, cause rapid temperature increases inside the structure

Fig. 3.2. Fan and pad evaporative cooling wall.

and create physiological issues such as leaf burn. Short days combined with continual cloud cover can reduce incoming solar radiation to levels where crops may become uneconomical to produce through low-light seasons. Not only do crops grow much slower under low radiation but produce quality can be negatively affected as well. Low light and a significant reduction in photosynthesis result in less assimilate produced for importation into developing flowers and fruits, resulting in low sugars and overall flavour in protected crops such as tomatoes. Leaf cuticles become thinner, under low light, plants become elongated and weaker and many diseases become more common.

Controlling light levels inside a protected cropping structure involves measurement and logging of year-round radiation levels at plant level. Light may be measured using a range of meters and units as PAR (photosynthetically active radiation in the 400–700 nm wavelength band) in watts per square metre (W/m^2) or per square foot (W/ft^2), or less commonly as milliwatts per square metre/foot (mW/m^2 or mW/ft^2); it may also be measured as the amount of light accumulated over a 24 h period as 'daily light integral'

(DLI). Researchers have found that fruit yield of greenhouse tomatoes is directly dependent on the solar energy received and on the total hours of bright light: a 10% reduction in sunshine hours results in 10% less yield (Warren Wilson *et al.*, 1992).

One of the most important aspects affecting light interception of a greenhouse is the design of the structural components and the cladding material. The more rafters and other solid structural components within the roof and walls of the greenhouse, the greater the amount of light that will be lost by these materials and the greater an increase in the shading effect on the crop. Some cladding materials transmit light better than others and there is always a compromise between double-skinned claddings to retain heat and the amount of light available for the crop. Covering the greenhouse floor with reflective material such as white plastic and painting structures white assists with redistribution of light which is reflected back on to the lower leaves of the crop, thus aiding photosynthesis.

While optimizing the amount of light transmission into a greenhouse is important, it can still be necessary to supplement this with artificial light to ensure a year-round supply of horticultural products (Hemming, 2011). Use of artificial or supplementary lighting is a common feature in many protected cropping situations based in low-light winter climates and also in indoor systems where all the plants' radiation requirements must be met with the use of lamps. Supplementary lighting is used to increase light levels when natural sunlight is limiting growth and production; this may be used to extend the natural day length and/or intensity of light during the day, thus increasing the DLI over a 24 h period. Indoor crops require artificial lighting to provide all the radiation required for growth and development, thus a higher intensity is required than with crops which are only receiving light to supplement natural sunlight. Some protected crops may use artificial light during the germination and seedling phase only, when radiation requirements are lowest.

There is a wide range of artificial lighting types and systems used in protected cultivation. Supplementary greenhouse lighting has traditionally been in the form of fluorescent lamps for smaller, low-light crops and propagation and high-intensity discharge (HID) lamps such as high-pressure sodium (HPS) and metal halide (MH) lamps for winter lighting of a wide range of plants including tomatoes, capsicum, cut flowers and flowering nursery plants. Lighting may also be used to prevent or induce flowering in certain day-length-responsive species as well as for increasing overall productivity. Depending on the type of lamp, artificial greenhouse lighting requires additional equipment including reflectors, ballasts, timers and metering equipment. Use of HID lamps providing supplementary or complete illumination for a crop can be a considerable running cost in many protected cropping situations and growers must weigh up the benefits of increased yields and harvest earliness against the capital costs of both lamps and equipment and energy consumption.

More recently, use of light-emitting diodes (LEDs) has become increasing popular for both supplementary and complete lighting sources due to their energy efficiency. The development of effective LEDs is a relatively new advancement in the greenhouse industry which significantly reduces the energy required for artificial crop lighting. LEDs are more energy efficient than HPS lighting systems as they do not generate heat, are durable and have considerably longer working lifespans than HPS and other forms of lighting. Recent advances in LED technology for crop lighting applications have improved these lighting systems considerably from early models which often did not provide sufficient light for crop growth. Ebinger (2016) reported that a 43% energy saving can be realized when controlled crop environments are lit with LED lamps compared with conventional HPS lamps, while Cuce et al. (2016) stated that LEDs can provide lighting-related energy savings of up to 75% per year.

3.5 Shading

While maximizing light interception by protected crops is often the focus in many climates, shading for light reduction is another aspect of climate control in protected cropping structures. Use of shading serves two purposes: it reduces excessive light falling on to sensitive plants, thus reducing plant stress; and it helps lower heat build-up inside the structure, thus having a cooling effect. Shading may be carried out via a number of different methods depending on the cladding, structure and whether the requirement for light reduction is temporary or permanent. Modern, high-technology structures use automated, computer-controlled thermal screens made from either woven or knitted material which are drawn across the top of the crop when light and/or temperature increases outside the optimal range; these are drawn back as required and may operate many times per day when light levels and temperatures are fluctuating. Other forms of shading include removable cloth covers of varying levels of shade which are installed either on the outside of the greenhouse structure or inside, above the crop. Shade-cloth materials may range from light shading of 15–30% to heavier shade of up to 85% depending on crop and natural radiation levels. Tunnel structures may have shade covers installed for the summer months which are removed in winter to maximize light transmission and warming of the internal environment. Shade paints or other liquid compounds may be applied to plastic films when light requires reduction and for assistance with temperature control, and whitewash-type liquids may be sprayed or painted on to glass and other rigid panelling.

3.6   Ventilation and Air Movement

Ventilation of protected cropping structures is vital for plant production in both small- and large-scale operations. Air movement around the plants, into and out of the environment, removes the stale 'boundary layer of air' which forms directly above the photosynthesizing leaf surface, ensures sufficient CO_2 for assimilate production and removes excess humidity, stimulating transpiration which is essential for normal growth and development. Ventilation also provides an effective way of cooling the air inside the greenhouse, removing excess heat build-up, and assisting with humidity control and thus with the prevention of many fungal and bacterial disease pathogens. Ventilation area inside a protected cropping structure is normally expressed as a percentage of the floor area and with naturally ventilated structures under temperate conditions, a ventilation area of 30% is recommended.

The two main types of ventilation used in greenhouses and other protected cropping structures are natural ventilation and fan or forced ventilation. With natural ventilation the warm air inside the greenhouse is replaced with cooler air from outside; the rate of cooling is then dependent on the temperature gradient, the outside wind velocity and its direction with respect to the vents, and the amount of ventilator opening. With fan or forced air ventilation, air is drawn through a vent and across the greenhouse to an extractor fan. Fan ventilation has become increasingly popular in plastic-clad structures and tunnel houses and is automated using thermostats. In a greenhouse with

sufficient ventilation, 60 complete air changes per hour should occur in order to supply large, mature crops with sufficient water vapour removal and CO_2 for rapid photosynthesis. An 'air change' is a term used to describe the replacement of all of the air in the greenhouse with fresh air from outside.

Many greenhouse designs incorporate both side and ridge ventilators, which run the length of the greenhouse and are capable of independent operation (Fig. 3.3). Under cooler conditions, this allows opening of the ridge ventilators only to reduce the rate of air exchange and allow air to mix and warm before reaching crop level. Under warmer conditions, all ventilators are fully opened to cool the air and lower the internal temperature. With natural ventilation, as warm air rises, this moves into the upper levels of the greenhouse and out the top vents, cooler air is then naturally drawn through the side vents and across the crop. Many ventilators are adjustable according to wind speed and direction and under computer control in commercial greenhouses.

Inside the greenhouse structure, air movement across leaf surfaces is vital for the promotion of transpiration and photosynthesis, processes which provide assimilate for growth and aid in the transportation of water and minerals from the roots and up into the canopy. Air movement within a crop should be sufficient to gently move leaves, which is an indication that the stale boundary layer of air is replaced regularly, supplying sufficient CO_2 to the stomata for photosynthesis. If there is insufficient air movement through the crop and over leaf surfaces, diffusion of water vapour out of the leaf and CO_2 into the leaf begin to fall as the boundary layer mixes too slowly into the rest of the environment. This air flow is not only essential for photosynthesis, but also assists in the prevention of physiological disorders such as blossom end rot and tipburn, as well a number of fungal pathogens. Horizontal airflow (HAF) fans installed above the crop canopy should be positioned so that they correctly circulate air within the greenhouse (Fig. 3.4). Airflow patterns should also be considered during greenhouse design so that optimal placement of vents, fans and air mixers creates good air

Fig. 3.3. Greenhouse top and side vents.

Fig. 3.4. HAF fans positioned to circulate air within the greenhouse.

movement in through vents, over and under the crop and out again. Small mixer fans can be installed in problem areas where moist, still air collects.

3.7 Humidity

Moisture content of the air inside protected cropping structures affects plant processes such as transpiration as well as the development and occurrence of a number of fungal and bacterial pathogens. Water vapour in the air is usually expressed as percentage relative humidity, or more commonly in greenhouses as vapour pressure deficit (VPD). VPD is a more accurate way of determining the moisture status of the air surrounding a crop as it takes account of the effect of temperature on the water-holding capacity of the air. VPD is the difference between the amount of moisture in the air and how much moisture that air can hold when saturated. The water-holding capacity of the air tends to double for every 10°C increase in temperature. A greenhouse at 80% relative humidity thus holds double the amount of water vapour per unit volume at an air temperature of 20°C as at 10°C. When the air in a greenhouse structure becomes fully saturated with moisture this is called the 'dew point' or 'saturation vapour pressure', and this is directly related to temperature. When the dew point is reached, free water forms on plants and greenhouse structures as condensation which growers aim to avoid as it is a significant disease risk under protected cultivation. At air saturation VPD, plants cease to transpire and physiological conditions may occur. Higher VPD values mean that the air has a higher capacity to hold more water and this stimulates plant transpiration from the foliage. Lower VPD means the air is at or near saturation, so the air cannot take more moisture from the plants under this high-humidity condition. Greenhouse humidity management aims to provide the correct VPD for the crop being grown and stage of development so that transpiration is assisted, which is essential for optimal growth, while avoiding conditions of high or low humidity which favour disease or excessive drying (Fig. 3.5). Humidity control inside protected cropping structures is largely to remove the water vapour created by the crop through the process

Fig. 3.5. Vertical air-movement fans assist air movement and humidity control.

of transpiration; however, in some systems, there may also be additional water vapour created by evaporation from the growing substrate surface. Protected cropping in certain arid, low-humidity climates may require the air surrounding the crop to be 're-humidified' to prevent desiccation, tipburn and physiological disorders associated with overly dry air.

Methods of humidity or VPD control inside protected cropping structures may be as basic as venting out warm, moist air from around the crop and drawing fresh, drier air from outside, to usage of dehumidification and re-humidification equipment in closed systems. Large-scale commercial greenhouses control humidity with use of a number of sensors positioned throughout the crop which are linked to a central controller. The greenhouse can then be programmed to vent out the warm humid air to the outside before the dew point is reached and condensation begins to form. The incoming air is then warmed with heating and later vented as it becomes saturated with water vapour. While this automated process works well in preventing condensation forming at night, it is not effective in climates where the outside air is just as humid as the air inside the growing environment. In more basic protected cropping structures such as high and low tunnels, crop covers and open-sided plastic houses, humidity control is extremely basic and relies on natural air movement into and out of the cropping area, and in these cases humidity may build where transpiration rates are high and air movement low. In dry, arid climates the air vented into the protected cropping structure during the day may have humidity levels lower than optimal for crop growth; in this case humidification can be carried out via damping down or spraying water on to plants, greenhouse floors and growing beds, or more sophisticated greenhouses may use automatic fogging or misting equipment connected to sensors. Fogging and misting from above, using evaporative coolers over air intakes or fans, or using large evaporative cooling walls all increase the humidity of the air coming into the protected cropping structure.

3.8 Carbon Dioxide Enrichment

CO_2 enrichment is highly beneficial for a range of species grown under protected cultivation and can increase crop yields, earliness and quality. Ambient CO_2 levels are slowly increasing and are now approximately 402 ppm (Dlugokencky, 2016), these may be rapidly depleted by the process of photosynthesis in a tightly closed growing structure with a large and healthy crop. Sufficient ventilation to replace the CO_2 taken up by the stomata for use in photosynthesis is essential to maintain crop growth rates. However, the rate of photosynthesis and assimilate production may be further increased by enriching the growth environment with CO_2 at higher than ambient levels. CO_2 enrichment to levels of 800–1200 ppm is used in various crops where it is considered economically viable to do so. The use of CO_2 enrichment is somewhat dependent on the requirement for ventilation in protected cropping structures. Ventilation lowers the levels of CO_2 in an enriched environment as high levels of CO_2 are lost to the outside; for this reason enrichment is often carried out more in winter, when less ventilation is required and greater benefits from enrichment are obtained. Crops vary in their response to CO_2 enrichment depending on species, light levels, temperatures, level of CO_2 maintained, stage of growth and other factors. Plants also have the ability to acclimatize to elevated CO_2 levels, so that the beneficial effects on photosynthesis become less over time, and for this reason enrichment may not be used continually throughout the growth cycle. Studies have shown that enriching the growing environment with 1000 ppm of CO_2 can increase winter growth of lettuce by 30% and result in a reduction in the time to harvest of 10 days, while under high light levels, CO_2 enrichment to 900 ppm has been found to increase canopy net photosynthesis by 40% (Hand, 1982).

The two most common methods of CO_2 enrichment of protected cropping structures are burning of hydrocarbon fuels such as natural gas or propane and use of bottled, liquid CO_2. Generation of CO_2 in commercial greenhouses burning fuels creates heat as a by-product, this may be used or stored, often as hot water, for heating under cooler conditions or at night or discharged as a waste product. When burning fuels to generate CO_2 for enrichment, only complete combustion must occur. Incomplete combustion produces a number of harmful by-products including carbon monoxide, sulfur dioxide and the plant hormone ethylene which can cause considerable damage to crops. Ethylene gas has the effect of inducing fruit and flower drop as well as leaf twisting and distortion; in less severe cases it can speed up fruit ripening and flower opening as it increases the rate of senescence in plants. Bottled CO_2 is a safer option for plant enrichment as there is no potential for toxic by-products to be produced. Liquid CO_2 is supplied in cylinders stored under high pressure (1600–2200 psi) and requires equipment such as pressure regulators, flow meters, solenoid valves and timers to run the system correctly. Some greenhouse installations may make use of waste CO_2 from adjacent industries for enrichment of the crop environment.

3.9   Greenhouse Automation

While simple, low-technology crop protection structures may rely solely on manual operation of any vents, heaters and basic timers for irrigation programmes, larger-scale commercial greenhouses are making increasing use of sophisticated computer control systems. Automated control systems consist of a number of different sensors, placed around the growing structure and crop, which continually record environmental factors such as relative humidity or VPD, temperature, light, CO_2 and moisture status in the root zone. The accuracy and efficiency of automated control systems are highly dependent on the quality of the information fed in from the environment. Sensors used require regular routine maintenance and careful placement to ensure accurate readings are taken, particularly for variables such as temperature and light which may not be uniform across all positions in the protected cropping

structure. These data recorded by sensors are fed back to a computer control system, which uses the information to compare to set points, pre-programmed by the grower, and to make regular adjustments to equipment settings to optimize growing conditions.

With irrigation systems in hydroponic culture, automation may measure not only moisture levels in the root zone, but also volume of leachate/drainage solution, electrical conductivity (EC) and pH (feed and drain solution), dissolved oxygen (DO) in nutrient solutions and root-zone temperatures, as well as integrating the irrigation programme with incoming solar radiation to determine the frequency and amount of nutrient solution to be applied. Computer-controlled systems also allow for the recoding and graphing of environmental and irrigation parameters so that historic data can be reviewed in terms of crop earliness, overall yields and other plant performance factors; this also allows growers to determine and compare the effectiveness of new crop production technology, practices or changes in environmental control levels. Automation also commonly employs a range of emergency alarms and backup generators should any variable fail to be controlled or a power failure occur. Recent technological advances have seen the increasing use of new sensors such as those which measure aspects of plant growth and physiology, including stem diameter, sap flow rates, expansion of fruit and leaf, and fruit surface temperatures; these all aim at an early detection of plant stress, thus optimizing growth rates and yields with precise control over the environment. Automation and computer controls are also an essential component in energy efficiency, allowing reductions in heating and running costs based on crop models of plant growth rates under different growing conditions and efficient use of thermal screens, lighting and CO_2 enrichment.

3.10   Energy Use and Conservation in Protected Cropping

The energy required to produce a crop under protected cultivation is a growing concern worldwide as fuel prices escalate, demand for energy increases and potential energy shortages develop. It has been reported that for the production of tomatoes, the minimum energy input was from soil-based unprotected field production in California and the highest energy requirement was for early crop production in heated glasshouses in England, and that the differences between protected heated crops in Northern Europe and unheated crops in the Mediterranean region approximately equalled the energy cost of transporting them by air (Stanhill, 1980). It is estimated that in temperate-zone greenhouses, the typical annual energy usage is 75% for heating, 15% for electricity and 10% for vehicles, indicating that efforts and resources to conserve energy should be utilized where the greatest savings can be realized (Bartok, 2015). While the majority of modern greenhouse heating is in the form of natural gas fuels and burning of other energy sources such as in oil-, wood- or coal-fired furnaces, there is an increasing trend in using waste heat from other industries to reduce energy requirements in protected cropping. Studies have found waste-heat utilization systems for greenhouses are significantly more economical to operate than purely natural gas systems (Andrews and Pearce, 2011). Lesser utilized forms of renewable energy in protected cropping also include geothermal for heating in certain regions of the world, use of solar energy and that generated by anaerobic digestion. Methane gas produced by landfill sites can be used to heat greenhouses as it is combustible in furnaces or boilers.

Energy conservation in protected cropping can be achieved with use of the correct structure design for the local climate, incorporation of improved glazing materials, more efficient heating and distribution systems and new technologies in climate control, thermal insulation and overall greenhouse management. Greenhouse design and materials are continually undergoing technological advances to become as energy efficient as possible. It has been reported that the lowest energy consumption occurs in multi-span greenhouse designs with an output–input energy ratio of 0.29 followed by gutter-connected greenhouses with a ratio of 0.21, while the highest energy consumption was found in single tunnel house designs (8 m × 15 m) with a ratio of 0.15 (Djevic and Dimitrijevic, 2009). Thus, energy productivity is higher where multi-span greenhouse structures are used rather than individual tunnel houses. The main reason for the greater energy efficiency of multi-span greenhouses is the fact these have 15–20% less surface area from which heat can be lost compared with several free-standing greenhouses covering the same area. Stand-alone greenhouses have a surface area to floor area ratio of 1.7 to 1.8 while gutter-connected greenhouses have a ratio of 1.5 or less (Sanford, 2011).

Use of modern insulation materials, energy-conservative claddings and thermal curtains for heat retention at night is estimated to save between 20 and 25% in energy costs. Use of double polyethylene films for greenhouse cladding reduces heating costs by up to 50% as compared with single poly coverings, while a single layer of film over glass can reduce annual heating costs by 5–40% (Latimer, 2009); however, there is also a corresponding loss of light transmission. Since up to 85% of the heat loss from a greenhouse occurs at night, thermal screens or blankets installed inside a greenhouse can result in a reduction in heating energy of up to 50% (Latimer, 2009). Locating greenhouses in regions with higher light, and in sheltered areas or installing windbreaks, can also assist with prevention of wind-induced heat loss and thus energy requirements for heating. Heating and cooling systems for protected cultivation also vary widely with regard to energy efficiency and the type of environmental control selected is dependent on the climate, crop and system of production. The varying efficiencies of different greenhouse heating systems have been estimated as being 90% for heated and insulated floors, to 85% efficiency for hot water pipes positioned near the floor, down to 60% for hot air heaters (Runkel, 2001). For protected crops in lower winter-light regions that use supplementary lighting, replacement of older HID lamps (HPS or MH lighting systems) with newer, highly energy-efficient LED technology can significantly reduce the cost of lighting by decreasing energy and maintenance costs and increasing longevity (Singh et al., 2015).

3.11 Energy Sources for Protected Cropping

Since protected cropping is an energy-intensive method of plant production, there has been a trend of moving away from the use of fossil fuels such as coal and oil for heating greenhouses towards more sustainable sources such as waste energy from industry, residual heat, and sustainable generation of heat and electricity (de Cock and van Lierde, 1999). As with many forms of horticultural production, greenhouse operation is still largely based on the use of non-renewable energy sources that will eventually be completely depleted and not replenished. These are fossil fuels such as coal, petroleum and natural gas which provide much of the energy used in machinery, electricity generation, transportation and heating on horticultural operations. Fossil fuels are a valuable source of energy, they are relatively inexpensive to extract, easy to ship and store, widely available and can be used for a range of purposes from transportation to the manufacture of fertilizers and plastics commonly utilized in greenhouse horticulture. Coal is currently used to generate electricity and directly for heating in greenhouses, with approximately half the electricity used in the USA coming from coal resources. Apart from being non-renewable energy sources, fossil fuels have a negative environmental impact and release CO_2 into the air.

3.11.1 Renewable energy sources

With the expected depletion in fossil fuel sources and raising costs, focus on renewable energy sources for use in greenhouse operations has grown rapidly over the last few decades. Renewable energy sources are those which can be used repeatedly, will not deplete over time and generally create less pollution than non-renewable energy sources. Renewable energy generation includes solar, wind, hydro, geothermal, biomass and biofuels.

3.11.2 Geothermal energy

Geothermal energy is generated and stored within the Earth's crust that originates from the original formation of the planet and from radioactive decay of materials. The molten core and surface of the Earth represent a considerable temperature gradient which drives continuous conduction of thermal energy into the upper layers of the Earth's crust. Here rock and water become heated and can be accessed as steam or hot water for a wide range of industrial and domestic uses including the generation of electricity. Geothermal energy is fully renewable and sustainable as only a small percentage of the Earth's geothermal resources can be exploited; it is also reliable and has little impact on the environment. However, the use of geothermal energy has largely been restricted to areas near tectonic plates where geological conditions permit a carrier such as steam to transfer the heat from deep hot zones to near the surface (Dickson and Fanelli, 2004). It is likely in the future as technological advances in geothermal energy drilling and usage occur that the cost of generating geothermal power will decrease further and become more readily available. Geothermal energy use in the horticultural industry has a long history which has largely been through direct heating of open-field cropping and within greenhouses. The use of low-pressure geothermal steam systems was developed to heat industrial and residential buildings as well as greenhouses in the period between 1920 and 1940 (Dickson and Fanelli, 2004). Greenhouse heating is now the most widely utilized application of geothermal energy in the horticultural industry and is used in many countries where it can reduce heating and operation costs considerably (Dickson and Fanelli, 2004). The most common method of using geothermal energy for greenhouse heating is to direct steam or hot water through a steel pipe system. However, in some locations, the corrosive nature of the steam or hot water means a heat exchanger must be used (Gunnlaugsson et al., 2003). One country which has made extensive use of geothermal energy in protected horticultural cultivation is Iceland. Since 1924, the greenhouse industry in Iceland has been based on the utilization of geothermal energy mainly for heating greenhouses, but also to some extent for soil disinfection (Gunnlaugsson et al., 2003). Worldwide, greenhouses account for 8.22% of the non-electric use of geothermal energy (Lund and Freeston, 2001) and the total geothermal energy used for greenhouse heating worldwide increased from 20,661 to 23,264 TJ/year in the period 2005–2010 (Lund et al., 2011). A total of 34 countries report the use of geothermal greenhouse heating, with Turkey, Russia, Hungary, China and Italy being the largest consumers with an estimated 1163 ha of greenhouse area heated geothermally worldwide in 2010 (Ragnarsson and Agustsson et al., 2014). In the Netherlands, which has an extensive, technologically advanced greenhouse industry, the horticultural sector uses extensive geothermal systems to heat greenhouses with nearly 30% of the total geothermal heat energy extracted used in protected cultivation heating (Richter, 2015).

3.11.3 Solar energy

Solar energy comes directly from the sun and is harvested via solar panels (termed 'photovoltaics') which convert radiant energy into electricity. PVs can be used to power a wide range of equipment including lighting, fans, pumps, irrigation systems and computer controllers or may be used to charge batteries. The main drawback of using solar energy to generate electricity is the current expense of PV systems; however, technological advances in the future are likely to see solar energy becoming more cost-effective to install. Solar panels used to generate energy for horticultural operations must be positioned correctly to achieve maximum sunlight interception. In many applications such as industrial and domestic use, PV panels are located on roofs, where they do not take up additional land. In rural areas, however, panels may be located on the ground where space allows, so a larger area of panel surface can be installed to generate higher levels of electricity. Solar energy utilization may often be the only source of electricity in remote areas which are off the general electricity grid or for sections of a property away from the central power supply. In particular solar energy is useful to power irrigation pumps in regions where no power supply is available, such as in many developing countries (Narale et al., 2013), and to supplement the large amounts of energy required in greenhouse operations.

Solar-powered greenhouses are a relatively new development with rapidly advancing technology. In the past, installation of PV cells on greenhouse roofs significantly reduced the amount of light able to penetrate to the crop below and required additional framework for support. However new PV systems have been developed specifically for installation on greenhouses which integrate the PV cells into the glass roof and allow sufficient light to enter for crop growth. PV cells are also installed above non-crop areas of greenhouses such as walkways, storage spaces, office areas and pack houses. The reduction in light available for crop use and thus the effect on yield and crop quality resulting from the use of greenhouse-roof solar panels is highly dependent on local sunlight levels and is most effective in high-light climates. In a study carried out in Almeria (Spain) which has high solar radiation levels, it was found that the installation of flexible solar panels mounted on greenhouse roofs so that they covered 9.8% of the roof area did not affect yield and price of tomatoes; however, fruit size and colour were negatively affected compared with the non-panelled greenhouse area (Urena-Sanchez et al., 2012). Recent developments include the use of semi-transparent PV modules for installation on greenhouse roofs. These semi-transparent PV systems or 'dye-sensitized solar cells' are proposed to be used as an alternative greenhouse glazing for suitable crop types (Allardyce et al., 2017). Semi-transparent PV modules contain a reduced area which is covered in cells – for example, 39% covered with solar collection cells and the remaining 61% of module area remains transparent to allow sunlight to enter the greenhouse for crop photosynthesis (Yano *et al.*, 2014). Other technologies for roof-based solar energy systems include those which have been developed to split the sunlight spectrum into the wavelengths used by plants for photosynthesis and those which are collected for electricity production. A greenhouse-roof PV system has been developed that allows penetration of PAR (wavelengths in the 400–700 nm range) for plant growth, but captures near-infrared radiation (wavelengths longer than 700 nm) for electricity production and heat storage (Sonneveld *et al.*, 2010a,b). Another development has been the use of Fresnel-lens greenhouse roofs which concentrate direct solar radiation on to PV and thermal collection modules, while diffused light remains unchanged and can be used by the plants below (Souliotis *et al.*, 2006; Sonneveld *et al.*, 2011).

The use of roof-installed PV cells to generate electricity for greenhouse operations is more suited to regions with high solar radiation levels, even if these only occur in the summer months. Many greenhouse operations experience solar radiation levels much higher than is required for crop production during at least part of the year, while others have year-round high solar energy levels which could be used for energy generation. Where high solar radiation levels occur, these are often reduced inside the greenhouse with use of thermal screens and shade-cloth-covered structures to protect the crop; however, the excess sunlight during these times could potentially be harvested to use to supplement the high energy requirements of greenhouse systems. Greenhouses which are sited in locations with additional unused land area at a close proximity can install land-based PV solar systems, or a large 'solar generation field', which prevent any loss of light for crop growth that occurs with roof panels. This, however, requires the use of additional land for the horticultural operation which in some areas may not be cost-effective.

Aside from PV solar systems, solar thermal collectors can also be used to generate energy. A solar thermal collector is simply defined as a unit that collects heat by absorbing solar radiation and is commonly termed a 'solar hot water module' (Cuce *et al.*, 2016). Concentrated solar thermal collections may also generate electricity as well as heating water; this is achieved by heating a working fluid to drive a turbine connected to an electrical generator (Cuce *et al.*, 2016). Solar thermal collectors are also widely used in greenhouses for space heating and drying purposes (Benli and Durmus, 2009).

3.11.4 Passive solar energy

Passive solar energy is widely used to heat greenhouse structures and requires no specialized PV cells. Passive solar heating is simply the collection of heat energy inside a greenhouse due to sunlight and this process can be effective for extending the growing season. A greenhouse which is optimizing passive solar energy for heating must be well insulated against heat loss and have glass panels angled to collect the maximum solar energy depending on the latitude or location of the greenhouse. Energy, thermal or heat banks may also be installed which absorb the heat of sunlight during the day and re-radiate this into the growing environment at night to keep temperatures higher than those outside. Thermal or heat-bank materials can be as simple as a concrete or terracotta brick floor or a wall of large black drums or containers containing water to collect heat during the day. Thermal screens which are pulled across inside the top of the greenhouse to retain daytime heat well into the night are also part of using solar heating efficiently. Heat-bank materials and thermal screens may not be sufficient to provide optimal temperature control in some climates, however they supplement the energy cost of other types of heating systems. More advanced methods of passive solar energy collection are being developed which not only store and release heat but can also contribute to the cooling required inside a greenhouse structure (Dannehl *et al.*, 2013).

3.11.5 Wind-generated energy

Wind power is one of the oldest forms of energy used in agriculture: for more than two millennia, windmills have been used to pump water to crops and livestock and to grind grain. While windmills convert the power in the wind to mechanical energy, wind turbines convert wind into electrical energy which may be used directly or stored. A more technological use of wind power is to generate electricity to feed directly into the local or long-distance grid system. Wind-generated energy is created by the flow of air through wind turbines which power generators to create electricity. Wind power is not only renewable, it utilizes minimal land and creates no emissions or by-products. The disadvantages of wind power are that its use is restricted to locations where there is sufficient and reliable wind to produce enough energy. Wind power generation is often combined with solar energy generation systems as wind is usually more prevalent in autumn, winter and spring with higher solar radiation available in summer, thus allowing a maximum year-round energy supply. Wind turbines vary in size from smaller models which power a single pump or irrigation system to larger systems designed to generate a significant amount of electricity for general use in a greenhouse operation.

3.11.6 Biomass and biofuels

Biomass includes natural renewable materials such as wood, organic wastes, manures and crops such as maize, soy or sugarcane which are used to produce energy. Most biomass materials are converted into energy by burning to provide direct heat, this may be used in greenhouses or where substrate warming is required. Another use of biomass is in the production of liquids or gases used to generate electricity or produce fuels for transportation and horticultural machinery and equipment. Biomass is converted into gas by heating it under pressure in the absence of oxygen, while manures can be converted in digester systems. Gas created from biomass can then be stored, transported and used to generate heat, steam or electricity. Fuels such as ethanol and biodiesel are produced from biomass via either fermentation or distillation (ethanol) or chemically converted in the case of biodiesel. Biodiesel can be produced from vegetable oils and animal fats, and waste cooking oil is being converted into biodiesel to fuel horticultural machinery as well as for transportation vehicles.

Wood still remains the largest biomass energy source and includes timber, wood chips and pellets, wood waste from manufacturing, particularly in the production of paper pulp, bark residue and sawdust. Industrial biomass is grown largely for the production of biofuels which may be used directly in horticulture. Biomass crops include maize, sugarcane, hemp, bamboo, switchgrass, poplar, willow, sorghum and trees species such as eucalyptus and oil palm, as these produce a higher biomass output per hectare with a minimal production energy input. Algae biomass is another non-food source which has the potential to replace many land-based biomass crops. Algae not only grow more rapidly than land-based crops, but also can be fermented to produce ethanol and methane as well as converted into biodiesel and hydrogen. Other sources of biomass include animal and human waste, food waste, rotting garbage and municipal waste, and general landfill waste; these all generate methane gas which is also termed 'biogas'. Biogas generated by old landfill sites has been tapped and used as a heating fuel and energy source for greenhouse operations in some countries.

While biomass is considered to be a renewable resource as it is obtained from easily and often very rapidly grown crops, it does produce CO_2 emissions into the environment when used as an energy source. Despite this, biomass energy is considered to be largely carbon neutral as carbon released into the atmosphere is reabsorbed by biomass crops during photosynthesis. This carbon cycling ensures that biomass fuels do not contribute to global warming in the same way fossil fuels have. Energy generated from some biomass is also considered less expensive than many types of fuel such as coal and oil and biomass sources are abundant globally, however extraction of some bio-mass fuels can be expensive. In the future, technological developments in biofuel processing are likely to bring production costs down, while increases in fossil fuel costs are likely to rise, making biomass a more economical proposition for energy generation.

References

Allardyce, C.S., Fankhauser, C., Sakeeruddin, S.M., Gratzel, M. and Dyson, P.J. (2017) The influence of greenhouse integrated photovoltaics on crop production. *Solar Energy* 155, 517–522.

Andrews, R. and Pearce, J.M. (2011) Environmental and economic assessment of a greenhouse waste heat exchange. *Journal of Cleaner Production* 19(13), 1446–1454.

Bartok, J. (2015) Greenhouse energy conservation checklist. *Farm Energy*, 17 November 2015. Available at: https://farm-energy.extension.org/greenhouse-energy-conservation-checklist/ (accessed June 2017).

Benli, H. and Durmus, A. (2009) Performance analysis of a latent heat storage system with phase change material for new designed solar collectors in greenhouse heating. *Solar Energy* 83(12), 2109–2119.

Carlsson-Kanyama, A., Ekstrom, M. and Shanahan, H. (2003) Food and life cycle energy inputs: consequences of diet and ways to increase efficiency. *Ecological Economics* 44, 293–307.

Cuce, E., Harjunowibowo, D. and Cuce, P.M. (2016) Renewable and sustainable energy saving strategies for greenhouse systems: a comprehensive review. *Renewable and Sustainable Energy Reviews* 64, 34–50.

Dannehl, D., Schuch, I. and Schmidt, U. (2013) Plant production in solar collector greenhouses – influence on yield, energy use efficiency and reduction in CO_2 emissions. *Journal of Agricultural Science* 5(10), 34–45.

De Cock, L. and van Lierde, D.V. (1999) Monitoring energy consumption in Belgian glasshouse horticulture. *Agriculture Engineering International: CIGR Journal* vol. 1. Available at: http://www.cigrjournal.org/index.php/Ejounral/article/view/1042/1035 (accessed 2 September 2020).

Dickson, M.H. and Fanelli, M. (2004) *What is Geothermal Energy?* Istituto di Geoscienze e Georisorse, Pisa, Italy. Available at: http://users.metu.edu.tr/mahmut/pete450/Dickson.pdf (accessed 2 September 2020).

Dimitrijevic, A., Blazin, S., Blazin, D. and Ponjican, O. (2015) Energy efficiency of the tomato open field and greenhouse production system. *Journal on Processing and Energy in Agriculture* 19(3), 132–135.

Djevic, M. and Dimitrijevic, A. (2009) Energy consumption for different greenhouse constructions. *Energy* 34(2), 1325–1331.

Dlugokencky, E. (2016) *Annual Mean Carbon Dioxide Data*. Earth System Research Laboratory, National Oceanic and Atmospheric Administration, Boulder, Colorado.

Ebinger, F. (2016) Year-long study of LED greenhouse lighting yields illuminating results. Available at: https://www.cleanenergyresourceteams.org/blog/year-long-study-led-greenhouse-lighting-yields-illuminating-results (accessed 2 September 2020).

Gunnlaugsson, B., Agustsson, M.A. and Adalsteinsson, S. (2003) Sustainable use of geothermal energy in Icelandic horticulture. *Presented at International Geothermal Conference*, Reykjavik, September 2003. Available at: https://citeseerx.ist.psu.edu/viewdoc/download?doi=10.1.1.612.3672&rep=rep1&type=pdf (accessed 2 September 2020).

Hand, D.W. (1982) CO_2 enrichment, the benefits and problems. *Scientific Horticulture* 33, 461–479.

Hemming, S. (2011) Use of natural and artificial light in horticulture – interaction of plant and technology. *Acta Horticulturae* 907, 25–35.

Latimer, J.G. (2009) Dealing with the high cost of energy for greenhouse operations. *Publication 430-101*. Virginia Cooperative Extension, Virginia Tech, Blacksburg, Virginia. Available at: http://pubs.ext.vt.edu/430/430-101/430-101.html (accessed 1 September 2020).

Lund, J.W. and Freeston, D.H. (2001) World-wide direct uses of geothermal energy 2000. *Geothermics* 30(1), 29–68.

Lund, J.W., Freeston, D.H. and Boyd, T.L. (2011) Direct utilization of geothermal energy 2010 worldwide review. *Geothermics* 40(3), 159–180.

Narale, P.D., Rathore, N.S. and Kothari, S. (2013) Study of solar PV water pumping system for irrigation of horticulture crops. *International Journal of Engineering Science Invention* 2(12), 54–60.

Ragnarsson, A. and Agustsson, M. (2014) Geothermal energy in horticulture. *Presented at Short Course VI on Utilisation of Low- and Medium-Enthalpy Geothermal Resources and Financial Aspects of Utilisation*, organized by UNU-GTP and LaGeo in Santa Tecla, El Salvador, 23–29 March 2014. Available at: https://rafhladan.is/bitstream/handle/10802/5443/UNU-GTP-SC-18-23.pdf?sequence=1 (accessed 2 September 2020).

Richter, A. (2015) Netherlands ten-folds use of direct use geothermal heat production 2009–2014. http://www.thinkgeoenergy.com/netherlands-ten-folds-use-of-direct-use-geothermal-heat-production-2009-2014/ (accessed 2 September 2020).

Runkel, E. (2001) *Michigan State University Greenhouse Alert* issue 1, 16 January 2001.

Sanford, S. (2011) *Reducing Greenhouse Energy Consumption – An Overview (A3907-01)*. University of Wisconsin-Extension, Cooperative Extension, Madison, Wisconsin. Available at: https://learningstore.uwex.edu/Assets/pdfs/A3907-01.pdf (accessed 2 September 2020).

Singh, D., Basu, C., Meinhardt-Wollweber, M. and Roth, B. (2015) LEDs for energy efficient greenhouse lighting. *Renewable and Sustainable Energy* 49, 139–147.

Sonneveld, P.J., Swinkles, G.L.A.M., Bot, G.P.A. and Flamand, G. (2010a) Feasibility study for combining cooling and high grade energy production in a solar greenhouse. *Biosystems Engineering* 105(1), 51–58.

Sonneveld, P.J., Swinkles, G.L.A.M., Campen, J., van Tuijl, B.A.J., Janssen, H.J.J. and Bot, G.P.A. (2010b) Performance results of a solar greenhouse combining electrical and thermal energy production. *Biosystems Engineering* 106(1), 48–57.

Sonneveld, P.J., Swinkles, G.L.A.M., van Tuijl, B.A.J., Janssen, H.J.J., Campen, J. and Bot,

G.P.A. (2011) Performance of a concentrated photovoltaic energy system with static linear Fresnel lenses. *Solar Energy* 85(3), 432–442.

Souliotis, M., Tripanagnostopoulos, Y. and Kavga, A. (2006) The use of Fresnel lenses to reduce the ventilation needs of greenhouses. *Acta Horticulturae* 719, 107–114.

Stanhill, G. (1980) The energy cost of protected cropping: a comparison of six systems of tomato production. *Journal of Agricultural Engineering Research* 25(2), 145–154.

Urena-Sanchez, R., Callejon-Ferre, A.J.C., Perez-Alonso, J. and Carreno-Ortega, A. (2012) Greenhouse tomato production with electricity generation by roof-mounted flexible solar panels. *Scientia Agricola* 69(4). Available at: http://www.scielo.br/scielo.php?script=sci_arttext&pid=S0103-90162012000400001 (accessed 2 September 2020).

Warren Wilson, J., Hand, D.W. and Hannah, M.A. (1992) Light interception and photosynthetic efficiency in some glasshouse crops. *Journal of Experimental Botany* 43(248), 363–373.

Yano, A., Onoe, M. and Nakata, J. (2014) Prototype semi-transparent photovoltaic modules for greenhouse roof applications. *Biosystems Engineering* 122, 62–73.

4 Greenhouse Operation and Management

4.1 Introduction

Apart from directly producing crops, a greenhouse operation has a wide range of other factors which must be managed and controlled to both protect the profitability of the business as well as meet a range of legal, compliance, safety, hygiene and environmental regulations. Inputs into the production system such as water, fertilizers, energy, labour, growing substrates and other resources must be managed and handled correctly, while outputs such as wastewater, used plastics and spent media require the correct method of disposal to prevent environmental issues and meet the requirements of local regulations.

4.2 Greenhouse Sanitation and Hygiene

Hygiene and sanitation are vital aspects of greenhouse production which serve to not only protect consumers from foodborne illness, but also to prevent the spread of crop pests and diseases while maintaining produce quality and shelf-life. Horticultural sanitation and hygiene regulations and standards have come under increased pressure in recent years due to the occurrence of a number of serious foodborne pathogen outbreaks originating from fresh produce. The development of highly resistant and devastating crop diseases has also seen a more focused approach to the prevention of pathogen spread via sanitation, quarantine and improved hygiene procedures.

Growers have access to many tools and techniques to maintain high levels of sanitation and hygiene, with most crops and production systems having specific guidelines which outline these processes. Control over sanitation and hygiene in greenhouse production systems begins with the early stages of crop establishment during propagation and carries through to postharvest grading, packing and storage so that the entire production chain has a lowered risk of microbial or other contamination. A significant part of control over sanitation includes the safe and effective disposal of wastewater, organic material and other refuse created during crop production and processing. Reducing waste and optimizing waste management not only provide an economic advantage to a horticultural operation, often lowering the cost of disposal, but also serve the role of protecting the environment. Excessive and uncontrolled waste, disposed of using inappropriate methods, poses a risk not only to staff and public health, but also may contaminate soil, water and air and have an ongoing effect on wildlife and natural ecosystems.

4.2.1 Hygiene and sanitation for food safety

One of the most important reasons for maintaining a suitable level of greenhouse sanitation and hygiene is for food safety (Fig. 4.1). Fresh produce can harbour and transmit a number of foodborne pathogens, some of which can cause significant illness. Apart from the use of postharvest washes, dips and sprays of water containing sanitizer compounds to kill any surface bacteria, a number of other sanitation and hygiene measures must be followed to ensure food safety is maintained through the harvest, packaging, storage and distribution chain. Produce which is consumed fresh, without peeling or cooking, is a particular risk for

DOI: 10.1079/9781789244830.0004

Fig. 4.1. Maintaining a clean and hygienic greenhouse is an important aspect of food safety.

food safety pathogens. This includes salad greens, baby leaf or mesclun products, micro greens, fresh herbs, berries, fruit and sprouts. Sprouts pose a particular issue with food safety as bacterial pathogens present on the seeds used for sprouts production may multiply rapidly during the sprouting process; these may also be present in the water used for processing or washing. Contaminated seed sources of alfalfa, clover, radish and bean sprouts have been identified as causes of food safety outbreaks and regulations now require sprout producers to sanitize seed with high levels of chlorine prior to sprouting and to also test spout production water for *Salmonella*, *Escherichia coli* O157:H7 and *Listeria* prior to harvest. Outbreaks of foodborne illness linked to fresh fruit and vegetables can be potentially serious, while many milder cases are often misdiagnosed and go unreported.

The main food safety risk factors with greenhouse-produced fresh produce come from issues with water quality, organic material, human handling and staff hygiene, animals and insects, chemical contamination and postharvest storage. Untreated water supplies can carry pathogens which cause food safety issues and human illness. Only treated (potable) water should be used for crop spraying operations, irrigation of micro greens and baby salad greens, and washing harvested produce. Greenhouse water sources can be tested for food safety pathogens such as *E. coli* and keeping water testing records are part of good agricultural practices (Larouche, 2012). Where produce is to be washed postharvest, such as cut lettuce or in the production of salad mixes, water requires chlorination (100–200 ppm activated chorine maintained). Another potential source of food safety pathogens is organic matter such as composts and manure which are not generally used as part of hydroponic production but may still be used under protected cultivation in soil-based systems. Only fully composted organic matter should be used in production systems and foot baths of sanitizer solution installed at entrance ways of greenhouses.

Staff handling of fresh produce is considered to be a major risk for the transfer of many food safety pathogens. Basic recommendations incorporate washing hands before handling crops/use of gloves, avoiding eating, drinking or smoking when working with plants, general personal hygiene and wearing protective clothing where appropriate (Fig. 4.2). These risks are higher where produce is handled multiple times. Chemical contamination can be another risk to food safety and only safe and correct use of registered pesticides, fungicides and other spray/drench compounds is permitted on food crops. Withholding periods for each spray product must be strictly adhered to and records kept of spray application dates. Under protected cultivation and in all food-producing hydroponic systems, control over animals and insects is required. This includes insects such as flies and cockroaches which require exclusion from the growing area and from produce

Fig. 4.2. Hygiene and sanitation signage on greenhouse entrance: 'Cultivation area, you must use net, gloves and mask' (left); 'Forbidden to enter greenhouse without authorization, keep door closed' (right).

processing and packing areas as these insects are vectors of disease. Other pests including birds, wildlife, pets, domestic animals and rodents should also be excluded from these areas as they pose a significant food safety/contamination risk. Once harvested, packaging of high-risk products such as salad mixes and micro greens enhances food safety and helps prevent contamination during transport. Packaging should contain storage information and 'best before' dates.

4.2.2 Food safety and compliance programmes

Under commercial greenhouse production, food safety compliance has become an increasingly important aspect of business operation and management. Serious, widespread and significant outbreaks of foodborne illness originating from fresh produce have occurred in recent years, with crops such as spinach, sprouts, lettuce, melons and berries being common vectors. Because of this, many supermarket chains, wholesalers, restaurants, exporters/importers and other outlets require that fresh produce growers become certified with independent third-party agencies. The most common of these programmes is GAP (Good Agricultural Practices), which may also incorporate sanitation procedures and Hazard Analysis and Critical Control Points (HACCP). GAP is a quality assurance programme that provides a traceable, accountable system from crop to consumer for the production of fruit, vegetables and flowers. GAP-approved growers are able to demonstrate to their customers that their produce is of an acceptable quality, produced in a sustainable manner and is safe to eat. Greenhouse producers may become GAP certified through local programmes in their region if these exist, which require a high level of compliance, record-keeping and regular inspection.

4.2.3 Hygiene and sanitation for crop protection

Greenhouse, nursery and protected cropping sanitation and hygiene is based around the principle of preventing the occurrence and introduction of pests and diseases into the production system. This takes a number of forms and an understanding of how pests and pathogens are transferred is an important aspect of prevention. Basic hygiene starts with good greenhouse design and the incorporation of double-door entries, foot baths, insect screening, floor surfaces which can be easily sterilized or replaced, and sufficient facilities for workers to wash hands and equipment are vital for most hygiene programmes. Many common greenhouse insect pests such as whitefly, aphids, mites, thrips, leaf miners, tomato psyllids and others can enter greenhouse cropping systems through vents, air intakes, on workers, equipment and with seedling transplants. Sanitation and hygiene can help reduce the occurrence of these infestations through screening of air intakes and vents which are regularly cleaned as required.

Workers' clothing, hair and shoes can also transmit not only insect pests but also disease spores, particularly where staff have been working with outdoor crops before entering the greenhouse or nursery area. Foot baths containing a sanitizer solution at the entrance of all greenhouses or a change in foot coverings is essential in high-technology protected cultivation for continued sanitation. Foot baths need frequent maintenance; this includes mats always being wet, with an active sterilizing (disinfectant) solution in the mat, and being relatively clean (Tesoriero et al., 2010). Foot baths need to be regularly rinsed and the solution replaced to ensure they do not become clogged with plant debris and soil. An increasing number of greenhouse operations are now using a change in clothing or coveralls for greenhouse workers to assist with hygiene procedures and crop protection. Many disease spores can not only come in on workers, equipment and planting material, but also be blown in by wind and arrive via ventilation systems or door entries. Regular cleaning and sanitation of floor surfaces, particularly entrance ways, and occasional surface cleaning of air intakes are all part of routine greenhouse sanitation procedures.

Apart from the prevention of introduction of pests and diseases from outside, sanitation and hygiene procedures within protected cropping systems and nurseries are also aimed at prevention of the spread of these issues from plant to plant within the crop. A small outbreak of pests or diseases can rapidly infect larger areas if hygiene practices are not put in place to reduce the severity and spread of these. This includes general staff procedures such as working from new crops to older ones, which helps reduce transmission of pests and diseases to more recent plantings. Use of disposable gloves and/or frequent hand washing when moving from one crop to another – particularly when carrying out sap operations such as pruning – are also an important aspect of worker hygiene. Pruning equipment blades require regular disinfection so that disease is not passed from one plant to another during the pruning process. Greenhouse staff carrying out pruning operations to remove lateral stems and other unwanted stem growth on tomatoes, cucumber and other crops typically dip cutting tools in a solution of bleach, alcohol or other disinfectant between plants to prevent the spread of a number of diseases from plant to plant including viruses carried in plant sap.

Handling of plant material in greenhouses and nurseries must also be a component of general sanitation and hygiene procedures. This includes careful screening and inspection of any incoming plant material for the presence of pests and diseases which could be spread to existing crops. Since many greenhouse and hydroponic vegetable operations make extensive use of transplants grown off-site, clean planting material becomes important for the overall health of the resulting crop. Planting material may be treated with disinfection agents, fungicides or other compounds where required or held in quarantine before being planted into the field or greenhouse. During crop production ongoing sanitation and hygiene procedures include removal of senescing, damaged or diseased leaves and fruit. This material should be removed immediately from the greenhouse. Plant prunings that are heavily infested with spores of diseases such as *Botrytis* should be placed into large plastic bags and sealed to limit the release of spores into the air which may infect other plants. Greenhouse and nursery crops should be regularly scouted and any dead or heavily diseased seedlings or plants removed to prevent cross-infection of healthy plants. All greenhouse prunings and other waste plant material then need to be removed from the cropping area and disposed of in a way which will not allow pests and diseases to escape and migrate back into the crop. For heavily infested plant material, this may require disposal to landfill, burning, burying or being placed into a sealed container. Open composting areas close to a greenhouse or other cropping areas can result in infection as pests and disease spores are dissipated into the immediate area from waste material.

Weeds and other vegetation inside and surrounding greenhouses, shade houses and other protected cropping structures require routine control as these can also act a host and source of disease inoculum and crop pests. Regular removal of weeds within and around crops can reduce pest pressure and reduce virus incidence, which is particularly important with tomato spotted wilt virus and western flower thrips (Tesoriero *et al.*, 2010). Where herbicides cannot be used, plastic mulches and weed matting can be installed under greenhouse benches, on pathways and in outdoor areas. Weed- and plant-free zones may be maintained between cropping areas to act as buffer zones and help prevent the spread of pests and diseases. Algae control is also important for continued sanitation and hygiene in greenhouses, nurseries and outdoor areas where this may become a problem under damp conditions. Algae can act as a food source for fungus gnat pests which may spread spores of plant pathogens, damage plants via larvae feeding and may also act as a buffer, preventing effective sterilization (Tesoriero *et al.*, 2010).

After crop completion, greenhouses and other protected cropping structures require cleaning out and disinfection between planting. This is one of the most effective methods of reducing pest and disease carryover from one crop to the next (Tesoriero *et al.*, 2010). Cleaning and sanitation procedures require the complete removal of all old plant material and debris, algae, weeds and substrates,

followed by cleaning and disinfection of surfaces. This includes descaling irrigation systems, pipes, pumps and other equipment as plant diseases can be harboured in the biofilm which forms on these. Irrigation systems can then be flushed with sanitation chemicals to remove pathogens. Benches, floors, pathways, hydroponic channels and other surfaces can be washed down and sprayed or rinsed with disinfectants or a high-pressure hot water unit with detergent. As a final part of the sanitation programme, the greenhouse should be left empty for a period of time before replanting.

Hydroponic systems may have a reduced risk of certain soilborne diseases; however, sanitation and hygiene are still essential to prevent the dissemination of certain pathogens. In recirculating hydroponics systems, the nutrient solution can aid the spread of certain pathogens, most notably those that produce zoospores which are disseminated through water and can swim towards host root systems. In recirculating systems, treatment of the nutrient solution with sterilization agents such as UV radiation, ozone, chemical sterilants, membrane filtration, surfactants, heat or slow sand filtration may be used to help control the spread of root disease pathogens (Stanghellini *et al.*, 1996).

4.3 Source Water Quality and Treatment

Water used for greenhouse irrigation can be obtained from a number of sources, from collected and stored rainwater, to wells, aquifers, springs, rivers, lakes, collection of dew trapped at night in certain climates and, increasingly more common, the reuse of treated or untreated wastewater. The requirement for high-quality, safe irrigation water has become a global concern as dissolved minerals, sediment, chemicals, organic matter and pathogens contaminating water used in crop production can pose a number of production issues. The main issues with irrigation water quality are the presence of dissolved concentrations of different mineral ions. The presence and levels of dissolved minerals vary considerably across the globe, with the main concern often being high salinity (high EC) particularly where unwanted salts such as sodium are present in high levels. Hard water sources, high pH and alkalinity may also be issues with some crops as this influences the uptake of certain nutrient ions within the root system. In addition, hard or alkaline water can cause considerable 'scaling' or build-up of calcium carbonate deposits inside irrigation pipes, emitters, pumps and other equipment, requiring acidification or other treatment to keep these clear and prevent irrigation blockages. Irrigation water containing human pathogens that may originate from sewage or animal waste is a considerable risk for human health, particularly in crops such as fruits and vegetables which may be eaten without cooking.

Disinfection of irrigation water used in greenhouses, hydroponics or for propagation and other sensitive crops may be carried out on water supplies which originate from rainwater, dams and reservoirs, bore/wells, rivers and streams where contact with plant or food safety pathogens is likely to have occurred. Water may be treated by a number of different methods to effectively destroy microbes such as waterborne plant pathogens including *Pythium*, *Phytophthora*, *Fusarium*, *Rhizoctonia solani* and *Verticillium dahliae* (Orlikowski *et al.*, 2017) and food safety bacteria such as *E. coli*. The most commonly used methods of irrigation water treatment are filtration, heat, ozone, UV radiation and chemical compounds such as chlorine and hydrogen peroxide. Slow sand filtration has been reported to be one of the most effective, safe and low-cost methods of water disinfection (Orlikowski *et al.*, 2017). This method involves water slowly flowing down through a large filter composed of substrates such as sand of various size grades; however, other materials such as volcanic lava, charcoal and mineral wool may also be used. As the water flows through the filter material, beneficial microorganisms resident in the filter material that are antagonistic to pathogens provide a biodisinfection treatment.

4.4 Greenhouse Biosecurity

Biosecurity at a producer level has a number of objectives, the main one being reducing the threat of new pests (invertebrates, pathogens and weeds) entering, establishing and spreading on a property. Other objectives are the prevention of further spread of damaging organisms to either surrounding areas or further afield via infected plant material, harvested produce or other means, and the rapid detection and control of any new pests should they occur onsite. If a new pest, either exotic or previously established in the country, is introduced to a production operation it will affect the business through increased costs for monitoring, cultural practices, additional chemical use and labour, as well as reduced productivity and quality and a potential loss of markets.

All horticultural producers as well as pack-house operations need to be fully aware of biosecurity risks. This includes training staff in the identification of pest and disease problems so that outbreaks may be diagnosed and treated as quickly possible. Apart from education in identification, staff should be fully aware of the required hygiene practices for people, equipment, vehicles and organic materials on the property. One of the most important aspects of individual grower biosecurity is to ensure pests and diseases do not enter the property. Supplies and production inputs such as growing media, fertilizers and organic amendments, composts and in particular new planting stock or propagation material, including seeds, should only be purchased from reputable suppliers (AHDB, 2020). Greenhouse producers who are importing plant material or propagation stock from other countries need to be particularly vigilant, even after border inspection, as some pests and diseases may be overlooked or not visible at the time of assessment. In particular, viruses, viroids and phytoplasmas may not display symptoms and some pests may only be present as difficult-to-detect eggs, all of which can develop into significant plant health outbreaks until sufficient quarantine measures are taken. Keeping such imported plant material in a specifically designed quarantine area with regular monitoring and inspection is advisable until it can be fully cleared of carrying any unwanted organisms.

Other practices and procedures which should be implemented on individual grower properties to help prevent the introduction and spread of unwanted organisms include the correct management of water and production waste. Water sources which become contaminated can spread pests and diseases not only throughout production areas, but also into the surrounding environment and waterways. All wastewater from production areas such as greenhouses, nurseries, packhouse operations and wash sites should be contained in a catchment area and treated before discharge. Areas around water-storage locations such as dams should be kept free of plant waste and other potential sources of infestation. Stored water may also be regularly tested for the presence of certain pests and pathogens and treated where appropriate before use on horticultural crops.

Management of crop and processing waste is another aspect of onsite biosecurity which is often overlooked by producers. Maintaining good hygiene and correct disposal of organic material can help minimize cross-contamination and breeding environments for pests. All plant waste including used growing substrates should be disposed of away from the production area and any water sources. Many insect pests contained in crop resides can rapidly spread to other crops and newly planted areas; some are strong fliers and may travel some distance to infect new crops, while many diseases can produce spores that may be spread by wind and rain to new locations. Appropriate disposal mechanisms for plant waste include hot composting, burning or removal to a waste-management facility or landfill.

4.4.1 People movement, human activity and biosecurity measures

Human activity and movement within and between greenhouse properties is another method of transfer of unwanted pests, diseases and weed seeds. These may be vectored in a number of ways, including on vehicles, tools and equipment, footwear and clothing. Biosecurity signs should also be placed at all entrances to greenhouse operations and in visitor car parks which inform visitors, contractors, transportation agencies and others of the requirement to formally register their presence with the manager or appropriate staff member before entering any production areas. This is particularly relevant to greenhouses, plant factories and nursery operations. Biosecurity signage at the entrance to and around locations on production properties should also contain contact details such as the office and mobile numbers for appropriate staff or the site manager.

Greenhouse and nursery operations usually maintain a visitor register onsite which records visitor movements and contact details; this is not only a safely measure, but in the case of a biosecurity outbreak can help trace the possible source of the problem. Any employees or visitors who have recently returned from overseas pose a significant risk if they visited production nurseries, greenhouses, farms or markets where plant material and produce were sold. Much of this risk is with soil contamination on shoes; however, pests and disease may be brought in with many items including clothing and headwear. All employees and contractors should be fully briefed on biosecurity measures such as the use of foot baths, cleaning of footwear, hand washing and change of clothing where contamination is a possibility before entering sensitive areas such as greenhouses, propagation areas, indoor growing facilities and nurseries. Foot baths are an essential component of biosecurity in many protected cultivation operations such as greenhouses, shade houses and indoor plant factories. Shoes should be free of organic matter and soil before using foot baths and remain in the solution for at least 30 s to be effective. Providing disposable overalls, foot and hair coverings and gloves is becoming standard practice under high-technology protected cultivation as a means of biosecurity to control the entry of pests and diseases. Under greenhouse and in nursery production, staff need to take additional precautionary measures to prevent the plant-to-plant spread of pests and disease and in particular viruses which are transmitted via plant sap. Pruning operations require not only regular treatment and sanitation of tools and knives while working within a crop, but also of workers' hands. The prevention of virus spread by washing hands in hot soapy water followed by the use of alcohol gel has been shown to be most effective (AHDB, 2020). Alcohol foam or gel is also the most effective disinfectant for hands to control fungal spores and should be used prior to entering the production facility. Rinsing hands in water alone or with soap and water is not as effective at preventing disease spread within a growing operation.

Greenhouse businesses and nurseries which are open to the public are prone to an increased occurrence of biosecurity risks. These opportunities may include field days, equipment demonstrations, visits by grower groups and students, onsite 'pick your own' operations and gate sales of produce. Public parking areas should be fully signposted and located away from production sites. Restricting access by the public to highly sensitive areas, such as propagation nurseries and greenhouse production areas, is often more effective than using foot baths and providing coveralls for all visitors. Produce sales should be in a designated area well away from production facilities and maintained with a high standard of hygiene.

4.4.2   Staff, visitors and hygiene

One of the largest sources of potential contamination, whether for diseases and insects or foodborne illnesses, is the people inside production facilities (Currey, 2017). Enforcement of good worker hygiene is essential; this includes clean hands, clothing, footwear, and procedures such as employees using hand sanitizer or wash stations before entering and handling plant material. Sanitation stations may be placed inside each greenhouse entrance and in convenient locations around larger greenhouse facilities and may also dispense disposable gloves. Some greenhouse operations, high-technology facilities such as tissue culture laboratories

and plant factories may also provide other items such as lab coats or disposable coveralls, hair and beard nets, and disposable shoe covers or booties (Currey, 2017). Visitors to greenhouses, nurseries and plant factories should follow similar hygiene practices as workers and staff because pests and diseases can be carried in on clothing and even on mobile phones (Tesoriero et al., 2010). Visitors should minimize moving between growing operations on any one day to assist with the prevention of pests and diseases transferring from one site to another and always use foot baths provided at greenhouse entrances and other sensitive production areas.

4.5   Waste Management and Disposal

Waste management is an environmental problem which has long been a concern for horticultural producers and the public. Greenhouse operations can generate large volumes of not only bulky organic waste but also plastics and other materials which may be difficult to economically recycle. Methods of waste prevention, recycling and disposal that reduce environmental harm are continuing areas of research which are likely to see further innovative methods of waste recycling within the greenhouse industry.

Removal and disposal of waste material from protected cropping operations must meet local regulatory standards for plastics, growing media/soil and organic refuse and also for discharge of wastewater and nutrient solutions. In many greenhouse operations which utilize hydroponics, particularly in non-recirculating or drain-to-waste systems, disposal of spent nutrient solution is often the main concern. In countries such as the Netherlands all nutrient leachate must be re-collected and cannot drain into the soil or wastewater systems. In many other countries, nutrient solution drainage may be collected and reused on outdoor soil-grown crops such as pasture, forestry, fruit and ornamental crops to provide additional nutrition; in other situations it may simply drain into the soil beneath the greenhouse floor or into wastewater drains. Plastic disposal is also another concern for the protected cropping industry as many materials such as temporary row covers, frost cloths, mulches, drip-irrigation tape and tunnel covers may only be used for a short period of time before disposal. Some of these plastics are recyclable and may have a collection system set up by the supplier or manufacturer; however, at the present time much of this is disposed at landfill or by incineration. Organic growing substrates and crop waste such as pruning, trimmings, reject produce and other vegetation may be composted on- or offsite, or ploughed into fields by some protected cropping operations, while others dispose of this as general waste.

4.5.1 Disposal of greenhouse wastewater

Management of wastewater from greenhouse operations has come under increased scrutiny in recent years as the environmental effects of food production become a growing concern. Wastewater may take a number of different forms, the most common being nutrient-rich water from greenhouse operations, drain-to-waste hydroponic systems and discharge from pack-house operations that are sources of wastewater, which need to be managed and disposed of correctly. Wastewater from crop production is often unnoticed as it may drain away through the soil, evaporate or flow away from the site as runoff. An increasing number of greenhouses and other intensive horticultural operations now attempt to contain this wastewater runoff and may reuse this as irrigation after treatment. Wastewater runoff is typically directed to a holding pond, dam or tanks where it is permitted to settle so that sediment can be removed and then undergoes further treatment if required.

Wastewater or 'runoff' from greenhouse operations usually contains a high level of nutrients, often some sediments and plant pathogens, and is typically described as nutrient loaded; this wastewater is defined as 'effluent' (Badgery-Parker, 2002). Wastewater which enters the environment can have a number of detrimental effects such as nitrification of groundwater supplies. Wastewater which enters natural waterways causes an increase in nutrients which in turn causes algae blooms and may kill fish species. In natural water systems total nitrogen levels of 0.1–0.75 mg/l contribute to algae blooms, while phosphorus is also a key factor with phosphorus levels of 0.01–0.1 mg/l also seen as a contributor (Badgery-Parker, 2002).

4.5.2 Treatment of wastewater

Water being discharged from a closed horticultural system such as a greenhouse or hydroponic crop that has been used for irrigation may contain contaminants and plant pathogens which require removal or treatment before either being discharged to waste or reused for further irrigation. This type of sanitation procedure is aimed at protecting the environment into which the water is to be discharged and crops where diseases may be spread via water reused for irrigation purposes.

Heat treatment of drainage water to lethal temperatures is one of the most reliable methods of disinfection (Postma et al., 2008). A temperature set point of 95°C for a minimal time of 10 s is sufficient to kill most disease-causing organisms present in wastewater or recycled fertigation solutions. Both source water and wastewater may be treated with oxidation chemical compounds such as sodium hypochlorite and hydrogen peroxide. Disinfection with sodium hypochlorite is widely used for water treatment and is effective against most pathogens. Concentration rates of 1–5 mg/l with an exposure time of 2 h are recommended for plant pathogenic fungi and bacteria (Postma et al., 2008).

Wastewater from onsite greenhouse pack-house operations is one of the major sources of waste from many horticultural processing facilities. Every effort should be made to recycle washing water wherever possible, after the appropriate treatment has been carried out. Water that has been used for washing fruits and vegetables may require treatment before it can be discharged to land or into the public sewer system depending on the regulations of the region where processing is occurring. Such pack-house wastewater may require solids removal, screening, sedimentation or filtration and biological treatment before it is discharged in order to meet trade effluent consents from local water authorities (FERA, 2008). High-risk wastewater may need further treatment before it can be recirculated or discharged in order to eliminate pathogens; this may include treatments such as UV irradiation, heating, microfiltration, oxidation and use of appropriate disinfection chemicals.

Wastewater from hydroponic systems, which is typically nutrient rich but lower in sediment and organic matter, can be treated and reused in soilless production systems. Treatment methods include the use of sterilization techniques to control plant pathogens including slow sand filtration, heat, chemicals such as chlorination and hydrogen peroxide, UV, ozone, or a combination of any of these. Nutrient levels may then be adjusted and the water reused in the hydroponic system. The final or 'end-of-pipe' purification of discharged drain water' from greenhouse or hydroponic systems must still meet environmental standards in many production countries. Environmental awareness and legislation (e.g. the EU Water Framework Directive 2000/60/EC) restrict the emissions of nutrients and plant production products (PPPs), so that wastewater treatment is required before disposal (van Ruijven et al., 2015). Such treatments as heat, filtration, ozone and UV are commonly used for wastewater disinfection and pathogen removal; however, only ozone treatment has been shown to destroy both pathogens and PPPs in water. Ozone has been found to remove up to 98% of PPPs from greenhouse discharge water so that it meets environmental standards for disposal (van Ruijven et al., 2015). Where treatment is not feasible wastewater or nutrient solution leachate may simply drain into the soil beneath the crop; however, this option is becoming less frequently used due to environmental concerns. Wastewater or drainage may also be

discharged into the municipal sewage or wastewater system if available, provided resource consent is obtained where necessary for large volumes of disposal.

If wastewater is collected and contained onsite, it must be managed in such a way that will clean or treat the water to prevent environmental issues. This usually means removal of sediment or dissolved solids as well as nutrients. The main pollutants in horticultural wastewater are phosphates and nitrates. Phosphates may be removed by sedimentation since phosphates easily attach to sediments such as clay particles (Badgery-Parker, 2002). Nitrates, however, are highly soluble but will be readily taken up by plants or may go through the biological conversion to nitrogen gas which is released into the atmosphere. Irrigation of the wastewater on to green waste crops can be used to absorb excess nitrates, the vegetation produced is then mown or harvested and converted into green waste which may be composted. Solid compost is a more environmentally safe product to dispose of than wastewater high in nitrates and may be incorporated back into the cropping system with minimal concern for pathogen carryover if composted correctly.

Other systems used to clean and dispose of greenhouse and pack-house wastewater and runoff are passive systems which continually operate with a low cost and high degree of efficiency. The most common of these are wetland-based systems which slowly filter wastewater through beds of vegetation. As the wastewater moves through the wetland system, it travels through four different components: the first is a sediment trap which removes heavy sediment and large particles of organic matter from the water before entering the wetland system, this helps prevent sediment build-up and blockages further down the system. Sediment is periodically removed from the trap and this is where the majority of the phosphates are held attached to the sediment particles. After the sediment trap, a filter bed containing vegetation such as grass traps the finer sediments and a large proportion of the dissolved nitrates. In man-made wetland systems, this area is typically lined with plastic to prevent excessive leaching of nitrates into the water table. Grass growth absorbs much of the nitrates in this zone. The third stage of a wetland wastewater treatment system is the planted constructed or artificial wetland consisting of either soil-based or aggregate beds lined with plastic to prevent leaching. This wetland stage removes the remaining nutrients via uptake by wetland plant species. The final stage is the retention pond or tank, a deep water area holding the treated water ready for reuse or further treatment if required. Some plant pathogens may survive the wetland treatment system and may need further removal before the water can be irrigated back on to crops (Badgery-Parker, 2002). Wetland water treatment systems require careful consideration during the design and planning phase because to be effective in sediment and nutrient removal, the system must be sized correctly to allow for the correct retention time for water in the system. Retention time is the time taken from when the water enters the system to when it exits and is based on the nutrient load of the wastewater.

4.5.3 Disposal of and reduction in organic waste

Waste management in horticultural operations is based on the basic principles of prevent, reduce, reuse, recycle, recover and finally, when all other options have been considered, dispose (Drury, 2017). Preventing and reducing waste generation is a key objective which greatly improves the efficiency of any waste management plan. A large proportion of the waste generated from greenhouse crops postharvest is due to produce not meeting strict size, visual and quality guidelines specified by retailers such as supermarkets, which are claimed to be led by consumer preferences. Over 380,000 tonnes of vegetable waste is produced annually in the 'Campo de Dalias' zone (Almeria, Spain), one of the most intensive horticultural areas in the world, deriving from the diverse range of crops grown under plastic cover (Parra, 2008).

Prevention of such waste generation should be forefront in greenhouse production as disposal is not only costly to the grower, but results in less efficient use of fertilizers, water, fuel and all crop inputs into a final marketable product. Using cultivation practices which minimize quality losses due to cosmetic damage, under- or oversized produce and other factors is one aspect of waste minimization. However it has been suggested that retail markets revising regulations and standards on visual appearance for fruit and vegetables sold in supermarkets and educating consumers while promoting the consumption of produce with cosmetic defects would greatly reduce waste levels (Ghosh et al., 2016).

Waste may be prevented or reduced by strategic production planning so that crops are produced at the correct time and in suitable volumes for the market; this reduces the amount of unwanted produce due to a lack of demand. Some waste produce may be recycled in a number of ways – culled fruit and vegetables and some trimmings may be used as stock feed for pigs or other livestock or incorporated into vermicast (worm farm) systems. Higher-quality culls or reject produce may be suitable for processing and may be used for the production of other products.

Where solid waste is considered high risk due to the presence of serious plant pests and pathogens, recycling may not be an option and disposal is required. Disposal of solids can be via landfill; however, this may require the appropriate approval and licence from the relevant environmental agency (FERA, 2008) in some circumstances. Incineration may be permitted in some regions and certain types of organic horticultural waste such as wood, prunings and waste oil may be burnt to generate heat for greenhouse operations.

Solid waste plant material that is a risk factor for transmitting pests and pathogens thus must be treated and disposed of correctly. Treatment of solid waste which exposes it to elevated temperatures for several weeks generally minimizes the risk of pathogen carryover and is widely used to dispose of organic material from horticultural operations. The most common form of heat treatment is composting, which is an aerobic and exo-thermic process during which temperatures rise to at least 50°C, often higher, for several weeks. Composting not only creates a valu-able by-product, but also significantly re-duces the volume of waste; however, where this cannot be carried out onsite, bulky waste organic material must be transported to a suitable composting site or commercial operation which carries out this process.

4.5.4   Disposal of plastics

The disposal of plastics from horticultural operations has become of increasing con-cern over recent years as, unlike organic waste, these cannot be disposed of onsite or composted to reduce volume and create a viable by-product. Plastics used in green-houses incorporate a wide range of products from nursery plant punnets and flats, trays, planter bags and pots to thin plastic films, greenhouse floor coverings as well as green-house claddings and cloche coverings. Shade cloths, bird netting, insulation materials, plant supports, ties and tags are all typically composed of plastic materials. The first ob-jective with the use of horticultural plastics is to reduce or recycle these materials. Many plastic containers such as plant pots can be reused but require sorting and cleaning which in the past has been considered un-economical by many horticultural oper-ations. The issue with recycling used plant pots and containers is that even with man-ual scrubbing, commonly transmitted dis-eases resulting from pathogens could still survive and currently cleaning of plastic containers is often accomplished by home gardeners rather than large-scale commer-cial nurseries (Meng et al., 2015). While pot recycling schemes are a good starting point, they do not address the large volumes of plastic waste still produced by the green-house industry.

The wide range of plastic types used in greenhouses and nurseries can make sort-ing and recycling a non-viable economic op-tion. In the horticultural industry the most widely used plastic types are LDPE, HDPE,

polypropylene and closed-cell polystyrene. Those plastics which are recyclable can be ground up and the resulting granules or flakes sold for further plastics manufactur-ing. However, despite the possibility of re-cycling certain types of horticultural plastic, in the past, large quantities have ended up either in landfill or incinerated. Disposal of these plastics to landfill creates an issue due to the large volume of space required and the fact that these do not readily decom-pose and may remain intact for decades. Incineration of plastics poses issues with at-mospheric emissions which increase envir-onmental pollution and is often no longer permitted in many areas.

4.5.5   Disposal of pesticides and agrochemical containers

Pesticide and other agrochemical containers are a type of plastic that can be difficult to dispose of safely as these are considered to be toxic waste which can cause significant environmental impacts if residues enter waterways. It is usually impossible to remove all traces of toxic chemicals from pesticide containers and these should not be reused for storing and transporting fuel or any other substances (FAO, 2018). Often empty pesticide and other chemical containers are disposed of at landfill where residues can contaminate soil and be leached into water-ways. Under the International Code of Con-duct on the Distribution and Use of Pesticides, manufacturers and distributors of pesticides are expected to provide facil-ities that allow pesticide users to dispose of empty containers and pesticide-related waste safely (FAO, 2018). Manufacturers or re-sellers of pesticides may set up schemes for re-collection of empty containers and ar-range for the safe and correct disposal of these.

4.6   Occupational Health and Safety

While most horticultural activities take place in the field, enclosed structures such as greenhouses provide additional risks associated with environmental factors such as temperature, humidity, air quality, increased exposure to spray compounds and dust. One study reported that half of the greenhouse workers tested reported musculoskeletal pain (Coumbis and Anderson, n.d.), this was often back pain associated with greenhouse operations and lifting. The installation of waist-height benches was identified as being the single most important preventive measure that a greenhouse owner can implement to reduce the incidence of back-related disorders (Coumbis and Anderson, n.d.). Contact dermatitis is another occupational occurrence among greenhouse workers who may be exposed to a wide variety of plants as well as growing substrates, pesticides and other chemicals. High temperatures and humidity inside greenhouse structures can lead to heat exhaustion, respiration problems and fatigue, and workers may be required to carry out activities during the cooler times of day to prevent these issues. Repetitive greenhouse operations such as seed sowing, planting, harvesting and packing, filling containers with growing substrates, fertilizing, pruning and trimming may result in repetitive strain injury (RSI) in the arm, wrist and hand. These issues can be reduced by automating repetitive activities wherever possible, rotating jobs among staff, providing sufficient rest and break times, as well as comfortable seating, bench heights and suitable tools for the task.

References

AHDB (Agriculture and Horticulture Development Board, UK) (2020) Biosecurity in protected edibles. Available at: https://ahdb.org.uk/knowledge-library/biosecurity-in-protected-edibles (accessed 3 September 2020).

Badgery-Parker, J. (2002) Managing waste water from intensive horticulture: a wetland system. *Agnote DPI-381*, 2nd edn. New South Wales Agriculture, Australia. Available at: https://www.dpi.nsw.gov.au/__data/assets/pdf_file/0005/119372/horticulture-waste-water-wetland-system-eng.pdf (accessed 3 September 2020).

Coumbis, J. and Anderson, R.G. (n.d.) Assessment of occupational health of greenhouse workers.

Floriculture Research Report 13-04. Agricultural Experiment Station, University of Kentucky, College of Agriculture, Lexington, Kentucky. Available at: https://www.uky.edu/hort/sites/www.uky.edu.hort/files/documents/OccupationalHealth.pdf (accessed 3 September 2020).

Currey, C.J. (2017) The importance of sanitation. *Produce Grower Magazine*, December 2017. Available at: https://www.producegrower.com/article/the-importance-of-sanitation/ (accessed 3 September 2020).

Drury, S. (2017) Waste management. *HorticultureWeek*, 3 February 2017. Available at: https://www.hortweek.com/waste-management/products-kit/article/1422665 (accessed 3 September 2020).

FAO (Food and Agriculture Organization of the United Nations) (2018) Prevention and disposal of obsolete pesticides. Available at: http://www.fao.org/agriculture/crops/obsolete-pesticides/what-dealing/containers/en/ (accessed 3 September 2020).

FERA (The Food and Environment Research Agency) (2008) Code of practice for the management of agricultural and horticultural waste. Available at: http://webarchive.nationalarchives.gov.uk/20141203185052/http://www.fera.defra.gov.uk/plants/publications/documents/copManagementWaste.pdf (accessed 3 September 2020).

Ghosh, P.R., Fawcett, D., Sharma, S.B., Perera, D. and Poinern, G.E.J. (2016) Survey of food waste generated by Western Australian fruit and vegetable producers: options for minimization and utilisation. *Food and Public Health* 6(5), 115–122.

Larouche, R. (2012) Four food safety and handling considerations for greenhouse vegetables. *Greenhouse Grower*, 16 October 2012. Available at: https://www.greenhousegrower.com/crops/vegetables/four-food-safety-and-handling-considerations-for-greenhouse-vegetables/ (accessed 3 September 2020).

Meng, T., Klepacka, A.M., Florkowski, W. and Braman, K. (2015) Determinants of an environmental horticulture firm's recycle process in terms of type and quality: the case of Georgia. *Selected paper prepared for presentation at the Southern Agricultural Economics Association 2015 Annual Meeting*, Atlanta, Georgia, 31 January–3 February 2015. Available at: https://ideas.repec.org/p/ags/saea15/196761.html (accessed 3 September 2020).

Orlikowski, L.B., Treder, W., Ptaszek, M., Trzewik, A., Kowalczyk, W. and Lazecka, U. (2017) Necessity of disinfection water for crop irrigation. *Polish Academy of Sciences* no. IV/1/2017, 1387–1400.

Parra, S. (2008) Protected horticulture and the environment. An integral decision model for greenhouse waste management in Southeastern Spain. *Paper prepared for presentation at the 107th EAAE Seminar 'Modeling of Agricultural and Rural Development Policies'*, Seville, Spain, 29 January–1 February 2008. Available at: https://www.researchgate.net/publication/23508953_PROTECTED_HORTICULTURE_AND_ENVIRONMENT_AN_INTEGRAL_DECISION_MODEL_FOR_GREENHOUSE_WASTE_MANAGEMENT_IN_SOUTHEAST-ERN_SPAIN (accessed 3 September 2020).

Postma, J., van Os, E. and Bonants, J.M. (2008) Pathogen detection and management strategies in soilless plant growing systems. In: Raviv, M. and Lieth, J.H. (eds) *Soilless Culture: Theory and Practice*. Elsevier, Oxford, pp. 505–543.

Stanghellini, M.E., Rasmussen, S.L., Kim, D.H. and Rorabaugh, P.A. (1996) Efficacy of non-ionic surfactants in the control of zoospore spread of *Pythium aphanidermatum* in a recirculating hydroponic system. *Plant Disease* 80(4), 422–428.

Tesoriero, L., Jelinek, S. and Forsyth, L. (2010) On-farm hygiene and sanitation for greenhouse horticulture. *PrimeFact 1005*. Industry and Investment NSW, Australia. Available at: https://www.dpi.nsw.gov.au/__data/assets/pdf_file/0003/340284/On-farm-hygiene-and-sanitation-for-greenhouse-horticulture.pdf (accessed 3 September 2020).

Van Ruijven, J., van Os, E., Stijger, I., Beerling, E. and de Haan, C. (2015) Double use of water treatment in soilless growing systems: disinfection of recirculating solution and removal of plant protection products from discharge water. *Acta Horticulturae* 1170, 571–588.

5 Hydroponic Systems – Solution Culture

5.1 Introduction – Solution Culture Systems

Solution culture or 'hydroculture' systems are methods of crop production which do not employ the use of substrates to contain the root system and hold moisture between irrigations. 'Hydroponics' is the term correctly used to describe solution culture systems; however, in recent times hydroponics has also come to refer to substrate-based systems reliant on nutrient solution application for crop growth as well. There is a range of solution culture systems in both commercial and small-scale use for a diverse range of crops. These include NFT, DFT, float, raceway or raft systems, aeroponics and various modifications on these techniques. While not using a soilless substrate to fully contain the roots has advantages such as avoidance of handling and disposal of solid substrates, lower costs of growing media and less monitoring of moisture levels between irrigations, soilless culture can have disadvantages. The main concern is the requirement for a high degree of grower skill to manage, monitor and adjust the composition of the nutrient solution which continually bathes the root system. Root systems in solution culture must also be able to obtain sufficient oxygen, either as DO in the nutrient solution or by roots maintained above the flow of the liquid. Since liquid can only contain a small percentage of the oxygen that is contained in the air surrounding roots, oxygenation in some solution culture systems can become an issue, particularly under warmer growing conditions.

5.2 NFT – Nutrient Film Technique

NFT or the 'nutrient film technique' was the world's first method of crop production which did not use a solid rooting medium. Developed in the 1960s by Dr Allen Cooper at the Glasshouse Crops Research Institute in England, the system was initially met with much scepticism in the early stages until studies proved that the system was not only viable, but had considerable potential for a number of crops.

NFT is a system where a thin film of nutrient solution (2–3 mm depth) continually flows along the base of channels in which the root systems sit. Channels are often constructed of rigid PVC, specifically designed and manufactured for hydroponic crop production, or formed from thick plastic film which is folded up to form a triangular-shaped channel. In this system the objective is that part of the developing root system is within the nutrient flow, but many of the other roots are positioned up above this in the moist air, accessing oxygen without being submerged. In NFT systems, the nutrient solution is usually continually recirculated so that roots do not dry out; however, intermittent flow systems may also be used. A central nutrient reservoir holds the nutrient solution from where it is pumped up into growing channels, flows past the root systems and is piped back to the reservoir for monitoring and adjustment either manually or automatically with use of electronic dose and controller units (Fig. 5.1).

5.2.1 NFT crops

NFT systems have gained worldwide acceptance, mostly for the production of small framed, rapid-turnover crops such as lettuce, herbs, strawberries, green vegetables, fodder and micro greens, although longer-term plants such as tomatoes will grow in these

© L. Morgan 2021. *Hydroponics and Protected Cultivation* (L. Morgan)
DOI: 10.1079/9781789244830.0005

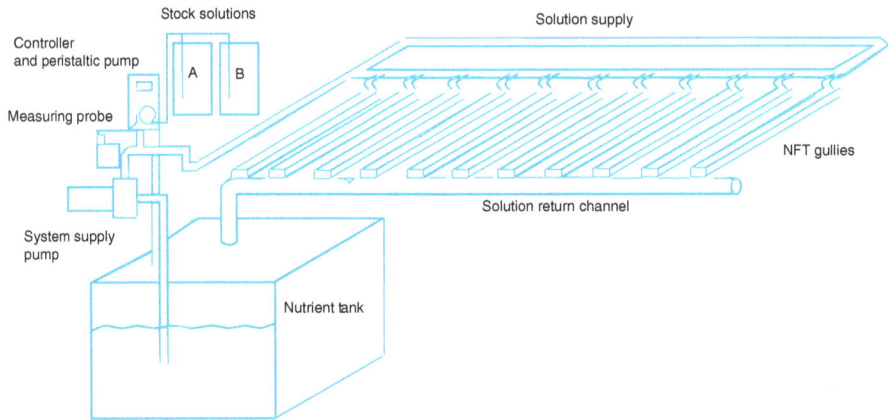

Fig. 5.1. Layout of an NFT system showing positioning of pump, irrigation pipes, controllers and nutrient stock solutions.

systems in larger channel sizes. NFT allows for the production of 'living greens' crops: those harvested and sold with roots intact and wrapped in plastic sleeves to prolong shelf-life postharvest. NFT also allows production of salad greens, lettuce and herbs without the risk of contamination from grit which can occur in soilless substrate systems and crops are usually sited on benches at a comfortable working height. Small plants such as lettuce and herbs may be grown in NFT by a number of different methods (Fig. 5.2). For lettuce head production and many herb crops, plants are introduced into the NFT system as seedlings at the correct density, grown to maturity and the entire plant harvested as required (Fig. 5.3). However, an increasingly common use for NFT is now in the production of mesclun or baby leaf salad greens where successional harvests are carried out while the plant remains in the system. Under these production systems, small pots or cubes containing a number of seedlings are placed into the NFT system and cut on a regular basis once leaves have obtained a sufficient size. For micro greens and some mesclun production, wide, open NFT channels are utilized which are lined with a porous material such as micro green matting or other thin sheets of material which does not overly impede the flow of the nutrient solution down the NFT channel. These mats not only hold moisture, but also

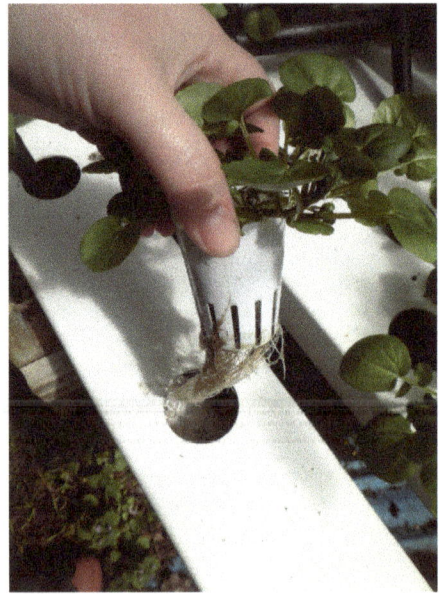

Fig. 5.2. Lettuce, herbs and other salad greens are grown in NFT using small pots to contain the plant and root system.

contain the seed in place as germination occurs. The solution flow in these types of NFT systems may be continual or intermittent and aims to avoid oversaturation of the young seedlings, which are grown at a high density, while at the same time providing

Fig. 5.3. NFT system growing a range of salad greens.

sufficient nutrients, oxygen and moisture for rapid growth. Micro greens and baby leaf products grown in this way are harvested within 2 weeks and the NFT channels are then cleaned and replanted immediately.

The main constraint with using NFT systems for all hydroponic crops is the size of the root system which may develop during the cropping cycle. NFT channels are characterized by a highly restricted root-zone volume, within which the shallow flow of nutrient solution must be maintained in order to maximize oxygenation levels. Longer-term crops such as tomatoes, cucumber, melon, capsicum, aubergine and others are often grown for a number of months and during this time develop the extensive root mass required for high rates of nutrient and water uptake that maintain plant growth and transpiration. Root systems, which increase in length and density over a longer period of time, slow the rate of nutrient solution flow down each channel and may result in stagnant areas and even overflowing of the channels. Increasing the slope of the growing channels as longer-term

crops develop can assist in maintaining flow rates and adjustable-slope systems can be designed for this purpose. Longer-term crops can be accommodated by utilizing larger sized, flat-based channels with an increased capacity for root system expansion as well as systems which increase the oxygenation of the nutrient solution, both within the nutrient reservoir and within the channels themselves. Crops such as cucumbers that have a high root-zone oxygen requirement require greater attention to DO levels as these can deplete along the length of NFT channels, particularly under warm growing conditions. Studies have found that the oxygen content of a flowing nutrient solution becomes more depleted in NFT cucumber systems than for other crops such as tomatoes and a marked loss of oxygen occurs between the inlet and outlet ends of growing channels (Gislerod and Kempton, 1983). Oxygen levels are typically lowest during the day and in the afternoon when solution temperatures and root respiration rates are highest and should be regularly monitored in NFT systems.

5.2.2   Types of NFT systems

Although NFT systems may seem more complex than substrate-based methods of hydroponics, they are extremely flexible with regard to size, design and equipment. NFT growing channels with a flat base profile can be purchased from hydroponic suppliers; however, over the decades many innovative materials have been used where these are not available. These include rainwater downpipes, large-diameter bamboo, channels formed from timber, metal or concrete and lined with plastic, and long rolls of plastic film itself folded into channels of various sizes. These types of basic NFT systems require manual checking, adjustment of the EC and pH of the recirculating nutrient solution and regular water-level adjustments.

Large-scale commercial NFT systems are more automated than small, backyard systems and make use of electronic and computer-controlled EC, pH adjustment and float valves to replace water used in the system. Some of the more advanced commercial NFT systems may also utilize DO and solution temperature measurements to monitor these root-zone variables. An NFT-based growing operation often has both seedling and finishing NFT benches through which the plants progress during production. Seedling systems have plants at a much higher density, which are moved to a wider spacing in the finishing system to maximize the growing area inside greenhouses. Other types of NFT systems include moveable channel systems where the distance between adjacent channels is increased as the plants grow in size. Channels may be completely removed from the system at harvest, cleaned, replanted and then replaced at the opposite end of the system to maintain a rapid crop throughput. Vertical NFT systems are also popular and maximize growing space for lower-light crops such as lettuce and some herbs and NFT systems can also be installed outdoors or under basic shade covers.

Constructing and running a successful NFT system are reliant on a few vital factors. These include slope, flow rate, nutrient reservoir size, channel volume and pump capacity. The channels need to be positioned on a slope so that the nutrient solution flows steadily down each at a suitable flow rate. Some systems allow the slope on the channels to be adjusted and increased as the root system develops and expands inside the channels, thus ensuring that nutrient ponding does not occur. NFT channel slope recommendations vary depending on the crop and system as slope and flow rate influence the oxygenation levels in the root zone. For tomato crops grown in NFT it has been found that a channel slope of 4% compared with 2% increased the DO content in the nutrient solution and also improved tomato yields (Lopez-Pozos et al., 2011). Nutrient solution flow rates also vary between systems with a flow rate of 1 l/min per channel often used. It has been found the fresh weight of NFT-grown crisphead lettuce was highest at flow rates of 1–1.5 l/min per channel as compared with 0.75 l/min per channel (Genuncio et al., 2012). Higher flow rates are implicated in greater

oxygenation levels in the root zone which is likely to account for increased fresh weight at harvest.

The dimensions of the NFT channels are dependent on the type of plants being grown. Small lettuce and herbs may be grown in small NFT channels 10–15 cm wide and 5–8 cm high made from UV-stabilized, Food and Drug Administration (FDA)-approved HDPE. Larger plants such as cucumbers and tomatoes require a wider and taller channel to accommodate the considerably larger root system of these plants. NFT channels may be moulded as a single piece in a range of different profiles and designs. Some channels have a number of grooves formed on the base which assist in spreading and widening the nutrient flow. Other designs which may be used in outdoor systems have a sloped profile on the top of each channel which helps shed rainfall from the channel surface and assists with keeping the base of the plants dry. Larger channels may have removable clip-on lids as an optional feature, these greatly improve accessibility to the inner channel surfaces for cleaning and sanitation between crops and also allow for viewing of the developing root system and nutrient flow depth.

Reservoir size is also important and many smaller NFT systems are set up with undersized nutrient tanks (Fig. 5.4). A buffer capacity volume of at least 50% is advisable when selecting a reservoir. This means that when all the channels are carrying nutrient solution at the desired depth, sufficient nutrient remains in the reservoir to

Fig. 5.4. Nutrient reservoirs need to be correctly sized for NFT systems.

half fill it or the reservoir is large enough to hold twice the amount of nutrient contained in the channels. This is considered a minimum and many systems have considerably larger tanks than this as a greater volume slows the rate of change in EC, pH and temperature in the nutrient solution held in the reservoir. Pumps for NFT are another piece of equipment which are often misjudged and may fail to provide the capacity required. Not only is it important to consider the capacity of the pump in terms of litres per hour but also the head height, which is the pump's ability to move nutrient solution up to a certain height (i.e. the top of the NFT system). Most NFT systems use centrifugal pumps which are capable of producing a constant flow of nutrient solution. These pumps will either pump a small volume of nutrient solution to a high head, or conversely a large volume of nutrient to a low head height. In between the two extremes the characteristics of the pump will vary and 'pump performance' curves should be used to select the pump with the correct head height versus flow rate characteristics for a particular NFT system.

5.2.3 Nutrient solution management in NFT

NFT is a system that continually recirculates the nutrient solution, thus it is more prone to nutrient imbalances: build-up of unwanted elements and changes in nutrient ratios over time as compared with 'open' substrate-based systems. For this reason, a high-quality, low-mineral water supply is required which does not contain levels of unwanted elements that may accumulate over time (such as sodium). Where plant-usable elements such as calcium, magnesium and some trace elements are present in the water supply, these can be adjusted for with a customized nutrient solution that takes these into account. A build-up of unwanted minerals or a significant change in the nutrient balance over time requires the NFT solution to be regularly changed, either partially or completely, to maintain good growth rates. NFT systems also require regular monitoring

and adjustment of pH and EC as changes in these can occur rapidly with some crops. For smaller systems once-daily checking and adjustment is usually sufficient for manually monitored systems. For systems with automated controllers for EC and pH adjustment, the system still requires occasional manual checks of these variables as a backup to equipment failures and errors.

NFT systems, as with all methods of hydroponics, are not without their potential issues. In the early days of commercial adoption of the technique there was concern that such a recirculating system would lead to the rapid spread and infection of root rot pathogens. Potentially one diseased plant could result in the loss of an entire crop as the pathogen was spread via the nutrient solution to all plants in the system. While disease outbreaks do occasionally occur in NFT, as they can in all hydroponic systems, careful management and following the principles of NFT with a high rate of oxygenation, minimal flow depth and solution temperature control all assist with minimizing these issues. An area of increasing research and understanding with NFT, particularly with the relatively new development of organic NFT systems, is the role that microbial populations, present in the root system, growing channels and nutrient solutions, play in disease prevention and crop health. Beneficial microbes develop in NFT systems just as they do in soil and soilless substrates and growers may choose to use methods to boost these such as slow sand filtration or the use of inoculant products designed specifically for hydroponics.

Nutrient management is also critical to success in NFT as it is possible for a rapidly growing crop to strip out an essential element from the recirculating system (often nitrogen or potassium) faster than it is being replenished via stock solution additions (Fig. 5.5). Use of nutrient solution analysis is a common practice with commercial growers which allows rapid adjustment of the recirculating solution as required so that imbalances are prevented. NFT nutrient formulations usually require adjustment for different seasons, crops and

Fig. 5.5. Rapidly growing crops can strip essential elements from a recirculating nutrient solution faster than they are replenished.

over time as plant uptake of the different nutrient elements is rarely consistent. Growth may be slowed when NFT solutions start to become out of balance and may go unnoticed until a severe and visible deficiency becomes obvious, so being proactive with solution monitoring and adjustment for larger growers is also recommended. Smaller growers who do not want to go to the expense of solution analysis testing can simply carry out regular partial or complete solution changes to ensure all nutrient elements remain in balance and also start with a good, well-matched nutrient formulation or product.

NFT systems may incorporate heating or cooling of the nutrient solution as is required for year-round cropping and allow a very 'clean' system to be run, with sterilization of equipment between crops if necessary. The main disadvantage of NFT is that power outages which halt the continual flow of nutrients can result in crop death within a short period of time, particularly under warm growing conditions.

5.3 Deep Water Culture/Deep Flow Technique – Float, Raft or Pond Systems

DFT or the 'deep flow technique' is similar to NFT in that a continual flow of nutrients is supplied to the root systems which are suspended in deep channels or beds (Fig. 5.6). In DFT, unlike the thin film of nutrients maintained in NFT, the solution flow is deeper and often relies on the introduction of oxygen along the entire length of each growing channel or trough so that oxygenation rates in the root zone are continually kept high enough for root growth. As with NFT, the nutrient solution is recirculated with EC, pH and temperature adjusted at the main nutrient reservoir.

Other types of deep water culture (DWC) systems consist of large tanks, narrow raceways or ponds of nutrients, upon which the developing plants float, supported by lightweight materials such as polystyrene (Fig. 5.7). Seedlings are usually raised in a separate nursery area, sown either into blocks of

Fig. 5.6. DFT system channels.

Fig. 5.7. Float system growing a mixed crop range.

inert propagation medium such as rock-wool, oasis or other foam material or into small tubes or lattice pots of substrate. Once the seedlings have reached an optimal size, they are inserted into holes cut into the floating rafts and set adrift on the nutrient pond. Roots develop down into the nutrient solution, which supplies not only mineral elements, but also DO for root respiration. The top of the plant develops supported only by the raft, which restricts crops

to those which are self-supporting in systems where the rafts float continually on the pool of nutrient solution. Similar systems have been designed for larger plants such as tomatoes, cucumbers and melons where the crops are stationary and supported by overhead wires. These types of systems for longer-term plants usually incorporate a solution which is flowing past the roots – otherwise known as DFT. The continual flow of the solution in DFT systems prevents stagnation and delivers fresh supplies of DO to the root system.

These types of floating hydroponic systems have two distinct advantages for the production of small framed crops: the nutrient pool is a virtually frictionless conveyor belt for planting and harvesting from the movable floats; and the plants are spread in a horizontal plane so that interception of sunlight by each plant is maximized. Young seedlings are transplanted into holes in floats in staggered rows at a high density, cutting or harvesting of the plants can be carried out at regular intervals and the floats replanted weekly (Fig. 5.8). As a crop of several floats is harvested from one end of the production tank or pool, new floats with transplants are introduced at the other end. The floats (polystyrene sheets) which support the plants are most efficient when they are at least 5 cm thick and crops

produced in nutrient pools with an average depth of 7–8 cm perform less consistently than those grown at greater depths (12–23 cm deep) (Morgan, 2012) (Fig. 5.9).

Float systems as a commercial option for plant production are not only used for the production of vegetable crops such as lettuce, herbs and other greens, but as a system for raising field transplants. These systems allow young seedlings to be evenly watered while receiving a well-balanced nutrient solution which can be adjusted in composition and strength during the critical stages of development before seedlings are transplanted out in the field. Under greenhouse production, large-scale float systems maximize growing area as the need for pathways is reduced and automation can be adopted where required.

The 'speedling float system' is an example of a DWC production method for young

Fig. 5.8. Lettuce production in a shallow DFT system.

Fig. 5.9. Greenhouse float system with seedlings.

plants and seedlings. The speedling foam trays contain a large number of small cells in a range of configurations for different crop species. A commonly used seedling float tray is 33 cm wide, 66 cm long, 4.4 cm deep and contains 336 cells (Mattson, 2016). These floating cell trays are filled with a hydroponic substrate, often perlite mixed with peat or coconut fibre, and one seed per cell is sown into each. After 2–3 days' germination, the speedling trays are floated on a nutrient pond until crop completion, which is typically less than 14 days. The speedling float system was initially used to grow transplants which were later established outdoors; however, it has been adapted for the production of a range of baby leaf species such as rocket, kale, lettuce, Asian greens and spinach. Yields for the speedling float system have been reported to be in the range of 3700–4600 g/ m^2 for rocket 'Astro', kale 'Red Russian' and lettuce 'Outredgeous' when germinated for 3 days followed by 10 days in a float system (Mattson, 2016).

5.3.1 Management of DWC and DFT systems

While float or DFT systems may appear to be a basic and easy-to-manage system of growing a wide range of crops, there are a number of principles of solution culture which must be adhered to. First, plant roots, whether in solution culture or substrate, require sufficient oxygen in the root zone. Roots use oxygen for respiration, which is either taken up directly from the air as oxygen (O_2) or as DO from moisture surrounding the root surface. Without this vital supply of root oxygenation, water and nutrient uptake is restricted, roots die back and opportunist pathogens such as *Pythium* may occur. While the air-filled pores in a growing substrate contain atmospheric oxygen at high levels (air is 21% O_2), depending on temperature nutrient solution can only maintain 6–13 ppm of O_2 at saturation, thus solution culture systems need adequate oxygenation for plant growth. Oxygenation in float or DFT systems can be provided via air pumps, injector systems or circulation of the nutrient solution whereby the solution cascades back into the pond or a reservoir, thus introducing air bubbles and replenishing DO levels. Systems which generate 'microbubbles', gas bubbles with a mean diameter of 50 µm or less in water, have been investigated as a way of increasing oxygenation in DWC, aquaponics and other solution culture methods. Microbubbles reside in the nutrient solution for a long period of time (Zheng et al., 2007) compared with the more easily formed macrobubbles which quickly rise to the surface of the water and burst. Thus, microbubbles have the potential to increase oxygenation in DFT systems and are an efficient method of delivering dissolved gas into a solution. Park and Kurata (2009) found that the application of microbubbles in a DFT nutrient solution increased fresh and dry weights of lettuce, which were 2.1 and 1.7 times larger respectively than with untreated nutrient solutions. This was partially attributed to the fact that microbubbles adhered to the root system which may have stimulated root growth by supplying oxygen directly to the surface of the root (Park and Kurata, 2009).

The second principle of float systems or DFT is nutrient management – the large volume of nutrient solution in the system acts as a buffer to slow changes in temperature, EC, pH and elemental levels; however, these still require monitoring and adjustment. As with any system which continually recirculates the nutrient solution, the ratio of elements can become unbalanced over time as the crop removes more of some nutrients than others and accumulation of unwanted salts such as sodium can occur (Fig. 5.10). Regular nutrient solution monitoring and adjustment is carried out in larger, commercial systems; however, for smaller growers, a float system can be managed by an occasional partial or complete replacement of the nutrient solution. EC and pH change slowly in the large volume of nutrient and allow for less frequent monitoring and adjustment for smaller systems. These systems may be utilized by small growers in situations where electricity is unreliable or not available to continually power the pumps and timers required in other types of hydroponic systems. In situations where basic raft systems have been set up without access to electricity or other forms of power, hand aeration by agitation or whisking of the nutrient solution may be applied in small systems to increase DO levels.

Management of the nutrient solution may, in some environments, include temperature

Fig. 5.10. Nutrient deficiency occurring in a fast-growing DFT brassica crop.

control. Under warm, tropical conditions, chilling of the nutrient solution pond is carried out to grow crops such as lettuce which would otherwise struggle to develop and yield well due to excessive heat. In cooler climates, the nutrient pool can be heated to improve nutrient uptake and growth rates. Deeper pools offer more buffering capacity to changes in solution variables, thus are likely to be more productive.

Float systems, as with other solution culture methods, have a risk of pathogens such as *Pythium*, which produces free-swimming zoospores, being transported around recirculating systems via the nutrient solution and can potentially infect a large number of plants. Ensuring temperatures are within range, EC levels optimal and sufficient oxygenation is provided for root health assists with prevention of disease outbreaks. If root disease is a concern, commercial growers may choose to construct smaller individual ponds which can be used to isolate a disease outbreak and minimize plant losses. Other approaches for disease prevention are similar to those used in all types of hydroponic systems where the nutrient solution is recirculated past many plants. These include use of microbial inoculants which can assist to suppress disease pathogens and use of nutrient sterilization such as ozone, heat, slow sand filtration or UV treatment.

5.4 Aeroponics

This is a method where the plant roots are suspended in the air with a fine mist of nutrient solution applied either continually or intermittently over the root surface. Intermittent misting of the root system has an advantage in terms of saving in running costs since pumps are only on for a short period of time while the roots are still contained within the nutrient-, moisture- and oxygen-rich environment between misting. A misting cycle of 1–2 min of misting followed by 5 min off may be used to ensure the root system does not dry out under most conditions. Continual misting may be preferable for larger plants with a higher evaporative transpiration rate to ensure moisture

is always sufficient for root uptake. Aeroponic systems may be further divided into those which have a separate nutrient reservoir and that pump the nutrient solution into the root chamber and those with a combined chamber and nutrient reservoir. Simpler aeroponic systems spray the nutrient solution up from the reservoir in the base of the root chamber where it falls back down after misting the root system. Larger, commercial aeroponic systems return the nutrient solution after misting to a separate nutrient reservoir where EC and pH can be corrected as required.

Aeroponics offers some advantages for crop production including high oxygenation in the root zone, potential for root-zone cooling and elimination of stagnation problems. However, the major drawback is the high energy cost of running the system, as large, high-pressure pumping systems are required, and a great deal of capital can be tied up in root chambers and equipment. The ideal droplet size for most plant species grown in aeroponics is 20–100 µm. Within this range, the smaller droplets saturate the air, maintaining humidity levels within the chamber; the large droplets, 30–100 µm, make the most contact with the roots; while any droplets over 100 µm tend to fall out of the air before contacting any roots. While most plants perform well in aeroponic systems, the capital expense and the maintenance required in running these systems mean this system of production has not been widely adapted for commercial crop production. As a modification on aeroponics, 'fogponic' systems have also been developed although these are largely used by small hobby growers rather than in commercial production. Fogponics, also known as 'mistponics', is a form of aeroponics where roots suspended in a chamber are supplied water and nutrients via extremely fine droplet sizes (5–10 µm) or 'fog' created by electric foggers.

While aeroponics is a capital-intensive method of crop production, it has been identified as a productive system for growing high-value medicinal crops, allowing roots to be harvested from accessible spray chambers while the plants remain actively growing (Hayden, 2006). High yields of plants that produce medicinal and phytop-harmaceutical compounds have been found using 'A frame' aeroponic systems which increase yield per unit area of greenhouse by maximizing vertical space (Hayden, 2006).

5.5   Vertical Systems

Vertical hydroponic systems aim to increase the number of plants per unit of floor area using a range of designs. These include the standard 'tube' systems, which are the most basic, consisting of a straight column into which plants are spaced at uniform intervals. As advancement on the basic tube, there are 'stack' systems which involve individual planting units stacked on top of one another, creating planting spaces or pockets for each plant. There are vertical systems which do not rely on circular 'column'-type structures but consist of tiers of growing channels, beds or chambers which can be constructed from the floor to almost the top of the growing area. Vertical or tiered tray systems create 'multiple environments' within the levels or layers of the growing area that need to be carefully managed and planned for. Plants growing at the top of a vertical system receive the highest light (assuming only an overhead light source is available); there is then a gradient of light from the top to the base of the vertical system along with shading from those plants on the upper levels. In basic systems, this shading and light fall-off effect is often what causes problems in both naturally lit and vertical systems with supplementary lighting. So, for this reason, successful commercial vertical systems are more common in climates with naturally high light levels which allow acceptable production on the lower, shadier levels of the system. Vertical soilless systems are often used in indoor or warehouse environments where each tier or shelf is lit individually with only artificial lighting. This allows maximum utilization of the vertical space within buildings, although production is often limited to small framed crops such as potted plants, lettuce, herbs, micro greens, strawberries and salad greens. Artificial lighting for vertical systems may

be supplied via compact fluorescent tubes, MH or HPS lamps, or more modern arrays of highly energy-efficient LEDs.

The main benefit of a successful vertical system is the number of plants that can be grown per unit area of floor space, which is many times more than can be supported in single-plane cropping (Fig. 5.11). Nevertheless, because of the much higher density, other factors such as the requirements for sufficient light and a greater degree of air movement and ventilation must be taken into account. The design of the vertical structure that holds the plants is important as it determines factors such as the root volume available to each individual plant, how the nutrient solution is delivered and the flow passage of moisture down the system, the collection system at the base of the stack and how air may move under and around each plant for humidity control. System design can also affect light interception by plants and the amount of space each has for upward growth and development. Many vertical systems rely on the use of a high-quality growing medium to support the plant and provide a reserve of moisture; however, solution culture and aeroponic vertical systems also exist.

5.6   Aquaponics

Aquaponics is a system of crop production that combines fish farming (aquaculture) with soilless crop production (hydroponics). In aquaponic systems the fish consume food and excrete waste primarily in the form of ammonia. Bacteria convert the ammonia to nitrite and then to nitrate (Khater and Ali, 2015) which is used for plant nutrition. Aquaponics is a system which is often considered sustainable due to the ability to maintain water quality, minimize consumption of fresh water and provide a dual crop of both fish and vegetables (Lennard and Leonard, 2005; Graber and Junge, 2009); however, it requires a reasonable degree of technical skill and experience to manage both interconnected systems efficiently.

Aquaponic systems may consist of a number of different designs of varying levels of

Fig. 5.11. Vertical systems increase the number of plants which can be grown per unit area of floor space.

technology. In the simplest systems, fish are produced in large tanks or ponds, while crops are grown suspended on floating rafts or platforms with their roots growing in the nutrient-rich fish production water. In more complex systems, the wastewater from the fish production operation is removed, solids are collected, and the remaining water transferred to tanks where microbiological processes carry out mineralization of the organic matter into plant-usage nutrients. The treated aquaponic solution is then incorporated into a hydroponic system to grow crops where additional nutrients may be added if required to obtain a balanced solution for growth. The design of an aquaponic system is an important consideration for the correct functioning, waste removal and eventual growth and yields of crops. Not only do the culture of fish and plant species need to be planned for, but the functioning

of the microbial populations within the system which carry out nutrient conversion must be considered as well. One study reported that using NFT channels as the crop production component of an aquaponic system was less efficient in terms of lettuce crop biomass gain than floating hydroponic systems or gravel beds (Lennard and Leonard, 2006; Maucieri et al., 2018). NFT treatment of the fish wastewater was 20% less efficient in terms of nitrate removal than either floating hydroponic subsystems or gravel beds with the gravel acting as a substrate for nitrifying bacteria (Lennard and Leonard, 2006). Where NFT channels are to be used as the crop production side to an aquaponic operation, the length and water flow rate down each channel are another important consideration. The optimal flow rates for NFT systems in aquaponics have been found to be 1.5 l/min for a 2–3 m length of gully and 2 l/min for a 4 m length of gully (Khater and Ali, 2015). Oxygenation of the aquaponic nutrient solution is an important determinant of crop yields as the system has a higher biological oxygen demand (BOD) than non-aquaponic nutrient solutions due to organic loading and larger microbial populations which compete with plant roots for oxygen (Suhl et al., 2019).

An alternative system, which is often termed 'total waste aquaponics' or 'complete aquaponics', is where all of the wastewater containing the solids is fed directly on to large, open media beds, often gravel, containing the plants and beneficial bacteria which carry out the waste conversion. The total waste system with large media-filled beds is often used for longer-term crops. In all aquaponic systems a compromise needs to be achieved between the needs of the crop and those of the fish species, mainly in terms of the nutrient strength and pH. This system relies largely on the health and functioning of a group of bacteria responsible for converting the fish waste into usable plant nutrients.

5.7 Organic Solution Culture

In recent years, the possibility of 'organic hydroponics' or organic soilless production has become a topic of much debate. In some countries, such as the USA, organic soilless systems, even those using NFT or solution culture, are certifiable as organic despite not making use of soil. However, in much of the rest of the world, soilless systems are currently not certifiable as organic due to the absence of soil which is considered to be the 'cornerstone' of organic production. Many of the organic nutrient solutions in use in solution culture are based on seaweed, fish or manure concentrates, allowable mineral fertilizers, processed vermicast or plant extracts, and other natural materials. Where organically allowable solution culture systems are in use, most growers make use of commercially bottled organic nutrient products which have been processed via a number of methods to create a nutrient solution that can be used in NFT and similar systems. Using organic fertilizers to provide a complete and balanced nutrient solution for soilless production is a difficult and technically challenging process and these systems are far more prone to problems with nutrient deficiencies, particularly in high-nutrient-demanding crops such as tomatoes. Aquaponics, which uses organic waste generated during fish farming processes to provide nutrients for crop growth via bacterial mineralization, is sometimes considered a form of organic hydroponics when no additional fertilizer amendments are added.

5.8 Hydroponic Fodder Systems

Hydroponic fodder systems producing fresh wheat or barley grass for animal production are essentially the same as those used for many other soilless crops such as lettuce, herbs and wheat grass for juicing. NFT is the preferred method of fodder production, although other systems using overhead misters which deliver water and nutrients to the trays of sprouting grains exist. Large shallow channels contain the pre-soaked grain while hydroponic nutrient solution is flowed through these channels at a shallow depth providing mineral elements and moisture. Within 24 h the barley or wheat grains begin to sprout and rapidly send up a fresh green shoot, and at the high density at which the grain is sown, a dense mat of fodder is ready to harvest within a week. Hydroponic fodder systems have been reported to use only 2–3% of the water that would be used under field conditions to produce the same amount of fodder, with barley being considered the best choice for hydroponic green fodder production (Al-Karaki and Al-Hashimi, 2012).

Most hydroponic fodder systems for small-scale feed production for a few animals do not need a lot of space. The growing trays or channels are usually stacked on to a tiered system of shelves from floor to roof allowing sufficient space for the sprouts to grow upwards and for air movement over the trays. Hydroponic fodder grown on a large scale is typically produced in specifically designed, climate-controlled sheds rather than greenhouses as high light levels are not required. However, many smaller-scale fodder producers modify existing buildings, use shaded greenhouses or propagation areas provided they are warm, rodent free and can be kept clean and dry.

5.9 Nutrient Chilling Systems

Root-zone chilling of the hydroponic nutrient solution is a technique used commercially by many growers in warm or tropical climates, most often with cool-season crops such as butterhead lettuce, herbs and other vegetables. In Singapore, NFT, aeroponic and deep flow culture systems are utilized with extensive nutrient chilling to grow butterhead and romaine lettuce, crops which otherwise do not grow or yield well at ambient air temperatures (He *et al.*, 2001). Chilling the nutrient solution down to as low as 16–18°C allows the cool-season vegetables to crop well at ambient air temperatures which are often well above optimal for these crops (28–36°C). Without nutrient chilling, the root zone usually warms to the temperature of the air and this causes numerous growth problems including slow growth, lack of heart formation, bolting, tipburn and low marketable yields. Chilling the nutrient tricks the physiology of the plant into growing in air temperatures which would otherwise not be economical.

5.10 Automation for Solution Culture Systems

Automatic control, monitoring and dosing equipment may be incorporated into recirculating solution culture systems such as NFT, DFT, float systems and aeroponics. Smaller systems often rely on growers manually checking pH, EC, DO and temperature at the nutrient reservoir and making adjustments as required, although these types of systems often incorporate automatic water top-ups. Automation of solution culture and other recirculating systems allows more frequent adjustment of the nutrient solution and this may be linked back to a computer program which tracks changes in EC, pH and solution temperature as well as stock solution and acid volumes. Automatic

dosing equipment requires good technical support for installation and backup as well as regular checks and maintenance to ensure correct readings are being obtained from sensor equipment; it also requires a reliable electricity supply. Automatic controllers for solution culture systems usually consist of solenoid valves or peristaltic pumps for nutrient and acid stock solutions which respond to sensors positioned in the nutrient flow from the main reservoir. While automation can provide a high degree of control over solution EC and pH, manual checks should still be carried out to confirm the accuracy of the dosing system.

References

Al-Karaki, G. and Al-Hashimi, M. (2012) Green fodder production and water use efficiency of some forage crops under hydroponic conditions. *ISRN Agronomy* 2012, 924672.

Hayden, A.L. (2006) Aeroponic and hydroponic systems for medicinal herb, rhizome and root crops. *HortScience* 41(3), 536–538.

He, J., Lee, S.K. and Dodd, I.C. (2001) Limitations to photosynthesis of lettuce growth under tropical conditions: alleviation by root zone cooling. *Journal of Experimental Botany* 52(359), 1323–1330.

Genuncio, G.C., Gomes, M., Farrari, A.C., Majerowicz, N. and Zonta, E. (2012) Hydroponic lettuce production in different concentrations and flow rates of nutrient solution. *Horticultura Brasileria* 30(3). https://doi.org/10.1590/S0102-05362012000300028

Gislerod, H.R. and Kempton, R.J. (1983) The oxygen content of flowing nutrient solutions used for cucumber and tomato culture. *Scientia Horticulturae* 20(1), 23–22.

Graber, A. and Junge, R. (2009) Aquaponic systems: nutrient recycling from fish wastewater by vegetable production. *Desalination* 246(1–3), 147–156.

Khater, E.S.G. and Ali, S.A. (2015) Effect of flow rate and length of gully on lettuce plants in aquaponic and hydroponic systems. *Journal of Aquaculture Research and Development* 6, 318. https://doi.org/10.4172/2155-9546.1000318

Lennard, W.A. and Leonard, B.V. (2005) A comparison of reciprocating flow versus constant flow in an integrated gravel bed, aquaponic test system. *Aquaculture International* 12, 539–553.

Lennard, W.A. and Leonard, B.V. (2006) A comparison of three different hydroponic sub-systems (gravel bed, floating and nutrient film technique) in an aquaponic test system. *Aquaculture International* 14, 539–550.

Lopez-Pozos, R., Martinez-Gutierrez, A. and Perez-Pacheco, R. (2011) The effects of slope and channel nutrient gap number on the yield of tomato crops by a nutrient film technique system under a warm climate. *HortScience* 46(5), 727–729.

Mattson, N. (2016) Growing hydroponic leafy greens. *Greenhouse Product News*, October 2016. Available at: https://gpnmag.com/article/growing-hydroponic-leafy-greens/ (accessed 3 September 2020).

Maucieri, C., Nicoletto, C., Junge, R., Schmautz, Z., Sambo, P. and Borin, M. (2018) Hydroponic systems and water management in aquaponics: a review. *Italian Journal of Agronomy* 13, 1012.

Morgan, L. (2012) *Hydroponic Salad Crop Production*. Suntec New Zealand Ltd, Tokomaru, New Zealand.

Park, J.S. and Kurata, K. (2009) Application of microbubbles to hydroponics solution promotes lettuce growth. *HortTechnology* 19(1), 212–215.

Suhl, J., Oppedijk, B., Baganz, D., Kloas, W., Schmidt, U. and van Duijn, B. (2019) Oxygen consumption in recirculating nutrient film technique in aquaponics. *Scientia Horticulturae* 255, 281–291.

Zheng, Y., Wang, L. and Dixon, M. (2007) An upper limit for elevated root zone dissolved oxygen concentration for tomato. *Scientia Horticulturae* 113, 162–165.

6 Substrate-based Hydroponic Systems

6.1 Introduction

The main purpose of the substrate in hydroponic systems is to provide plant support, allowing roots to grow throughout the medium absorbing water and nutrients from the nutrient solution. The nutrient solution may be applied either to the surface of the substrate or through the base of the growing container via sub-irrigation or capillary action. Early substrate systems ran on the drain-to-waste principle, which meant the systems were simpler and less sensitive to nutrient solution composition or water salinity issues. With the concerns over water and fertilizer usage and environmental/disposal issues of draining nutrient solution to waste, collection and recirculation of the solution has become standard practice. This type of system consists of either large, shallow, media-filled beds or individual pots, slabs or containers, positioned in or above collection channels (Fig. 6.1). Drainage percentages are at least 10% or more where the solution is recycled after each irrigation and sterilization may be provided through the use of UV, filters, heat or slow sand filtration to prevent the accumulation and spread of root pathogens.

6.2 Properties of Hydroponic Substrates

Since the volume of soilless substrate is highly restricted as compared with soil cropping, the medium must be of a suitable physical structure to not only hold sufficient moisture between nutrient applications, but also drain freely, thus supplying the root system with essential oxygen, the level of which is dependent on the porosity of the substrate. Both air-filled porosity and water-holding capacity are dependent on the physical properties of a substrate, which in turn are affected by the shape and size of the constituent particles. Other important properties of substrates under hydroponic production are that they are free of weed seeds, salinity, contamination from pests and diseases, of a suitable pH for crop production and do not unduly influence the composition of the nutrient solution which is applied for plant growth. Soilless substrates also need to be biologically stable and not rapidly decompose, break down or fracture during the cropping period, particularly since the medium may be used for many successive crops. Inert substrates used in soilless culture typically have a low cation exchange capacity (CEC), thus do not play a role in plant nutrition which is controlled solely with the use of a well-formulated and balanced nutrient solution. Worldwide, a vast array of soilless media has been tested, developed, blended and manufactured for use under hydroponic production. The type of soilless substrate selected often depends on the materials available locally, as shipping bulky media long distances is costly. However, many substrates such as rockwool, perlite and coconut fibre (Fig. 6.2) gained acceptance rapidly and are now shipped worldwide to high-technology greenhouse and hydroponic growers in many different countries.

6.3 Open and Closed Soilless Systems

Soilless system nutrient management can be divided into two categories: those that allow the nutrient solution to drain to waste once it has passed through the root system, termed

© L. Morgan 2021. *Hydroponics and Protected Cultivation* (L. Morgan)
DOI: 10.1079/9781789244830.0006

Fig. 6.1. Substrate system showing collection channels underneath troughs.

Fig. 6.2. Coconut fibre grow slab.

'open systems'; and those that recirculate the nutrient solution, termed 'closed systems'. Closed or recirculating systems are used in solution culture such as with NFT, DFT, float/raft systems, ebb and flow (flood and drain) and aeroponics, whereas hydroponic substrate systems may be open or closed depending on how the nutrient is managed. Open systems are more commonly used where the water may contain excess salts which prevent extended use or recirculating of the nutrient solution that

can cause these to accumulate or where disease spread may become an issue. In open or 'drain-to-waste' systems, growers aim to minimize the volume of leachate and the associated loss of water and nutrients while at the same time providing sufficient nutrient solution to the crop. In closed systems, the nutrient drainage is collected from the base of the substrate and channelled back to a central reservoir where it often receives treatment such as filtration, sterilization, EC and pH adjustment before being reintroduced to the irrigation system (Fig. 6.3). Recirculation of the nutrient solution in closed systems is more cost-effective and poses less environmental risk; however, recirculation can spread certain root diseases and treatment with UV, ozone, heat or slow sand filtration is often used to control these issues. Recirculating systems also require a higher degree of control over the composition and balance of nutrient ions in the irrigation solution to ensure deficiencies do not occur over time and this is achieved with use of nutrient solution analysis and adjustment. Recirculating systems

Fig. 6.3. A closed cucumber production system where the nutrient solution is collected and recirculated.

require a high-quality water source which is low in excessive and unwanted salts such as sodium as these will accumulate through the recycling process, can lower yields and cause crop damage. Source water with a high content of unwanted salts can be treated with reverse osmosis (RO) to remove these contaminants and create pure water as a base for the hydroponic nutrient solution. If the water supply is of medium quality it is advisable to use 'semi-closed' systems where flushing out or periodically discarding the concentrated recirculating solution is carried out (Castilla, 2013).

6.4 Common Hydroponic Substrates

Soilless substrates utilized in hydroponic production fall into two main categories: those which are purely organic materials and those which are mineral substrates, the latter may include naturally occurring media such as gravel and sand, or artificially manufactured substrates such as perlite and rockwool. Organic materials used in hydroponic

systems may be by-products of other industries such as sawdust, coconut fibre (coir), rice hulls, sugarcane bagasse, biochar, or other horticultural substrates such as peat and ground bark. Naturally occurring minerals such as scoria, pumice, sands and gravels are heavier, higher-density growing media, while manufactured substrates include artificially transformed minerals used to create mineral wool/rockwool, perlite, vermiculite and light expanded clay aggregates (LECA) or polyurethane foam, formed into a range of growing cubes and slabs. Under worldwide commercial production, the most commonly utilized soilless substrates are mineral wool, of which rockwool dominates, coconut fibre (coir) and perlite.

Hydroponic substrate-based systems used for crop production come in a diverse range of forms. Many are simply based on growing slabs of substrate (rockwool or coconut fibre) placed on a levelled floor, which allows the nutrient solution to drain away from underneath the plant. Grow slabs of rockwool or coco may also be contained within a rigid or plastic-film channel or support system or suspended up above floor height as in the 'hanging gutter system'. Loose substrates such as perlite, vermiculite, expanded clay, peat, sawdust, bark and others are placed inside growing bags, buckets, containers, trays, troughs or beds of a suitable volume for the crop to be grown. Drainage holes in the base of these containers allow nutrient solution to be channelled away after irrigation.

6.4.1   Stone wool (mineral wool, rockwool or glass wool)

Early substrate culture systems utilized naturally occurring media for plant support such as sand and gravel; however, by the 1970s Scandinavian and Dutch greenhouse growers were testing the use of stone wool, a man-made 'spun mineral wool' substrate that was originally produced for use as thermal insulation in the construction industry. Stone wool is manufactured by melting basaltic rock and spinning this molten mix into thin fibres which are then cooled by a stream of air. Grodan dominates the rockwool market worldwide and is the most common brand used by large and small hydroponic growers alike. Rockwool and other stone wool products developed for hydroponic cultivation are manufactured in a range of different products with varying degrees of density and water-holding capacity for different cropping uses. Stone wool of many brands is manufactured into a range of sizes from small propagation plugs for seeds to larger cubes for cuttings and fruiting vegetable transplants, as a wide range of slab sizes and as a granulated product.

Stone wool products are characterized by a moisture gradient within the substrate. At the base of the stone wool slab there is plentiful moisture, usually at media saturation levels, while in the upper layers of the slab, the roots are in drier conditions and hence have access to plenty of aeration and oxygen for root uptake and respiration. It is this moisture gradient from top to bottom of stone wool material that makes it a highly productive substrate for soilless production.

With tomatoes and similar crops, growers have the option of using the EC (strength of the nutrient solution) and moisture content of the stone wool/rockwool slab to help 'steer' the plants into either more vegetative or 'generative/reproductive' growth depending on what is required. Drying the slab back between irrigations and allowing the EC in the root zone to increase directs tomato plants into a more generative or reproductive state with less leaf growth and more assimilate being imported into the fruit. A higher level of moisture maintained in the stone wool/rockwool and a lower EC directs the plants towards more lush vegetative growth. Skilful growers use these techniques to direct their crop and control leaf, flower and fruit growth at different times and stone wool/rockwool is a substrate that allows this type of control via the root zone.

Stone wool/rockwool can be reused and commercial growers may grow many successive crops from rockwool slabs by steaming these after the plants have been removed and then replanting. Solarization is also possible, as is using chemical disinfectants, although the substrate must be rinsed well with water after using these. Commercial Grodan rockwool users have the option of the Grodan recycling service which picks up the used slabs and recycles them into new product. However, other growers may recycle the material by shredding it and reusing as a growing medium, as a component of potting mixes or incorporation into outside soils and gardens.

6.4.2   Coconut fibre (coir, palm peat, coco peat, coco)

Coco or coir is the outside layer of coconut husks (or mesocarp) which consists mainly of coarse fibres, but also finer material known as 'coir dust'. Harvested coconuts are first soaked in water, a process termed 'retting' which makes the fibre easier to remove. Usually the longer coarser fibres are removed for other uses while the coir pith then undergoes further processing and decomposition which makes it more suitable as a plant growth medium. Coir pith consists of mixture of shorter fibres and cork-like particles ranging in size from granules to fine dust.

'Coco fibre' is also the term often used to refer to the general-purpose grade of coco which is ideal for growing longer-term crops under hydroponic cultivation. Worldwide coco is used for soilless crops such as tomatoes, peppers, cucumber, melons, aubergines, ornamentals, cut flowers and many others because the structure of the coco does not break down over the time frame these longer-term crops are grown for. Thus, high rates of root-zone aeration and moisture retention are typical in both short- and long-term soilless crops and this results in high yields and good root health. Most coco fibre used in soilless systems has a water-holding capacity in the range 80–88% at container capacity and an air-filled porosity of 23–29% (Morgan, 2012).

Coco also comes in a range of different products – from small to large compressed 'bricks' to 'grow slabs' to pre-expanded, ready-to-use bagged product. Compressed bricks of coco fibre ensure the cost of shipment can be kept to a minimum; a typical 5 kg block of compressed coco can be expanded in water to over 65 litres of ready-to-use growing substrate. Pre-wrapped slabs of compressed coco can be less than 25 mm thick but when expanded with water within their plastic sleeve give a full-sized growing slab comparable in volume to rockwool (Fig. 6.4). The coco slabs only need be placed in position, slits cut in the plastic sleeve and water irrigated on – the coco expands and can be planted out with no further effort. The disadvantage of slabs is that they require a level surface to be placed upon so that drainage is even, and slabs only provide a relatively shallow depth of root volume compared with other bag- or container-based systems. Although coir fibre works well as a stand-alone medium, it can be blended with other substrates to improve the properties of the medium. Perlite can be mixed with coir fibre in a 50:50 mixture, which results in a medium that gives good support to the plants' roots and stems as well as holding more moisture than perlite alone.

6.4.3 Peat, bark and sawdust

Peat is used in many hydroponic systems and in seed-raising mixes mostly in Europe

Fig. 6.4. Grow slabs contain a substrate within a plastic sleeve.

where there are still reserves of good-quality peat available for use. The use of peat for soilless cultivation originated from the nursery industry where large volumes of peat-based potting media have been used for many decades for the production of container-grown plants. High-quality peat has good physical properties for soilless production; however, lower-quality peat can cause a number of problems if used in hydroponic systems as it can lose its structure and become waterlogged, resulting in root death. Peat is sometimes mixed with other substrates such as sand, pumice or media to 'open out' the structure and allow greater aeration and drainage in the mix.

Composted bark fines as a soilless substrate first came into use in the nursery industry in response to a decline in the availability and quality of peat resources. It was soon found that bark could also be utilized under hydroponic production and as a transplant-raising medium in soilless systems. To prepare a suitable substrate, bark of certain tree species such as *Pinus radiata* is composted with some form of nitrogen fertilizer to destroy any toxic substances or resinous materials and to prevent nitrogen drawdown once in use as a hydroponic substrate. If carried out correctly, the composting process also sterilizes the bark medium.

Fresh sawdust (i.e. uncomposted) of a medium to coarse grade is used as a short- to medium-term hydroponic substrate with reasonable water-holding capacity and aeration; however, problems arise with this medium when the sawdust starts to decompose and lose its physical structure. Once decomposition of the sawdust starts a number of problems can arise in the root zone, many of which the grower may not be aware of in the early stages. Growers using sawdust aim to produce one or two crops in this medium before it begins a process of rapid decomposition and needs to be replaced with fresh medium. Sawdust for hydroponic use needs to be free of toxins, plant resins and chemicals which are present in many tree species and in tanalized wood. *P. radiata* sawdust is the most commonly used in countries such as Australia and New Zealand, although other tree species are also suitable.

6.4.4 Perlite

Perlite is a siliceous material of volcanic origin which is heated to cause expansion of the particles to small, white kernels that are very light. The high processing temperature of this medium gives a sterile product which is often used for raising seedlings, particularly when combined with vermiculite, and for use in many types of hydroponic systems. Perlite is a free-draining medium that does not have the high water-retentive properties of many other substrates such as rockwool or vermiculite. It is essentially neutral with a pH of 6–7, but with no buffering capacity; unlike vermiculite, it has no CEC and contains no mineral nutrients. While this medium does not decay, the particle size does become smaller by fracturing as it is handled. Growers who handle perlite, particularly the finer grades, are required to wear either a respirator or dust musk to prevent inhaling the fine perlite dust particles or to wet down this substrate before filling growing containers or disposing of spent medium. Perlite is commonly used in the 'Dutch or bato bucket' system of soilless production that incorporates large rigid plastic growing containers with an interconnected drainage system which channels away the nutrient solution from the base of the bucket. Bato or Dutch bucket systems are typically used for larger fruiting plants such as tomato, capsicum and cucumber.

6.4.5 Pumice and scoria

Pumice and scoria (Fig. 6.5) are lightweight, siliceous, vesicular minerals of volcanic origin, characterized by high porosity and low bulk density. Pumice and scoria are generally free of pathogens and weed seeds and have a low CEC. The structure is stable and, being a natural product, can be used for many successive crops and disposed of with little environmental risk. Both substrates are available in a range of particle sizes from fine grades used in propagation to coarser grades which give a more rapid rate of drainage for long-term crop production.

Fig. 6.5. Scoria used as a hydroponic substrate.

6.4.6 Vermiculite

Vermiculite is a porous, sponge-like, kernel material produced by expanding certain minerals (mica) under high heat and it is thus sterile. Vermiculite is light in weight, with a high water-absorption capacity, holding up to five times its own weight in water, and a relatively high CEC – thus it holds nutrients in reserve and later releases them. Horticultural-grade vermiculite is available in a number of grades, with finer particle sizes used for seedling production and larger grades for hydroponic systems growing larger sized crops. Vermiculite is suited to hydroponic systems where the substrate is allowed to fully drain between nutrient applications, thus preventing waterlogging and allowing air to penetrate between the medium's granules.

6.4.7 Expanded clay

Expanded clay has a physical structure much like naturally occurring pumice or scoria and

is produced by baking specially prepared clay balls or pebbles in ovens at 1200°C. The clay expands and the final product is porous, allowing good entry of both water and air. Expanded clay is known by many names and produced by a number of different manufacturers: pebbles or balls can range in size grades from 1 mm up to 16–18 mm. LECA is light expanded clay aggregate which is usually not formed into uniform round balls, but is of similar structure to expanded clay balls or pebbles. All types of expanded clay products are sterile after baking, inert and well suited to many hydroponic systems due to their free-draining nature and attractive appearance. Expanded clay, however, does not have a high moisture-holding capacity between nutrient applications and salt accumulation and drying out can be common problems in systems that are not managed well to prevent these from occurring.

6.4.8 Rice hulls

Rice hulls are a lesser known and utilized hydroponic substrate in most parts of the world; however, they have been proven to be effective for the production of a range of hydroponic crops including tomatoes and strawberries. As a by-product from the large rice production industry in many warmer regions of the world, rice hulls have the potential to be an inexpensive and effective medium for hydroponic production. Rice hull is a free-draining substrate, with a low to moderate water-holding capacity, a slow rate of decomposition and typically low levels of nutrients in the raw product. Rice hull media can be sterilized before use in hydroponic systems by the application of steam or through 'solarization' in climates where bright sunlight is consistent.

6.4.9 Sand and gravels

Sand and gravels of various particle sizes are one of the oldest soilless substrates utilized in hydroponics. Both have the advantage of being locally available and inexpensive in many regions of the world. The ideal sand particle size for use in hydroponic substrate systems is between 0.6 and 2.5 mm in diameter, rather than fine sand which can compact down when wet, excluding oxygen from the root zone. Sand and gravels used in hydroponic systems need to be inert and not release minerals into the nutrient solution, as well as free from contamination from soil, weed seeds, pests and disease, and salinity.

6.4.10 New substrates

Less widely used substrates in soilless production include polyurethane grow slabs, which are durable, light, inert and recyclable, but have a low water-retention capacity (Castilla, 2013); and a number of organic products such as composts, sphagnum moss, vermicasts and natural waste materials such as sugarcane bagasse from other industries. New substrate materials are continually being developed and assessed for hydroponic crop production. Much of the focus of this research is not only to improve crop performance and optimize plant growth, but also to improve sustainability, recycle organic materials and provide solutions to the waste generated by protected cultivation which is recognized as a major challenge (Urrestarazu et al., 2003). While research into the use of various composts and vermicasts as soilless substrates for crop production has been ongoing, new materials are continually being evaluated. Biochar is one such substrate that has undergone investigation as both a stand-alone substrate and in combination with other hydroponic media. Biochar, a charcoal-like material produced by heating biomass in the absence of oxygen, can be produced from a wide range of organic materials including greenhouse crop waste and has shown potential as a hydroponic substrate. Dunlop et al. (2015) found that biochar created from tomato crop green waste could be used subsequently as a substrate for soilless hydroponic tomato production, thus creating a closed-loop system. Other investigations have found biochar combined with various

substrates resulted in a suitable hydroponic substrate for a wide range of applications. Awad et al. (2017) found that a blend of perlite and rice husk biochar produced a substrate which resulted in an increased yield of leafy vegetables compared with plants grown in perlite alone.

Other materials which have undergone investigation as new hydroponic substrates include sphagnum biomass formed into slabs, which proved to be a potential replacement for rockwool, while substrates based on sheep wool were less successful (Dannehl et al., 2015). The use of hemp (Cannabis sativa) and flax (Linum usitatissimum) bast fibre has also been assessed and while flax was unsuitable, hemp showed promise as a novel hydroponic substrate (Rossouw, 2016).

6.5 Substrates and Water-Holding Capacity

Many hydroponic substrates such as perlite, vermiculite, coconut fibre and others are available in a number of different grades, from fine, medium to coarse and mixed particle or fibre sizes. This allows growers to select different grades for propagation, smaller or larger plants, different hydroponic systems, longer-term crops and also for air-filled porosity and water-holding capacity. A good example of this is with horticultural coconut fibre used in hydroponic production. While orchids prefer a very coarse coco 'chip', using coco for propagation and germination of small seeds requires a much finer grade which will hold sufficient moisture as well as oxygen. While the high water-holding capacity of coir dust is required in some situations, it can create problems with oversaturation of the root zone, and grades of coco commonly used in grow slabs tend to consist of a mixture of longer coarse fibres or 'flakes' of coco, which keep the substrate open and aerated, and finer particles which hold more moisture. These grades of coco are ideal for longer-term hydroponic crops such as tomatoes, cucumbers, melons, peppers and cut flowers as the fibres assist in the prevention of the substrate 'packing down' over time. Other manufactured substrates such as stone wool, mineral wool or rockwool blocks/slabs also have a range of products with differences not only in overall moisture and air-filled porosity levels but also carefully calculated moisture gradients between the top and base. Many of these different products are aimed at different crops, growing climates and uses, and help provide the optimum levels of moisture in the root zone for different applications.

6.6 Substrates and Oversaturation

When nutrient solution is irrigated on to a substrate it displaces air in the open pores of the material; when draining subsequently occurs, more air is drawn down into the root system. If irrigation is too frequent the air-filled pores remain saturated and the plant has less access to the oxygen contained in air. Plants exhibit a strategy termed 'oxytropism' where roots will avoid growing into oxygen-deprived areas such as waterlogged soils, overwatered hydroponics substrates and stagnant nutrient solutions. This is most commonly seen inside the base of growing containers or slabs of substrate – areas devoid of any root growth, or small, thin brownish roots which have died back due to suffocation and oversaturation. If plants have been performing poorly observing root outgrowth into all areas of the substrate in the growing container or slab, particularly the base, will reveal if oversaturation has become a problem and prevented vigorous root growth. Other symptoms of root waterlogging include chlorosis (yellowing) or paleness of the new foliage, older leaves may also yellow and abscise, and flower and fruitlet drop become common. One of the more extreme symptoms of over-irrigation is epinasty where ethylene gas builds up within the plant causing the upper side of the leaf petiole cells to elongate whereas those on the lower side do not. The result is a severe bending downwards of the leaves which may be mistaken for wilting caused by a lack of moisture. Growers need to carefully check whether wilted plants are actually suffering from a lack of irrigation or epinasty due to waterlogging in the root zone.

Large amounts of algae may also grow on the surface of the medium if overwatering has been occurring. In seedling trays, high levels of moisture often lead to problems with 'damping off' caused by opportunist pathogens such as *Pythium* and *Rhizoctonia* that prey on young plants stressed by oversaturation and lack of oxygen. Cuttings and clones may suffer from stem rot and dieback as oversaturation cuts out much of the oxygen required for callus and root formation.

6.7 Matching Substrates to Crop Species

Some plant species are highly prone to problems with overwatering, while others are more tolerant. Strawberries are one common hydroponic plant that has no tolerance for a saturated substrate and many strawberry crop losses have resulted from overwatering the crown and the root rot that this causes (Fig. 6.6). Many cacti and succulents will also rot when over-irrigated and prefer a coarse and very free-draining substrate such as perlite or coarse sand. Other plants, more notably those that are grown under warm conditions, have large leaves and a rapid rate of growth and are better suited to highly moisture-retentive media that will hold sufficient water between irrigations. Cucumbers, tomatoes, squash and similar crops perform well in a medium that has a high water-holding capacity and also a good rate of air-filled porosity.

The rate of transpiration, temperature and humidity levels also play a role in substrate selection. Crops growing under warm, high-light, low-humidity conditions require frequent irrigation and benefit from a moisture-retentive medium which helps prevent drying out of the root zone and gives more of a safety buffer should failures with pumps or the power supply occur. Under cool conditions with slow growth and small plants, substrates which are highly free-draining and retain lower levels of moisture assist

Fig. 6.6. Hydroponic strawberries have little tolerance for oversaturated root zones, which can predispose the plants to crown and root rot pathogens.

with prevention of oversaturation under these conditions. These types of very free-draining, open substrates are also more for-giving of the application of higher levels of nutrient solution which may be needed, but at the same time not contributing to a water-soaked root zone when growth and transpir-ation rates are low.

6.8   Physical Properties of Soilless Substrates

The physical and chemical properties of a stand-alone medium for hydroponic produc-tion are an important component of crop performance. A growing medium is typically contained within the limited, fixed volume of a pot, container, trough, bed, grow bag or other system and must provide all of the oxy-gen and moisture requirements of the root system. The medium must also act as a physical

support for plants, especially if these are large specimens grown over long periods (Carlile *et al.*, 2015). Determination of the physical properties of growing media is car-ried out via a number of different analysis methods, these aim to measure variables such as water-holding capacity, air-filled por-osity, and weight or bulk density which affects handling of the substrate. Chemical properties commonly determined include pH, CEC, salinity, specific ion contents, microbiology and potential contaminants.

6.8.1   Bulk density

Bulk density is related to the weight of a sub-strate and is defined as its dry mass per unit of volume (in a moist state), measured in g/cm^3 or kg/m^3 (Raviv *et al.*, 2002). Most tests of bulk density are relatively simple to carry out: wet substrate is lightly packed or compressed under a given pressure into a container of a known volume, it is then dried down completely and weighed. Bulk density can vary considerably between different growing substrates and mixes; however, many producers of growing mixes seek media with a bulk density less than 300 kg/m³ (Carlile *et al.*, 2015). A low bulk density is often desirable in intensively grown greenhouse crops requiring a high air-filled porosity to meet the oxygen requirements of the root systems under frequent irrigation programmes.

6.8.2 Particle size distribution

The size and shape of substrate particles largely determine the air-filled porosity and water-holding capacity of a medium and vary considerably between different materials. Particle sizes in fine coconut fibre may range from 0.5 to 2 mm, while scoria, perlite and expanded clay can be several magnitudes larger. Peat may be screened into particle size grades ranging from 0–5 mm (fine) to 10–20 mm (coarse) or greater than 20 mm (very coarse). Substrates composed of one or more materials are often a blend of finer and coarser particle sizes, aimed to provide sufficient air-filled porosity but at the same time hold sufficient moisture between irrigations. Coarser materials such as perlite may be added to fine peat or coconut fibre to open out the structure of the growing medium, improving drainage and aeration where required. The particle size distribution of a specific medium is useful in estimating the hydraulic properties of the medium such as the water-retention characteristics and the hydraulic conductivity (Raviv *et al.*, 2002). Growing substrates with larger particle sizes are generally more free-draining, highly aerated with a lower water-holding capacity than those with a higher percentage of small particles. These larger particle media require more frequent irrigation as less water is held in the substrate between waterings and are often utilized on crops that are sensitive to saturation of the root zone or require high levels of aeration around the root zone. Fine particle-sized substrates are suited to seed

germination and seedling raising where there are no coarse particles to impede young root growth and sufficient moisture is held around germinating seeds.

6.8.3 Total porosity

Total porosity, or total pore space, is the combined volume of the aqueous and the gaseous phases of the medium (Wallach, 2008). The porosity of a growing medium is determined by the size, shape and arrangement of the particles and is expressed as a percentage of the total volume of the medium. Total porosity is an important variable as it can be further divided into the air-filled porosity and water-holding capacity which affect plant growth and root function. Most soilless substrates contain between 60 and 96% total pore space depending on particle size distribution. Total pore space includes those pores which are closed off and not accessible to water which tends to occur frequently in vascular materials such as pumice and scoria. 'Effective pore space' is a term which describes only those pores that can be saturated with water. Water is mainly held by the micropore space of a growth medium, while rapid drainage and air entry is facilitated by macropores (Drzal *et al.*, 1999). Substrates with a good physical structure for plant growth tend to have an optimal distribution of large and small pores to meet the requirements for both high aeration and water-holding capacity.

6.8.4 Air-filled porosity

Air-filled porosity is defined as the volumetric percentage of the medium filled with air at the end of free (gravitational) drainage (Raviv *et al.*, 2002). However the issue with air-filled porosity is that this value varies greatly with the height and shape of the container; for this reason, air-filled porosity is determined as the volumetric percentage occupied by air at a pressure head of 10 cm (or water suction of 1 kPa) (de Boodt and Verdonck, 1972). Most growing media have an air-filled porosity in the range 10–30% (Wallach, 2008), while highly aerated media such as those required for rooting cuttings have an air-filled porosity of greater than 20%. While air-filled porosity is a good indication of general oxygenation around the root system, problems can exist in the base of growing containers where a perched water table may form, thus excluding oxygen from this area. Growing container shape and size affect the moisture and air gradient from the top to the bottom of the container. Air content tends to decrease and moisture to increase from the top to base of growing beds and containers due to gravitational effects. Container geometry also influences air–water relationships within the medium, with air volume increasing by 25% and water volume decreasing by 13% in tapered pots compared with cylindrical types (Bilderback and Fonteno, 1987). Root systems, which respond to gravity, may also form a dense mat in the base of pots, beds or growing containers, which results in less than adequate root aeration despite the medium itself having suitable air-filled porosity.

6.8.5 Water-holding capacity or container capacity

The water-holding capacity or container capacity is dependent on the medium's particle size and container height: it is the portion of container pore space that retains water after drainage is complete. The water-holding capacity is comprised of both unavailable water, that which is not extractable by plants due to being held within very small pores or adsorbed on to particles, and available water, which is readily taken up by plant roots. To measure water-holding capacity, a container is filled with growing medium, fully saturated with water and then permitted to drain. Once draining is completed the damp medium is weighed, dried at 110°C until all moisture has been removed, then reweighed. The difference between the fresh and dry weights of the medium is the water-holding capacity. A water-holding capacity of 45–65% is considered optimum for most soilless growing media and container mixes.

Substrates with a lower water-holding capacity require more frequent irrigation than those with a higher water-holding capacity.

While porosity, air space and water-holding capacity can be measured before a substrate is used for crop production, these variables can change over time, particularly with organic materials which are prone to decomposition. The structural stability of the growing medium is an important factor as microbial degradation and compaction or shrinkage can reduce aeration and increase moisture levels in substrates. Substrates and container medium components such as composts, wood fibre and sawdust are particularly prone to a loss in physical structure over time due to the breakdown of cellulose and other easily hydrolysable substances (Carlile et al., 2015).

6.9 Chemical Properties of Hydroponic Substrates

6.9.1 pH

The pH of a substrate is a measure of the level of acidity or alkalinity of the medium, which affects nutrient ion availability and uptake. The ideal pH in the root zone of most crops in a soilless medium is between 5.5 and 6.5. At high pH levels, the uptake of some trace elements such as iron and zinc and macroelements such as phosphate may be reduced, resulting in an induced deficiency. pH values can vary considerably between growing medium materials and may need amendment to obtain the correct values for optimum plant growth. Peat typically has a low pH in the range pH 3–4 depending on the degree of decomposition and peat-based substrates are amended with lime to adjust the pH as required. Dolomitic lime also provides calcium and magnesium and is applied at rates of 2 to 3 kg/m³ for less decomposed (H2–H3) peat and 3 to 7 kg/m³ for more decomposed peat (H4–H6) (Maher et al., 2008). Coconut fibre including coir dust, fines and chips generally has an acceptable pH within the range 5.5 to 6.5 and requires no pH adjustment. Pine and other bark, after the composting process, typically has a pH range of 5.0 to 6.5 and lime may be added as required to increase the pH in some composted bark substrates (Jackson *et al.*, 2009).

6.9.2 Cation exchange capacity

The extent of cation adsorption by the surfaces within a growing medium is termed the 'cation exchange capacity' and may be expressed as meq/l, meq/100 g, meq/100 cm or mmol/kg; or, more commonly when referring to container mixes, CEC is related to volume unit in centimoles of charge per litre (cmol/l) or even as meq/pot. CEC refers to the medium's ability to hold and exchange mineral ions and varies considerably between soilless substrates. Growing media materials have electrical charges on the particle surfaces; negative charges contribute to the CEC and can bind positively charged ions (cations) to these sites. These cations can then be exchanged for another ion, releasing elements for plant uptake. The larger the number of negatively charged sites found on the substrate particles, the higher the CEC of the growing medium. A high CEC allows a soilless medium to hold nutrient ions in reserve and release these back later for plant growth, it therefore increases the buffering capacity or rate of change in fertility levels in the root zone. A high CEC can also buffer or resist changes in pH in a substrate; media with a high CEC require more lime to raise pH than those with a lower CEC. Soilless growing substrates can also have positively charged sites which attract negatively charged particles – this is referred to as the 'anion exchange capacity', which is far less significant than the CEC in a growing medium. While a moderate to high CEC value of 150–250 meq/l provides some good buffering capacity to a container or potting medium to pH changes and fertility, CEC is less of a concern in hydroponics or soilless crop production. Under hydroponic production a lower CEC is often preferable as plant nutrient requirements are fully met with frequent irrigations of nutrient

solution containing all of the elements required for plant growth. Thus, a medium which retains high levels of ions, which are later released, can create imbalances in the nutrient solution surrounding the plant roots. Commonly utilized hydroponic substrates such as perlite and stone wool have low CEC values of 25–35 and 34 cmol/kg, respectively (Argo and Biernbaum, 1997; Dogan and Alkan, 2004). Materials with high CEC values include peat, with a CEC of between 150 and 250 cmol/kg, and composts, which can vary considerably in CEC depending on source materials and degree of decomposition (Carlile *et al.*, 2015). Intermediate CEC values are found in other organic materials such as coconut fibre and composted pine bark. The CEC in some materials such as coconut fibre destined for use as a hydroponic medium can be adjusted before use with addition of solutions such as calcium nitrate which also help counter any nitrogen drawdown in this medium during the early stages of crop growth.

6.9.3 Specific ion contents, salinity and electrical conductivity

Salinity, which may include sodium as well as a number of other ions, can be an issue with certain materials used as soilless growing media. These are typically organic materials that can vary considerably with regard to EC, which is typically used to measure salinity levels. Some sources of coconut fibre can have high levels of sodium chloride (originating from coastal areas) as well as high levels of potassium present (50–200 mg/l) and require leaching or washing in water to remove these before use as a horticultural substrate.

6.9.4 Testing methods

While agricultural laboratories can carry out a full range of testing of soilless substrates, onsite testing by growers and manufacturers of growing mixes is a useful tool when evaluating media. The standard evaluation test for pH and EC (salinity) is the 1:1.5 medium/water extract. This involves mixing a representative sample of the growing medium with distilled water at a ratio of 1 to 1.5, stirring or shaking well, allowing to settle, filtering and then measuring the resulting pH and EC of the extract. The extract sample can then also be sent to a laboratory to analyse for specific ions to determine the fertility level of the substrate and for the presence of unwanted contaminants.

Conducting an onsite bioassay of new substrates is another rapid and inexpensive method of assessment of a hydroponic medium (Fig. 6.7). Since bioassay testing can also identify potential problems within a substrate that are not shown up by chemical physical analysis only (Kemppainen *et al.*, 2004), these types of evaluations play an important role in substrate quality assessment. A plant bioassay involves growing seedlings of a sensitive species such as cucumber, lettuce or tomato (Ortega *et al.*, 1996; Owen *et al.*, 2015) on a new substrate and comparing a number of growth variables against a standard or control substrate. Cucumber seedlings, which germinate rapidly, are a highly sensitive bioassay species and are commonly used for such assessments (Fig. 6.8). Common uses for bioassay tests are often with organic-based substrates such as peat, bark, sawdust, wood chips, coconut fibre, composts and similar materials, which may contain traces of phytotoxic compounds under certain circumstances (Ortega *et al.*, 1996). However, bioassays can be used for any substrate to compare the growth rate and plant performance to standard substrates. Once seedlings are grown for a limited period of time as part of the bioassay, they are assessed for variables such as germination count, coloration, height, leaf width, fresh and dry weight, root mass and overall appearance. Use of simple bioassays gives a close correspondence to actual growing conditions and should be part of the evaluation of new substrates.

Fig. 6.7. Bioassays or growth comparison is an inexpensive way of assessing a hydroponic medium.

Fig. 6.8. Cucumber seedlings are a highly sensitive bioassay species.

6.10 Nutrient Delivery in Substrate Systems

6.10.1 Drip irrigation

Drip irrigation, also termed 'micro-irrigation', 'trickle irrigation' or 'low-volume irrigation', is an efficient method of supplying nutrient solution slowly and directly to the root zone.

Flexible drip systems incorporate the use of a wide range of emitters which operate at low pressure and allow adjustment of nutrient flow rates. Drip irrigation systems consist of either a pressurized water supply with nutrient injectors, or a central nutrient reservoir with a pump to provide low pressure to the system. Filtration is also commonly included to remove sediment and other material which may block components and emitters, and in recirculating drip systems a filter may also be included on the nutrient return system. Drip systems then feed nutrient solution through a main irrigation line or ring out to the cropping area which feeds smaller-diameter lateral pipes carrying

the solution along the rows of plants. Small-diameter drip tubing (also termed 'microtubing' or 'spaghetti tubing') may then be installed into the length of lateral lines, these each have a dripper or emitter fixed to the end; this is then staked around the base of each plant to hold the dripper in position. These types of systems give maximum flexibility, with placement of the emitter allowing for differences in plant spacing to be easily accommodated.

Drippers or emitters control the flow of nutrient to the plant, come in a wide range of different options and are sized by the volume of nutrient solution they deliver (Fig. 6.9). Drippers are then further divided into pressure compensating and non-pressure compensating. Pressure-compensating drippers are more commonly used in large-scale commercial hydroponic systems as they are designed to discharge nutrient solution at a very uniform rate under a wide range of system pressures, so they deliver the same flow rate irrespective of pressure. These types of emitters are particularly useful in hydroponic systems as all drip emitters will start dispensing solution at the same time and prevent drainage of the solution after the irrigation has been switched off. Pressure-compensating emitters often incorporate a turbulent flow design which helps keep sediment and other particles in motion to help reduce clogging. One of the issues with pressure-compensating drippers is that they do not perform well on very low pressures such as with gravity-fed systems. Non-pressure-compensating drippers are the second type of emitter which also has applications for hydroponic systems. These emitters consist of two parts: a central body which is installed into the microtubing and a screw-on top which can be used to adjust the flow of nutrient. By winding the top further down on to the body of the emitter the flow of nutrient can be slowed and by winding this up, the flow is increased. This allows for emitters in the same system, under the same pressure, to deliver different rates of nutrient flow depending on what is required by individual plants.

The main issue encountered with drip irrigation is clogging of the emitters with salts, sediments or other material resulting

Fig. 6.9. Drip irrigation is an efficient method of nutrient delivery in substrate systems.

cleared by tapping the dripper to loosen any sediment; accumulated salts can be removed by submerging the dripper in hot water for a few minutes to dissolve any deposits. Between crops, or at least once per year, drip irrigation systems can be cleaned with acid which removes not only salt deposits, but algae and bacteria as well.

6.10.2   Drip-irrigated systems – design and layout

Irrigation layout and design affect the distribution of nutrient solution to individual plants around the drip-irrigated system. One of the main issues in drip irrigation is differences in the volume of nutrient solution received by plants in different parts of the system. Some plants may end up underwatered while others are constantly too wet, and these issues become difficult to remedy once the irrigation system is in place. To achieve uniform and constant nutrient flow rates to all emitters in the system, a 'ring' or 'loop' layout can be installed. This consists of emitters placed into lateral irrigation pipes connected at both ends to a ring main system, which evens out the flow and pressure round an irrigation system. Each ring main is supplied by a main irrigation pipe directly from a pump or pressurized water supply (Fig. 6.10). This largely prevents the issue of plants furthest away from the pump receiving the lowest volume of nutrient solution at each irrigation.

in uneven flow rates. High-quality water and prefiltration assists with some water supplies, particularly where sediment, sand or other organic matter might be present. Iron minerals in some water supplies are also a major contributor to blockages of irrigation equipment and are best removed before using to make up nutrient solutions. Drippers which are fully exposed to direct light can result in salt deposits accumulating and algae can also grow around emitter outlets causing blockages. In recirculating hydroponic systems where the solution drainage is redirected back to the nutrient tank for further irrigation, particles of growing substrate, pieces of root system and other organic material can all result in emitter blockages unless suitably sized filters are installed on the system. To avoid these issues resulting in plant growth problems, drippers should be regularly monitored while the irrigation is on to ensure all are working correctly. Blockages can often be

Selecting irrigation pipe diameter and pump size is another aspect of hydroponic drip-irrigated system design which is often overlooked. A large-capacity pump will not compensate for irrigation pipes that are too small to carry the nutrient solution flow rate required. If the flow of nutrient appears too low from some emitters, it is often more effective to increase the diameter of the delivery pipes rather than invest in a more powerful pump. The type of irrigation emitters, the number of emitters and their flow rate determine the size of irrigation lateral pipe required in terms of flow rate and pressure needed.

Fig. 6.10. An irrigation design which evens out flow and pressure ensures equal volumes of nutrient solution are delivered to each plant.

6.10.3 Ebb and flow (flood and drain) nutrient delivery systems

In ebb and flow (flood and drain) systems, nutrient solution is pumped up into the base of the growing area where the plants are positioned in pots, beds, containers or troughs of soilless substrate. The nutrient solution ebbs for a given period of time, re-wetting the growing medium from the base, before it is drained back to a reservoir tank below floor level. The ebb and flow process takes place several times per day depending on water and nutrient demands of the crop. Ebb and flow systems are more commonly used for potted plants and young nursery crops such as vegetable transplants being raised in rockwool propagation cubes. While allowing uniform delivery of nutrient to each plant, ebb and flow systems can have the disadvantage of pushing excess nutrient salts to the surface of the growing substrate particularly under conditions of high evaporative water loss.

6.10.4 Capillary watering systems

Capillary watering systems provide a continual shallow volume of nutrient solution around the base of the growing container, pot or slab of substrate, allowing the natural capillary action of the medium to draw up water into the root system area. The success of capillary systems depends very much on the natural structure of the growing medium and the ability of moisture to be 'wicked' up at sufficient rates to supply the plant. Capillary systems can cause problems with uneven moisture levels around the plant roots and salt build-up in the upper layers of the substrate.

6.10.5 Gravity-fed irrigation

Gravity-fed irrigation systems are another form of nutrient dispersal which rely on irrigation water flowing downwards to cropping areas under gravity rather than being pumped or manually moved. Rainwater collected from rooftops and stored in irrigation tanks may be used in gravity-fed systems in areas where no other source of mechanical water movement is available. Gravity-fed systems may also use pumps to fill reservoirs or tanks at a higher level, which then use gravity to irrigate crops as required on a low-pressure system of dispersal.

6.10.6 Nutrient dosing and injectors

There are two main methods of dosing nutrients into drip-irrigated substrate systems, these are 'batch feeding' and direct fertilizer injection. Batch feeding involves a mixing tank where the nutrient concentrates are dosed into water, EC and pH adjusted, and the working-strength nutrient solution is then pumped out into the irrigation system for application to the crop (Fig. 6.11). With direct injection, separate A and B stock solutions are injected directly into the water supply line, this system does not require a mixing tank (Fig. 6.12). The injection of the nutrient stock solution occurs via a pressure drop in the main supply line which draws the fertilizer concentrates into the flowing solution. The EC of the nutrient solution is controlled by the dilution rate which may vary from 1:50 to 1:200, with a 1:100 dilution rate being commonly used. Separate tanks of acid for pH control and other nutrient stock solutions such as calcium or potassium supplements may also be added to the injector system where required (Fig. 6.13).

6.11 Irrigation and Moisture Control in Substrates

Irrigation methods and frequency determine moisture and oxygen levels in the root zone, meaning that nutrient application rates and timing are often different for the same crop growing in substrates with different physical properties. This means that irrigation programmes need to be worked through by each grower based on their particular crop, system, substrate, climate and regular observations of moisture levels, rather than just following some predetermined irrigation settings. Between irrigations of nutrient solution, the soilless substrate must retain sufficient moisture for the roots to extract to maintain maximum growth rates, while excessive irrigation frequency and volumes must be avoided as these exclude essential air and oxygenation

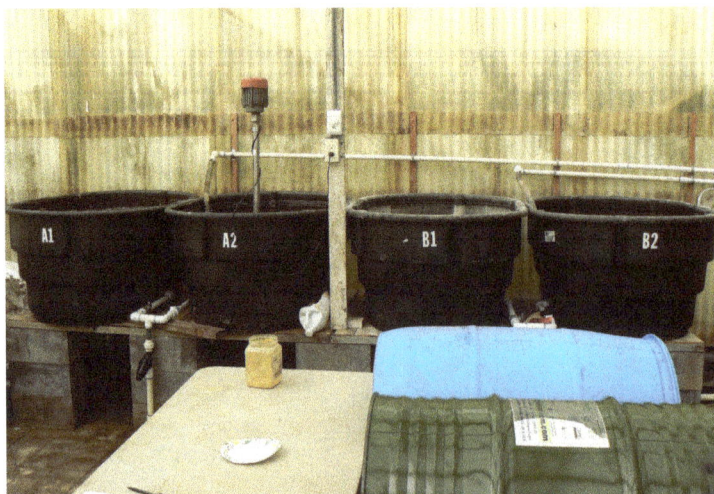

Fig. 6.11. Batch feed nutrient tanks.

Fig. 6.12. Nutrient tanks and Dosatron injectors.

Fig. 6.13. Computerized nutrient injector system.

from the root zone and can lead to infection by root rot pathogens. Irrigation determination and scheduling in hydroponic substrates are more critical than in the field as the root-zone volume is severely limited and requires more frequent application of water. Irrigation determination in hydroponic systems may be based on grower observation of the moisture status of the substrate, measurement of the volume of leachate after each irrigation (nutrient solution draining from the base of the growing slabs or containers), with use of substrate moisture sensors, by weighing the substrate to determine when sufficient moisture loss has occurred to trigger subsequent irrigations or through measurement of incoming solar radiation and computer models which estimate crop transpiration based on this value.

6.11.1 Substrate moisture, growth balance and deficit irrigation

Plant growth for fruiting crops such as tomato, capsicum and cucumber is divided into two parts: 'vegetative' (leaves, stems) and 'generative' (buds, flowers, fruits). Many plants such as indeterminate tomatoes, capsicum and cucumbers have both vegetative and generative growth occurring at the same time, so the balance between these two is referred as 'growth balance'. A highly vegetative crop will have large, thin, soft leaves, minimal flowering, poor fruit set and may drop small fruitlets, or set fruit may end up undersized. Generative plants may look vigorous and healthy; however, much of the plant's energy goes into flower/fruit production with little vegetative growth to produce more shoots and foliage to support the developing fruit with a photoassimilate supply and continue crop development. One way of manipulating plant growth balance is with the use of moisture control, both via substrate selection and use of careful irrigation programming which controls the supply of nutrient solution. High levels of moisture in the root zone tend to have a vegetative effect on many crops so 'controlled deficit irrigation' is used on a wide range of crops, both hydroponic and those grown in soil, to help direct growth in a more productive or generative direction with controlled levels of stress. Deficit irrigation may include reducing the volume applied at each irrigation, allowing more time between irrigations, and allowing the medium to dry slightly overnight by restricting early-morning and evening irrigations. Along with restricting moisture levels via a reduction in nutrient application, growers can select substrates which are coarser, more free-draining with a higher air-filled porosity and lower water-holding capacity to help with controlling overly vegetative growth. This is a useful tool for certain crops or cultivars which are naturally vegetative and benefit from some growth control or those growing under environmental conditions which may favour vegetative over generative growth. For plants that may tend to be overly generative and set large numbers of fruit at the expense of foliage growth, use of a substrate that has a naturally higher water-holding capacity such as fine-grade coconut fibre is preferable for favouring vegetative growth.

Along with influencing growth balance, mild deficit irrigation can be used to help improve fruit quality, shelf-life and even volatile concentrations in some plants. Tomato fruit flavour is one such aspect that can be improved with restricting moisture in the root zone which has the effect of increasing the dry matter percentage of the fruit and reducing water content, thus giving more concentrated flavour and often improved keeping qualities as well. Mild deficit irrigation can also be used to 'harden off' transplants before they are planted out into a different environment to reduce stress and improve plant survival rates.

6.12 Microbial Populations in Substrates

Microflora develop rapidly after planting a crop in a hydroponic system and consume plant exudates, compounds in the nutrient solution and dead plant materials, with the composition of microbe species affected by environmental factors and the source of nutrients. Some of the microbe species may be pathogenic; however, these are generally outnumbered and outcompeted by populations of non-pathogenic organisms, unless conditions change which favour infection by disease-causing fungi or bacteria. In most hydroponic systems the species of beneficial resident microflora most commonly found are *Bacillus* spp., *Gliocladium* spp., *Trichoderma* spp. and *Pseudomonas* spp.

Studies into the effects of various species of beneficial bacteria have found a number of positive results on yield and quality of hydroponically grown fruits and vegetables (Sambo *et al.*, 2019). Some studies have shown that growth-promoting rhizobacteria may provide a direct boost to plant growth by providing crops with fixed nitrogen, phytohormones, iron that has been sequestered by bacterial siderophores and soluble phosphate (Sambo *et al.*, 2019). Other species may protect the plant from potentially highly damaging pathogens which would otherwise limit plant growth, quality and yields. *Bacillus amyloliquefaciens* has been shown to increase vitamin C content and water-use efficiency in tomatoes, while *Bacillus licheniformis* increased fruit diameter and weight of tomatoes and peppers and promoted higher yields (Garcia *et al.*, 2004). A strain of *Pseudomonas* sp. (LSW25R) has been found to promote growth of hydroponic tomato crops and increase the uptake of calcium which in turn reduced blossom end rot of tomato fruit (Lee *et al.*, 2010). In hydroponic strawberries, inoculation with plant-growth-promoting rhizobacteria (*Azospirillum brasilense*) resulted in a higher sweetness index and a greater concentration of flavonoids and flavonols in the fruit as well as increasing the concentration of micronutrients (iron) (Pii *et al.*, 2017). In hydroponic tomato studies evaluating a number of different plant-growth-promoting rhizobacteria, it was found that a *Bacillus* sp. (strain 66/3) was effective in increasing tomato yield significantly – this increase in marketable yield was 37 and 18% compared with untreated control plants in autumn and spring crops, respectively (Kidoglu *et al.*, 2009). Lee and Lee (2015) even state that the application of potential beneficial microorganisms could lead to an improvement in the nutraceutical properties of hydroponic crops. These effects of increased yield and improved compositional quality in hydroponic crops from microbial interactions are likely to have occurred via complex processes, some of which are still not fully understood. Some species of beneficial rhizospheric bacteria in particular are known to improve plant performance under stressful environments and thus improve yields either directly or indirectly.

The inoculation of hydroponic systems with specific plant-growth-promoting rhizobacteria using commercially available products is possible, however the species available are still somewhat limited. While naturally occurring beneficial microbes do typically self-inoculate into new hydroponic systems, this can be a slow process and species diversity may be limited. Well-established hydroponic systems tend to have a greater diversity of beneficial microbial species than newer systems. However, microbes can be introduced

via a number of different methods. The plant-growth-promoting rhizobacteria commonly used in commercial inoculant products often include species of *Bacillus* and *Trichoderma* as these have been shown to have positive effects on a number of hydroponic crops. Such inoculant products are often designed to be added directly to the nutrient solution; however, some are in more widespread use as substrate inoculant incorporated into the growing medium before planting. If using inoculants, such as *Trichoderma*, it is simpler to establish beneficial microbes into a new substrate as little competition exists from microorganisms already present. However, if the substrate is relatively inert such as rockwool or other synthetic growing media, this is initially a difficult environment for microbial life to take hold. Once plants have established, carbon exudates from the roots and sloughed-off root material begin to provide organic substances for microbes to grow and population numbers then begin to build over time. Some microbial inoculants may also be applied as seed coatings and commercially obtained seed lots may be treated with inoculants aimed to improve germination and seedling establishment rates through a range of different processes including rot pathogen prevention and root growth promotion.

Alongside commercial mixes of inoculants, microbes may be introduced in other ways. The use of a 'slow sand filter' system for disease suppression and inoculation with beneficial microbes is one of the most effective ways of obtaining a diverse population of beneficial microflora. The sand filter system acts as a continuous source of inoculation with beneficial species and is particularly useful for solution culture systems where microbial life can be more limited than in substrate-based hydroponic systems. Other methods whereby beneficial microbial species may be introduced to a hydroponic system include the incorporation of composts and vermicasts, which naturally contain high levels of microbes. Where plant-growth-promoting rhizobacteria are being inoculated or actively encouraged, methods of nutrient sterilization should be avoided as these will not only destroy potential pathogens, but beneficial species as well.

Application of residual chemical agents such as chlorine to the nutrient solution should also be avoided as these can also negatively affect beneficial microbial life in the hydroponic substrate.

References

Argo, W.R. and Biernbaum, J.A. (1997) The effect of root media on root zone pH, calcium and magnesium management in containers with impatiens. *Journal of the American Society of Horticultural Science* 122, 275–284.

Awad, Y.M., Lee, S.E., Ahmed, M.B.M., Vu, N.T., Forooq, M., *et al.* (2017) Biochar, a potential hydroponic growth substrate, enhances the nutritional status and growth of leafy vegetables. *Journal of Cleaner Production* 156, 581–588.

Bilderback, T.E. and Fonteno, W.C. (1987) Effects of container geometry and media physical properties on air and water volumes in containers. *Journal of Environmental Horticulture* 5, 180–182.

Carlile, W.R., Cattivello, C. and Zaccheo, P. (2015) Organic growing media: constituents and properties. *Vadose Zone Journal* 14(6), vzj2014.09.0125.

Castilla, N. (2013) *Greenhouse Technology and Management*, 2nd edn. CAB International, Wallingford, UK, pp. 174–178.

Dannehl, D., Suhl, J., Urichs, C. and Schmidt, U. (2015) Evaluation of substitutes for rock wool as growing substrate for hydroponic tomato production. *Journal of Applied Botany and Food Quality* 88, 68–77.

De Boodt, M. and Verdonck, O. (1972) The physical properties of the substrates in horticulture. *Acta Horticulturae* 26, 37–44.

Dogan, M. and Alkan, M. (2004). Some physiochemical properties of perlite as an adsorbent. *Fresenius Environmental Bulletin* 13, 252–257.

Drzal, M.S., Fonteno, W.C. and Cassel, D.K. (1999) Pore fraction analysis: a new tool in substrate analysis. *Acta Horticulturae* 481, 43–54.

Dunlop, S., Arbestain, M.C., Bishop, P.A. and Wargent, J.J. (2015) Closing the loop: use of biochar produced from tomato crop green waste as a substrate for soilless, hydroponic tomato production. *HortScience* 50(10), 1572–1581.

Garcia, J.A.L., Probanza, A., Ramos, B., Palomino, M.R. and Gutierrez Manero, F.J. (2004) Effect of inoculation of *Bacillus licheniformis* on tomato and pepper. *Agronomie* 24, 169–176.

Jackson, B.E., Wright, R.D. and Seiler, J.R. (2009) Changes in chemical and physical properties of pine tree substrate and pine bark during long term nursery crop production. *HortScience* 44(3), 791–799.

Kemppainen, R., Avikainen, A., Harranen, M., Reinikainen, O. and Tahvonen, R. (2004) Plant bioassay for substrates. *Acta Horticulturae* 644, 211–215.

Kidoglu, F., Gul, A., Tuzel, Y. and Ozaktan, H. (2009) Yield enhancement of hydroponically grown tomatoes by rhizobacteria. *Acta Horticulturae* 807, 475–480.

Lee, S. and Lee, J. (2015) Beneficial bacteria and fungi in hydroponic systems: types and characteristics of hydroponic food production methods. *Scientia Horticulturae* 195, 206–215.

Lee, S.W., Ahn, I., Sim, S., Lee, S., Seo, M., *et al.* (2010) *Pseudomonas* sp. LSW25R, antagonistic to plant pathogens, promoted plant growth, and reduced blossom end rot of tomato fruits in a hydroponic system. *European Journal of Plant Pathology* 126, 1–11.

Maher, M., Prasad, M. and Raviv, M. (2008) Organic soilless media components. In: Raviv, M. and Lieth, J.H. (eds) *Soilless Culture: Theory and Practice*. Elsevier, Oxford, pp. 459–504.

Morgan, L. (2012) *Hydroponic Salad Crop Production*. Suntec New Zealand Ltd, Tokomaru, New Zealand.

Ortega, M.C., Moreno, M.T., Ordovas, J. and Aguado, M.T. (1996) Behaviour of different horticultural species on phytotoxicity bioassays of bark substrates. *Scientia Horticulturae* 66(1–2), 125–132.

Owen, W.G., Jackson, B.E. and Fonteno, W.C. (2015) Pine wood chip aggregates for greenhouse substrates: effect of age on plant growth. *Acta Horticulturae* 1168, 269–276.

Pii, Y., Graf, H., Valentinuzzi, F., Cesco, S. and Mimmo, T. (2017) Influence of plant growth-promoting rhizobacteria (PGPR) on the growth and quality of strawberries. *Presented at International Symposium on Microbe Assisted Crop Production: Opportunities, Challenges and Needs (miCROPe 2017)*, Vienna, 4–7 December 2017.

Raviv, M., Wallach, R., Silber, A. and Bar-Tal, A. (2002) Substrates and their analysis. In: Savvas, D. and Passam, H. (eds) *Hydroponic Production of Vegetables and Ornamentals*. Embryo Publications, Athens, pp. 25–89.

Rossouw, S. (2016) *A novel organic substrate based on hemp (Cannabis sativa), or flax (Linum usitatissimum) bast fibre for NFT hydroponic systems*. Master's thesis, McGill University, Montreal, Canada.

Sambo, P., Nicoletto, C., Giro, A., Pii, Y., Valentinuzzi, F., *et al.* (2019) Hydroponic solutions for soilless production systems: issues and opportunities in a smart agriculture perspective. *Frontiers in Plant Science* 10, 923.

Urrestarazu, M., Salas, M.C. and Mazuela, P. (2003) Methods of correction of vegetable waste compost used as substrate by soilless culture. *Acta Horticulturae* 609, 229–233.

Wallach, R. (2008) Physical characteristics of soilless media. In: Raviv, M. and Lieth, J.H. (eds) *Soilless Culture: Theory and Practice.* Elsevier, Oxford, pp. 41–116.

7 Organic Soilless Greenhouse Systems

7.1 Introduction – Organic Greenhouse Production

Worldwide, organic produce is one of the fastest-growing niche markets with an increasing number of greenhouses growers looking to improve profitability by adopting organic systems. 'Organic greenhouse horticulture' has been defined as the 'production of organic horticultural crops using inputs only from natural, non-chemical sources, in climate-controllable greenhouse and tunnels (permanent structures)' (van der Lans *et al.*, 2011). In the EU the largest areas in organic greenhouse production are in Italy (1000 ha) and Spain (1500 ha), with France (450 ha) ranking third followed by the Netherlands (100 ha), with non-EU countries such as Israel (500 ha), the USA (195 ha) and Mexico (estimated 800 ha) also having sizeable organic production areas under protected cultivation (van der Lans *et al.*, 2011). The main crops produced in organic greenhouses are tomato, capsicum, cucumber, lettuce and culinary herbs. Tomatoes are the main organic crop in Mexico, with Mexico's production almost exclusively in Baja California Sur and largely exported into the USA rather than sold domestically (van der Lans *et al.*, 2011). Many of the larger-scale producers in the EU such as Spain and the Netherlands are also exporters of organic produce, mostly fruit and vegetables. In the USA, organic tomatoes are the leading crop, followed by European cucumbers, lettuce, peppers and culinary herbs (Greer and Diver, 2000). According to the 2016 census of organic agriculture reported by the US Department of Agriculture (USDA), the 'vegetables under protection' category accounted for US$89 million in sales which was a 22% increase from sales in the previous year (USDA, 2017). The highest-value crops were fresh market tomatoes (US$39 million in sales) with 583 certified organic farms and fresh-cut herbs/other vegetables with 666 farms (US$49 million in sales) (USDA, 2017).

Most countries where an organic greenhouse industry exists have specific organizations and regulations for organic production. These usually include certification agencies which oversee organic production practices, carry out certification and inspection, and provide guidelines for producers. Organic standards, allowable inputs, rules and regulations vary somewhat from country to country and between organic associations and certifying bodies. In the USA, the National Organic Program (NOP) implements legislation for the Organic Foods Production Act of 1990. While the majority of organic greenhouse crops are grown in soil, in some parts of the world soilless organic systems are used to produce potted plants, transplants and seedlings, and where certification allows, a range of fruit, vegetable and herb crops. Much of the organic potted plant production makes use of natural substrates such as peat and composts; however, organic regulations in some countries still require some soil to be used as part of the substrate for container-grown plants.

Under protected cultivation, organic crops may be heated to provide year-round growth, however this varies depending on the organic regulations in different countries. Heating and lighting require significant energy inputs which may be considered as inconsistent with the philosophy of organic and sustainable production. Greenhouse heating is used by organic producers in the Netherlands; however, heating is not permitted in Italy and other countries only allow heating if it is derived from renewable energy sources or limit this to certain times of the year (van der Lans *et al.*, 2011).

CO_2 enrichment of organic greenhouse crops is generally permitted in most countries, but in the Netherlands CO_2 may only be used if it is generated as a by-product of the heating system.

Since the majority of organic greenhouse production still largely takes place in the soil, this creates a number of important issues for growers. Soil disinfection with chemicals and fumigants is not permitted under organic production. Since allowing soil to remain fallow, converted to pasture or rotation of different crop classes is difficult in high-cost greenhouse structures, other methods of suppressing soilborne pests and diseases must be used. These often include steaming and solarization of the soil where covers allow heating of the soil via sunlight, usually over the summer months. Despite these methods, soilborne pests and diseases, maintenance of soil structure and fertility are major challenges to organic greenhouse production.

7.2 Organic Hydroponic Systems

In recent years, the possibility of 'organic hydroponics' or organic soilless production has become a topic of much debate. The interest in organic hydroponics has largely been driven by the higher returns organically certified produce receives on many markets and the growing public demand for more sustainable methods of production. From a resource recycling viewpoint, it is also important to develop methods capable of using organic fertilizer sources in hydroponics (Shinohara *et al.*, 2011). While organic hydroponic methods may not be organically certified for commercial production in most countries, backyard, hobbyist and smaller growers around the world may use the method to come in line with their own philosophy regarding organics.

In some countries, such as the USA, organic soilless systems, even those using NFT or different forms of solution culture, are certifiable as organic despite not making use of soil. However, in much of the rest of the world soilless systems are currently not certifiable as organic due to the absence of soil which is considered to be the 'cornerstone' of organic production. Where organic soilless systems are considered to be allowable, these typically incorporate the use of natural growing substrate such as peat or coconut fibre which may be amended with perlite, compost or vermicast (Fig. 7.1). These substrates are irrigated with liquid organic fertilizers which do not contain non-organic, synthetic or 'man-made' fertilizer salts such as calcium nitrate, potassium nitrate and chelated trace elements. Many of the organic nutrient solutions are based on seaweed, fish or manure concentrates, allowable mineral fertilizers, processed vermicast or plant extracts, and other natural materials. Using organic fertilizers to provide a complete and balanced nutrient solution for soilless production is a difficult and technically challenging process and these systems are prone to problems with nutrient deficiencies, particularly with highly nutrient-demanding crops such as tomatoes. Aquaponics, which uses organic waste generated during fish farming processes to provide nutrients for crop growth via bacterial mineralization, may be considered a form of organic

Fig. 7.1. Organic substrates may be blends of a range of materials.

hydroponics (mostly in the USA) when no additional fertilizer amendments are added.

7.3 Organic Hydroponic Nutrients

Conventional hydroponic systems use only inorganic fertilizers as few microorganisms are present in hydroponic solutions to mineralize the organic compounds (Shinohara et al., 2011). The use of organic materials in hydroponics was long considered to be phytotoxic; however, correct selection, treatment and microbial processing have made organic hydroponics a technical reality even if the method is not recognized as being fully organic due to the lack of soil in these systems. Large amounts of unprocessed organic material, particularly in solution culture systems, generally have phytotoxic effects leading to poor plant growth (Shinohara et al., 2011). For organic nutrient sources to be successful in hydroponic systems a rapid degree of mineralization needs to occur. Some of this process may be carried out on organic materials before they are incorporated into a soilless system; however, significant mineralization also needs to occur during production for the crop to have access to readily available mineral ions.

There are a number of different options for organic hydroponic nutrients, from commercial fully processed, bottled concentrate products designed especially for soilless systems to a large number of other materials which growers may process, blend and modify before application. In the USA where solution culture systems such as NFT, float/raft or raceway can be certified as organic, a market has developed for organic hydroponic liquid concentrate nutrient products which are Organic Materials Review Institute (OMRI)-registered. These bottled organic products are diluted and the nutrient solution run through mostly NFT-type systems, often with the addition of a microbial inoculant product to assist with mineralization within the system. For practical and successful organic hydroponics, it is necessary to efficiently generate nitrate from organic fertilizer, thereby allowing direct addition of organic fertilizer to the hydroponic

solution during cultivation (Shinohara et al., 2011). This process is complex however, and many factors influence the rate of nitrification within the system including temperature, pH and oxygenation, the amount of organic material, and the species and numbers of microorganisms.

Organic nutrients used in solution culture systems such as NFT pose an increased risk of anoxia in the root system as both microorganisms and root systems require DO, placing a high BOD on the system. Many organic growers using solution culture methods install additional methods of oxygenation of the nutrient solution to help counter this. Ensuring the nutrient solution temperature remains within the optimal range also assists with oxygenation. Growers using drip-irrigated organic substrate systems (again, allowable under many USA certification agencies) have more flexibility with organic nutrient sources. The same commercial bottled liquid organic concentrate products used by NFT and raft system producers may be applied to substrate systems, however many other options are used successfully in a number of crops. These include blending a range of organic liquid concentrates such as those derived from fish, blood and bone, seaweed and similar sources with allowable fertilizer salts such as magnesium sulfate, potassium sulfate, iron with organically allowable chelates and sulfate trace elements to create a nutrient solution which is as balanced as possible. The main issue with running these types of organic blends through a drip irrigation system is clogging of drippers and emitters which is common with organic materials.

7.4 Microbial Mineralization of Organic Nutrients for Hydroponics

Organic hydroponics systems, just as organic cropping in soil, are reliant on the presence and activity of microbial species to carry out the conversion from organic compounds to plant-usable nutrient ions. In some cases, this may be partially completed in the nutrient concentrate before being irrigated on to the hydroponic substrate. Use of 'biodigester' systems and reactor tanks is a form of pre-processing which provides ammonification and nitrification of organic nitrogen compounds (Shinohara *et al.*, 2011). Other methods of mineralization include more conventional methods such as composting raw materials and use of vermiculture (worm farm) systems to provide both a solid growing substrate and liquid nutrient extracts.

7.5 Anaerobic and Aerobic Processing of Organic Materials

Processing of organic materials before addition to a hydroponic system allows at least partial mineralization to occur and is an important step particularly for solution culture methods such as NFT production. Anaerobic digestion is a process where microorganisms break down organic materials in the absence of oxygen and is widely used in waste management systems and for the production of biogas. The digestion process undergoes a number of steps beginning with bacterial hydrolysis of the organic material, followed by conversion of sugars and amino acids to hydrogen, ammonia, organic acids and CO_2, and finally the production of methane which may be used as a source of renewable energy. As a by-product, the energy-rich digestate can be used as a crop fertilizer. While anaerobic digesters are largely used for waste management of materials such as sewage sludge and municipal waste management, these systems can be used to process a range of organic materials as the resulting nutrient levels are dependent on the substances fed into the digester (Ulusoy *et al.*, 2009). Studies on the use of anaerobic digestate effluent in hydroponics have produced mixed results. This is likely to be attributable to the different raw materials initially incorporated during the anaerobic digestion process and the hydroponic crops these were applied to. Anaerobic digestion produces significant levels of ammonia and a high pH of the resulting digestate which can cause growth issues with many hydroponic crops. Studies have shown that digestate applied to hydroponic lettuce

crops produces the greatest response, whereas that applied to ammonium-sensitive crops often show poor growth compared with controls (Neal and Wilkie, 2014). Anaerobic digestate from thermophilic digestion of poultry waste applied to lettuce in an NFT system was found to produce shoot fresh weight which was not significantly different from control plants produced in a commercial inorganic nutrient solution (Liedl *et al.*, 2004). In this case, the digestate was diluted to 100 mg ammonia-N/l and pH adjusted to within the range of 5.5–6.1 (Liedl *et al.*, 2004). Other studies have found similar results on a range of leafy vegetables and concluded that biogas digestate is an effective nutrient source for high-quality vegetable production based on its synergistic effects and effectiveness in yield and quality improvement, which included depressive effects on foliar nitrate accumulation (Liu *et al.*, 2009). Anaerobic digester systems used to process fish sludge from aquaponic systems have also been found to give greater growth of lettuce plants in hydroponic systems as compared with supernatant from an aerobic digester and a control nutrient system; this was largely attributed to the presence of ammonium ions, dissolved organic matter, plant-growth-promoting rhizobacteria and fungi, and humic acid (Goddek *et al.*, 2016). In other studies, and with different crops such as hydroponic tomatoes, the predominance of nitrogen in the ammonium form (ammonium-N) in anaerobic digestate produced from food waste led to lower performance than those grown with traditional fertilizer and was attributed to the sensitivity of this crop to ammonia-N (Neal and Wilkie, 2014). Anaerobic digestate was found to have not only high levels of ammonia-N but also nitrate was essentially non-existent while pH was greater than 8 (Neal and Wilkie, 2014).

For hydroponic crops that prefer nitrate over ammonium, nitrification after the initial anaerobic digestion phase is required for optimal growth. This may take place before the solution is applied to crops via aerobic digestion in separate digester tanks or systems with the addition of nitrification bacteria, or within the hydroponic production system provided certain conditions are met. These include the correct temperature of the solution for maximum nitrification to take place, sufficient DO to support both plants and nitrification bacteria, and pH control. Both systems benefit from the addition of initial inoculants of nitrification bacteria with the species *Nitrosomonas* and *Nitrobacter* which carry out the conversion of ammonium ions (NH_4^+) and nitrite ions (NO_2^-) into nitrate ions (NO_3^-). Few microorganisms are present in hydroponic solutions to mineralize organic compounds (Shinohara *et al.*, 2011), thus these may be introduced and maintained during crop production. Using an inoculant enriched with nitrifying microorganisms enables organic fertilizers to be directly added to the solution during crop cultivation (Shinohara *et al.*, 2011). During this process, pH correction with calcium carbonate has proven to be beneficial to stabilize pH levels (Saijai *et al.*, 2016), as nitrification activity is enhanced by pH maintained between 6.5 and 8.5. Along with initial inoculation of nitrification microorganisms, organic nutrient solutions require ongoing additions of organic fertilizer in the correct volumes as nitrification can be inhibited by excessive amounts of fertilizer additions (Shinohara *et al.*, 2011).

7.6 Vermicast and Vermicomposting

Vermicast is derived from organic material which has been digested and passed through the body of worms (composting worm species such as tiger worms) and excreted as castings. Vermicomposting has gained popularity as an alternative waste management option due to its environmentally friendly approach (Quaik *et al.*, 2012). Solid vermicast contains plant-available soluble nutrients as well as a high diversity of microbial populations originating from the gut of the worms the material has passed through. Alongside solid vermicast, liquid vermicast (vermicomposting leachate) which drains from worm farm systems may be collected and used as a liquid fertilizer. Vermicast systems are an effective way of processing organic raw materials into mineral-rich substrates and liquid fertilizers

for use in hydroponic systems. One of the unique features of vermicast is that during the process of conversion of various organic wastes by worms, many of the nutrients are changed to plant-available forms (Chanda *et al.*, 2011). This makes vermicast a rich source of available nutrients such as nitrogen in the nitrate form (nitrate-N) or ammonium form, exchangeable phosphorus and soluble potassium, calcium and magnesium (Buchanan *et al.*, 1988). As well as nutrients and microbial species, vermicast contains organic growth promoters such as humic and fulvic acids (Quaik *et al.*, 2012), which are known to stimulate growth and increase nutrient uptake (Han *et al.*, 1989).

A wide range of raw materials may be used as feedstocks for vermiculture systems, the most common being kitchen waste, green waste, manures, paper waste and other by-products of food processing industries. Fully processed vermicast can be a useful source of plant nutrients, however the mineral composition is highly dependent on the raw materials initially fed into the system. For the production of a balanced organic hydroponic nutrient solution and substrate, vermiculture systems need to process organic raw materials of a high compositional quality. A combination of raw materials such as fish meal, blood and bone, manures, green waste and seaweed gives a greater balance and concentration of mineral ions than if only kitchen waste is processed through the vermiculture system.

Vermiculture systems have been set up on both small or domestic and on larger, commercial scales as the resulting vermicast or solid castings are a product sold for both soil conditioning and as an addition to container mixes. Vermicompost has been reported to improve the physical conditions of substrates via decreased bulk density and increased aggregate stability, porosity and water-holding capacity (Adhikary, 2012). In organic hydroponic systems, fully processed vermicast is a valuable input as a component of soilless growing mixes. Vermicast provides not only soluble nutrient ions immediately available for plant uptake, but also a diverse range of microorganisms which play vital roles in further nutrient mineralization and as potential disease-suppression agents. Solid vermicast is not used exclusively as a growing medium but blended with other organically allowable substrates to create a more stable physical structure and maintain porosity over the life of the crop. Common substrate blends used in organic hydroponic production include vermicast mixed with coconut fibre, composted bark or peat, with smaller additions of high-quality composts or perlite where organically allowable. Vermicast may also be used in seed-raising mixes, container mixes for potted plants and during the production of organic transplants.

For longer-term hydroponic crops such as tomatoes, capsicum or cucumbers, bags or beds of organic substrate may be top-dressed with fresh vermicast at regular intervals during cropping, largely to reintroduce a diverse range of species of beneficial microbe populations. This has the benefit of re-inoculating the substrate with microorganisms and increasing the range of species present in the root zone because, over time, the substrate can become dominated by a limited range of microbial species. Vermicast may also be used to inoculate solution culture systems with microorganisms by introducing a liquid vermicast extract on a weekly basis. Vermicast extracts may also be used as a foliar fertilizer provided sufficient dilution and organic surfactants/wetting agents are used.

Vermicast leachate or 'vermiliquer' is produced during the vermicomposting process as the microorganisms release water during the decomposition of organic material; this liquid is drained from the system to prevent oversaturation (Quaik *et al.*, 2012). Collection of the vermicast leachate from the base of the system is ongoing during the vermicomposting process and this is often sold as an organic liquid fertilizer product. While this leachate does contain minerals and microorganisms, it has usually drained through the system which still contains unprocessed and partially processed organic material and is therefore not as effective to use in hydroponics as the fully completed solid vermicast. For use in organic hydroponic systems, vermicast which has been processed through to completion in the vermiculture system (in much the same way

as a mature compost) is more effective particularly in organic solution culture. High-quality, solid vermicast can be extracted into water for use as an organic nutrient solution and has been successfully used in NFT systems to grow low-nutrient-requiring crops such as lettuce. Vermicast liquid or leachate collected directly from vermicast systems may be used as an organic hydroponic nutrient solution if it is collected from a mature system or one that has been processing for at least 8–10 weeks to minimize the presence of unprocessed organic material. Studies have shown that the use of such fully processed vermiliquer, when pH adjusted, can produce growth and yields in crops such as pak choi that are comparable with those treated with a conventional inorganic fertilizer (Churilova and Midmore, 2019). Other essential factors which contribute to the success of vermiculture solutions as an organic hydroponic nutrient are temperature and DO. As well as inorganic material dissolved in water, vermicast liquids contain diverse organic compounds and microbiota, the metabolism and dynamics of which are greatly regulated by temperature and the concentration of DO (Churilova and Midmore, 2019).

7.7 Using Vermiculture Liquids in Hydroponics

The properties of vermicast solution used as a nutrient source for hydroponic production vary depending on the type of organic raw material fed into the vermiculture system and the processing time. It has been reported that the resulting mature vermiliquer typically has an initial EC between 1.5 and 2.6 dS/m with pH values ranging from 7.6 to 9.2, total levels of nitrogen between 120 and 227 mg/l, potassium values of 81–236 mg/l and phosphorus values of 40–87 mg/l (Churilova and Midmore, 2019). While total nitrogen levels are usually sufficiently high in mature vermicast solutions, this is present in both the organic and inorganic forms with the former being more slowly available (Churilova and Midmore, 2019).

pH buffering with acid may be required with vermiculture leachates as the optimal pH for worm productivity is higher than that typically run for hydroponic crop growth. One potential issue with vermicast solutions, as with many organic nutrient sources, can be the presence of sodium, the levels of which are highly dependent on the initial raw materials fed into the vermicast system. High levels of sodium have been found in vermiculture solutions originating from chicken manure vermicast (Pant *et al.*, 2009) and other animal manures. Sodium levels can be managed by monitoring and solution replacement in recirculating systems but are rarely a problem with sodium-tolerant crops such as tomatoes.

7.8 Composting for Organic Nutrient Processing and Substrate Preparation

Compost is a material which has undergone aerobic and thermophilic decomposition (composting) of organic matter to produce a stable, humus-like product. A diverse range of materials may be composted including plant debris, prunings, food waste, wood waste, manures, sewage sludge, municipal solid waste, food industry and processing waste, and by-products of other industries such as sugarcane fibre, rice and peanut hulls, cotton gin waste, papermill waste, fish and shellfish waste, apple pomace, feather meal, blood and bone, maize cobs, seaweed, and olive or grape marc and other brewery or distillery waste. Compost is one of the oldest soilless substrates in horticulture and the use of composted organic wastes as growing media has been increasing globally over the last 40 years (Barrett *et al.*, 2016). Recently the use of composts as a growing substrate has developed more of an environmental incentive as composting allows for the reuse of many waste materials that would otherwise end up in landfill or incineration plants (Raviv, 2013).

Due to the wide availably and low cost of composts, these materials are a commonly used component of many growing mixes and are often combined with other materials such as peat, bark or coconut fibre to give a more stable structure and to improve physical properties. The main issue with the use of compost as a hydroponic growing medium is that composts vary considerably in physical, chemical and biological properties depending on the raw materials initially used in decomposition and the duration and nature of the composting process. Bulk density, air space, water retention, pH, nutrient levels, and microbial density and populations are highly variable with composts (Raviv, 2011), making standardization of container and soilless mixes difficult. Some countries, such as the UK and the Netherlands, have developed strict quality control procedures which are essential for the preparation of commercial quantities of compost for use in growing media (Wever and Scholman, 2011).

The main advantages for the use of compost as a hydroponic growing substrate are not only the utilization and safe disposal of waste organic matter via the composting process, but also the characteristics and performance of composts in containers, which are similar to peat with a considerably lower cost and also a high suppressiveness of many soilborne diseases (Fascella, 2015). Composts also naturally contain some plant-available nutrients; however, these can be variable and high salt content may be an issue with some materials. Disadvantages of composts as a growing substrate, apart from considerable variability, include the possibility of contamination principally from herbicides used on vegetation that is subsequently composted as well as other phytotoxicities, heavy metal issues or potential for the presence of human pathogens originating from raw materials or incomplete decomposition processes. Some compost may have a high salinity and require leaching before use and pH may also be outside the ideal range for plant growth. During the compost production process, identification of when compost is fully mature enough to be used as a growing medium can be difficult and lead to quality issues and biological instability. Young composts which are not fully stable contain organic compounds that undergo secondary degradation while in use as a growing medium. This leads to problems with oxygen and nitrogen (nitrogen drawdown) deficiencies in the root zone. Mature and fully processed composts are fully stable, and this is also a prerequisite for suppressiveness of many root pathogens (Fascella, 2015).

Due to the variability of composts and potential for secondary breakdown while in use as a hydroponic growing medium, composts are often used in combination with more stable materials to create blended growing media. The most commonly used blend is compost and peat. Such combinations aim to produce a substrate with a high hydraulic conductivity, air-filled porosity and easily available water content as well as an improved fertilizing capacity of the substrate. Composts with a naturally high pH also help counter the low pH of peat; however, blending ratios need careful consideration to create the optimum root environment for plant growth and avoid negative effects such as high soluble salt contents.

Mature composts that have undergone a sufficient curing stage have large and diverse microbial populations which have been reported to assist in the suppression of a wide range of phytopathogenic fungi (Hadar and Gorodecki, 1991). However, these beneficial microbial populations are destroyed if the compost is sterilized and the material then loses much of this disease-suppressive effect (Raviv *et al.*, 1998). The fact that mature composts are known to be suppressive of several soilborne pathogens to which peat is conducive has encouraged the use of compost/peat blends or the complete substitution of peat with composts (Raviv *et al.*, 2002).

7.9 Organic Materials for Vermicast, Composting and Biodigester Systems

7.9.1 Organic fertilizer/nutrient sources

The composition of raw materials fed into decomposition systems to create suitable hydroponic substrates and nutrient sources is vitally important to the quality of the final product. Organic raw materials are typically inconsistent with regard to mineral composition and often contain unwanted elements such as sodium, and for this reason several different materials may be blended in the same system to create a more optimal nutrient balance for crop growth. Organic fertilizers are those which are derived from naturally occurring materials including animal manures and guano, wastes and slaughterhouse by-products, plant materials, treated sewage sludge, peat, fish and seaweed materials. Under organic crop production standards other materials that are not carbon based but are naturally occurring, mined minerals such as limestone, rock phosphate and sodium nitrate may be considered to be organic fertilizers and used in these processing systems. While most organic fertilizers are relatively low in the concentration of essential plant nutrients, many have the advantage of increasing microbial diversity and populations.

7.9.2 Animal sources of organic fertilizers

Animal manures are commonly used organic fertilizer sources, often being relatively inexpensive when locally available or if produced onsite in farming operations. Animal manures may include poultry shed/chicken litter, farmyard manure (pig, cattle, sheep, goat, etc.), liquid manure or slurry. The mineral composition of manures varies considerably between sources, making accurate application of nutrients to match plant growth profiles difficult. Animal manures are a slow-release form of organic fertilizer: for nitrogen in farmyard manure, about 20–30% of the amount applied is available to the crop during the first year (Reetz, 2016). Organic fertilizers of animal origin used in the production of food crops must be fully and correctly composted or microbially processed via digester systems before application to destroy any pathogens present. Other animal-based organic fertilizers include by-products of slaughterhouses such as blood and bone meal, hoof and horn meal, hide and leather dust, fish meal and feather meal. These are dried and ground for application or may be compressed into pellets. The rate of release of nutrients from animal by-products is highly dependent on the speed of microbial action and factors such as moisture levels, oxygenation and temperature. Nutrient release from organic fertilizers is significantly faster under warmer temperatures which increase the rate of microbial breakdown.

7.9.3 Plant-based inputs

Plant-based organic fertilizer sources comprise a diverse range of materials including composts, crop residues, green crops, residues from other industries such as bagasse (sugarcane waste), rice hulls, oilseed waste, wood fibre and timber industry residues, seaweeds, dried algae, domestic and industrial green waste, and a wide range of vegetable materials from food processing industries. Processed organic fertilizers derived from plant materials include humic acid, amino acids and other enzyme-digested proteins.

7.10 Aquaponics

Aquaponics is a system of crop production that combines fish farming (aquaculture) with soilless crop production (hydroponics). In aquaponic systems the fish consume food and excrete waste primarily in the form of ammonia. Bacteria convert the ammonia to nitrite and then to nitrate (Khater and Ali, 2015) which is used for plant nutrition. In the USA, aquaponic systems using suitable inputs, including organic fish food, can become organically certified; however, in much of the rest of the world aquaponics is not considered to be organic as soil is not used for crop production.

Aquaponic systems may consist of a number of different designs of varying levels of technology. In the simplest systems, fish are produced in large tanks or ponds, while crops are grown suspended on floating rafts or platforms with their roots growing in the nutrient-rich fish production water. In more complex systems, the wastewater from the fish production operation is removed, solids are collected, and the remaining water transferred to tanks where microbiological processes carry out mineralization of the organic matter into plant-usage nutrients. The treated aquaponic solution is then incorporated into a hydroponic system to grow crops where additional nutrients may be added if required to obtain a balanced solution for growth. An alternative system, often termed 'total waste aquaponics' or 'complete aquaponics', is where all of the wastewater containing the solids is fed directly on to large, open media beds, often gravel, containing the plants and beneficial bacteria which carry out the waste conversion. The total waste system with large media-filled beds is often used for longer-term crops. In all aquaponic systems a compromise needs to be achieved between the needs of the crop and those of the fish species, mainly in terms of the nutrient strength and pH. This system relies largely on the health and functioning of a group of bacteria responsible for converting the fish waste into usable plant nutrients.

7.11 Organic Hydroponic Production Systems

Organic nutrients have been successfully used in a range of soilless systems from substrate culture to NFT and other solution culture methods. Since these systems must provide optimal conditions for both plant growth and microbial populations to carry out nutrient conversion, they are more complex than conventional hydroponics and often prone to nutrient disorders.

Drip-irrigated hydroponic systems which use a substrate to support plants are often more successful than recirculating solution culture for higher nutritionally demanding crops such as tomatoes, capsicum and cucumber with organic nutrition (Fig. 7.2). Organic substrates such as compost and vermicast typically contain some mineralized nutrients to provide early nutrition to the crop, while naturally occurring microbial

Fig. 7.2. Drip-irrigated organic substrate system.

species are present to start the process of nitrification. These are often blended with substrates such as coconut fibre, peat, perlite and similar materials which provide physical support and structure to the growing medium, helping to maintain aeration through the crop cycle. The high surface area of substrates provides a suitable environment for a biofilter to establish within the root zone where populations of microbial species reside and carry out the mineralization process. Organic liquid nutrient solutions are then drip irrigated on to the crop to provide the majority of the nutrients required for growth. Microbial inoculant products may be added to both the substrate pre-planting and top-dressed throughout the cropping cycle and also incorporated into the liquid nutrient solution for drip irrigation application. Substrates may have additions of small quantities of solid, allowable organic fertilizer inputs incorporated pre-planting for slow-release nutrient supplementation. Growing crops such as hydroponic greenhouse tomatoes using organic nutrient sources and substrates is carried out on a commercial scale in the USA where such systems are fully organically certified and often

highly successful (Fig. 7.3). Greenhouse tomatoes produced organically have been found to be comparable to those produced conventionally in regard to nutritional status, plant development and harvest yields (Rippy et al., 2004).

In recirculating hydroponic systems, using organic nutrients generates a number of challenges for growers. Obtaining plant yields and quality comparable to non-organic systems is more difficult, pH fluctuations are common and must be controlled using only organic methods. EC is not a reliable method to determine the concentration of nutrients and unwanted elements such as sodium are often high in organic fertilizer sources (Williams and Nelson, 2016). Despite these challenges, organic NFT and float/raft or pond systems are in existence and can produce high-yielding crops. The nature of soilless culture systems somewhat restricts the development of an extensive biofilter as no large volumes of solid substrate with a high surface area are present to house microbial species. However, biofilms containing nitrification bacteria and other microorganisms do develop on system surfaces such as NFT channels, tank walls and floors, floating

Fig. 7.3. Organic hydroponic tomato crop in the USA.

rafts, root systems and other components. Some solution culture systems incorporate additional surfaces for microbial species to reside on and these biofilters can assist the rate of nitrification. Use of slow sand filtration within recirculating systems is another efficient method of establishing a mature biofilter for the purposes of microbial colonization, nutrient mineralization and disease suppression.

Nutrient management with organic hydroponic production requires a different approach to the use of conventional nutrient solutions. Since many of the nutrient compounds present are organic, they do not conduct an electrical current, thus EC is not a precise guide to nutrient concentration in organic systems. Nutrient analysis may be carried out on organic nutrient solutions; however, this will only measure nutrient ions that have been mineralized and not organic compounds that are yet to undergo the nitrification process, so does not give a complete guide to nutrition levels. Nutrient solution analysis is useful to determine accumulation of unwanted elements such as sodium which are often present in recirculating organic systems and thus provides a guide to when

solutions need replacement. pH adjustment in recirculating organic nutrient systems is another issue which requires monitoring and adjustment to both maximize root uptake of nutrient ions as well as ensure nitrification bacteria are not inhibited. pH levels can change rapidly in recirculating organic nutrient solutions and in the root zone of substrate-grown plants depending on microbial activity and ion uptake. Low pH values can be corrected with solutions prepared from calcium carbonate and sodium bicarbonate while pH may be lowered with the use of organically allowable acids, with citric acid being the most commonly used for this purpose. Organic hydroponic systems are typically run at higher pH levels than conventional nutrient solutions, to allow for the higher pH preference of nitrification bacteria. In substrate-based organic systems, the medium's pH appears to play less of a role in nutrient availability than in conventional hydroponic systems. Good yields of organic tomato crops have been reported to be produced on substrates with a naturally high pH with yields not linked to pH levels in the same way they are in conventional hydroponic systems (Rippy *et al.*, 2004).

7.12   Biofilms in Organic Hydroponic Systems

Biofilms of microorganisms have been observed to develop on the root hairs of roots submerged in organic hydroponic solutions but were not found in conventional hydroponic systems (Shinohara et al., 2011). It is likely that nitrification bacteria reside in these root-system biofilms which are able to degrade organic nutrient solutions into ammonium and nitrate that are absorbed directly by the roots without diffusion into the hydroponic solution (Shinohara et al., 2011). This can mean that while mineral analysis of an organic hydroponic solution in recirculating systems such as NFT may show minimal nitrate levels, plants may be more than adequately supplied by nitrification occurring directly on the root system.

Biofilms may confer other advantages to organic hydroponic crops apart from nutrient mineralization. It has been claimed that organic hydroponics could be an effective method for both soilborne and airborne disease control (Chinta et al., 2015) through the development of biofilms in the root zone and subsequent induction of systemic disease resistance (Ramamoorthy et al., 2001; Saharan and Nehra, 2011). The presence of beneficial microorganisms contained in organic nutrient solutions has been shown to have an antagonistic action against the development of root rot by Fusarium oxysporum in lettuce grown in organic hydroponics (Fujiwara et al., 2012, 2013; Chinta et al., 2015) and against grey mould (Botrytis cinerea) in lettuce and cucumber (Chinta et al., 2015). Organic hydroponic nutrient solutions result in the development of biofilms in the rhizosphere rich in microorganisms, which has been found to lead to a reduction of disease symptoms in the above-ground plant parts due to the process of induced systemic resistance (Pieterse et al., 2003). Further to this, the use of organic nutrient solutions as foliar sprays has similar effects of disease suppression for some crops such as cucumbers (Chinta et al., 2015). It has been suggested that the jasmonic acid/ethylene signalling pathway is involved in induced systemic disease resistance whether organic solutions are used in the root zone or as a foliar spray. The potential benefits of organic nutrient solutions on plant root and foliar disease suppression may be a major advantage of organic hydroponic methods, even if the system cannot be certified as organic in the country of production.

7.13   Nutrient Amendments

Organic nutrient solutions may be applied to simply utilize waste streams from other industries in countries where organic certification is not allowable. In these circumstances, organic nutrients may be amended with conventional inorganic fertilizers to make up shortfalls in certain elements, typically nitrogen and calcium with the addition of calcium nitrate, as there are no organic certification regulations to comply with. The majority of nutrients may be supplied via organic fertilizers, with only minor additions required to boost elemental levels.

In countries such as the USA, where organic hydroponic systems are able to be organically certified, any additions to the nutrient solution must be allowable under the organic production standards and guidelines. Lists of allowable inputs provide information on which fertilizers may be used under organic crop production and a number of these are suitable for hydroponic production. Organic nutrient solutions may require amendment or blending with a number of different organic liquid concentrates and allowable fertilizer salts to provide sufficient and balanced nutrition, particularly for crops such as tomatoes, capsicum and cucumbers with a higher nutrient requirement. Depending on the certification agency, naturally occurring fertilizers such as potassium sulfate and magnesium sulfate can be used to amend organic nutrient solutions where required. Sulfate trace elements are also allowable under most production standards for supplementation and to correct deficiencies where they occur. While synthetic iron chelates such as Fe EDTA are not organically allowable, iron sulfate can be used and mixed with an organic chelation agent such as citric acid. In the USA, OMRI provides a list of all organically acceptable and restricted products and inputs which may be used in crop production. This makes selection of suitable inputs and nutritional additives more straightforward as products display their OMRI status on labels and are easily identified as suitable to use under organic production.

7.14 Organic Certification in the USA

While hydroponics or soilless cropping is not recognized as a form of organic horticulture in most of the world, in the USA, hydroponic growers using solution culture, container culture or aquaponics methods can become organically certified. At the current time, the NOP in the USA allows certification of many hydroponic systems provided the producer can demonstrate compliance with organic regulations. There are organic hydroponic operations in the USA that are certified by USDA-accredited certifying agents based on the current regulations and the operations' organic system plan (Organic Trade Association, 2020). In 2010, the US National Organic Standards Board (NOSB) defined hydroponics as 'the production of normally terrestrial, vascular plants in nutrient rich solutions or in an inert porous, solid matrix bathed in nutrient rich solutions'. In 2017 the acceptance of hydroponic and aquaponic methods for organic production came under review by the NOSB, an advisory body to the USDA. It was voted not to prohibit hydroponic and aquaponic farms from USDA organic certification and for them to remain recognized as organic; however, aeroponics was voted to become a prohibited practice under organic production (HortiDaily, 2017). In the USA it is likely that proponents of soil organic farming will continue to fight against the certification of organic hydroponic and aquaponic methods under the reasoning that these systems do not promote soil health and microbial populations.

In order to become organically certified in the USA under the USDA organic program, growers must undergo a certification process before being permitted to display the USDA organic label on produce. This process involves first contacting a NOP-accredited agent who will certify a soilless operation, and creating a suitable organic system plan for the hydroponic operation, before following through the USDA organic certification process. The organic certification agent reviews applications and if these are compliant will then carry out an onsite inspection to view the production and handling facility. This inspection will determine if the current site or growing operation is suitable for organic production including verification that no prohibited substrates will come into contact with the product. The certification process then involves a report being prepared from the onsite visit and this along with application files and interviews are reviewed for compliance. If approval is granted, the grower pays the required fees and is permitted to label the produce or company as USDA-certified organic. Ongoing records must be kept of all inputs and procedures carried out during crop production to maintain organic certification status and regular inspections are carried out. Organic hydroponic growers must only use allowable inputs for crop production, packaging and processing, and lists of these as well as production guidelines are provided by the organic certification agency the grower is working under. In addition to these, OMRI (www.omri.org, accessed 4 September 2020) provides detailed lists of organically acceptable products, their manufacturers or suppliers, and contact details of where these inputs can be obtained. Organically allowable inputs, including fertilizers, media, pest and disease control products, and many others, are listed by generic material, supplier and product name which is a valuable resource for organic hydroponic producers in the USA.

7.15 Organic Pest and Disease Control

Organic certification standards have a comprehensive list of products and compounds which are permitted for pest and disease control under organic greenhouse production. These usually include products such as those that contain natural formulations of *Bacillus thuringiensis* (Bt) for caterpillar control, botanical pesticides such as neem, pyrethrum, oil and soap-based smothering agents, microbial products and plant extracts. Compounds for disease control include bicarbonates, biological controls, copper- and sulfate-based products and microbial formulations. Lists of allowable organic pest and disease control options often change and are regularly updated as new technologies and products are developed. Organic pest and disease control relies heavily on prevention of problems via methods such as greenhouse screening and hygiene practices, regular crop scouting and early detection methods, environmental control for disease prevention, and the use of predators and parasites for integrated pest management (IPM) programmes.

7.16   Hybrid Systems

With an increasing focus on nutrient recycling and waste management, the development of 'hybrid' hydroponic/organic systems has become an area of research. While organic hydroponics may not be certifiable in most of the world, organic materials may still be incorporated into conventional hydroponic production systems. This may be as components of substrates, as nutrient solution additives or as foliar sprays. Recent research has often focused on the use of sustainable substrates for hydroponic production; these include composts, vermicasts and waste material from other industries such as rice hull. Hybrid systems typically utilize carbon-based growing substrates such as coconut fibre or peat which have been shown to develop stable and distinctive microbial communities in soilless systems (Grunert *et al.*, 2016). Incorporation of fully processed organic waste (municipal solid wastes compost and vermicast) into a peat/perlite soilless medium at a rate of 25% has been shown to enhance the growth and yield

of tomato in hydroponics (Haghighi *et al.*, 2016). It is likely that the presence of organic composts or vermicast may promote growth by the introduction of beneficial microbial species and by the presence of plant-growth-promoting substances (Zhang *et al.*, 2014). Other studies have found similar results for the use of liquid organic additives in conventional hydroponic systems. Priadi and Nuro (2017) found that a combination of organic and inorganic nutrient solutions (25%/75%) resulted in the highest seedling growth parameters of hydroponic rockwool-grown pak choi; however, the organic solution alone was not suitable for seedling production. Other studies have reported similar findings, with compost and vermiculture extracts increasing the growth and mineral content of pak choi growing in peat/perlite substrate, the results being attributed to the presence of nitrogen and gibberellin in the extracts (Pant *et al.*, 2012). Hybrid systems which incorporate some form of organic nutrients or additives in the root zone are still reliant on the presence of microorganisms to break down these compounds and the use of solution sterilization methods in recirculating systems is not recommended because of this.

7.17   Issues Commonly Encountered with Organic Hydroponic Systems

Apart of the lack of uniformity regarding what organic hydroponics actually encompasses on a worldwide scale, being a certifiable system in the USA but not in most other nations, the system has a number of technical challenges. With soilless systems not officially recognized as being certifiable as organic in much of the world, there has been limited research on the method and the development of effective commercial-scale hydro-organic nutrient concentrate products has been largely restricted to the USA. There is also limited technical information to support these systems and more research is needed to provide recommendations for appropriate substrate mixes and nutrient management (Rogers, 2017). Whether a hydroponic system is organically certified or simply using additions of organic material as substrates or nutrients, a number of issues commonly arise. Many of these originate from the process of microbial breakdown of organic materials, which places an increased demand on oxygen within the root zone and may not occur rapidly enough to supply plant-available nutrients to keep pace with rapid rates of growth. For this reason, the use of organic nutrients often leads to reduced growth and yields as compared with conventional hydroponic systems, nutrient deficiencies and toxicities, and accumulation of unwanted elements such as sodium. Use of organic nutrient sources also creates new issues with EC and pH measurement and control, with dramatic pH fluctuations being common (Williams and Nelson, 2016). The use of organic substrates such as composts in hydroponic systems can work successfully; however, there is considerable variation in the physical and chemical properties of composts and a lack of uniformity between batches. Immature and unstabilized composts result in the immobilization of nitrogen, may contain phytotoxic compounds and have high EC levels (Rogers, 2017).

One of the main issues faced by growers aiming to use organic inputs in soilless systems is the very wide range of materials to select from, many of which are proven in soil systems but may not necessarily be suitable for hydroponic production. With no universal and proven information or guidelines on how organic materials can be used in soilless systems, those who have successfully incorporated such materials have often done so through trial and error. These systems are often characterized by having a high rate of mineralization and oxygenation within the root zone, healthy and stable populations of microorganisms, correct timing of nutrient inputs and monitoring of nutrient levels. Many organic systems make use of pre-processing of organic materials so that at least partial mineralization has occurred before introduction into the production system. This may include nutrient processing such as vermicast and composting systems and using anaerobic and aerobic digester systems to break down organic raw materials into

plant-available nutrients which may be supplied via the growing substrate and as nutrient solutions. While both solution culture and substrate-based hydroponic systems have been used with organic nutrients, those incorporating a solid growth medium appear to be more reliable, possibly due to more favourable conditions for microbial species to colonize and proliferate and a larger overall root volume per plant. Soilless culture such as NFT and pond/raft/float systems are used with organic nutrients, particularly in the USA where these systems are organically certifiable, however nutrients must be more highly processed before being applied to such systems.

7.18 Conclusions

While soil-based organic greenhouse systems are rapidly growing in popularity worldwide due to premiums received for organically certified produce, organic hydroponics continues to be a controversial issue. Organic nutrient sources and substrates can be utilized in soilless systems with varying results, however acceptance of this method as 'organic' is not universal. With organic hydroponics not being a certifiable system in much of the world, research, technology and information on the use of organic materials are unlikely to be prioritized. Even in the USA where hydroponic and aquaponics systems can be organically certified at the present time, there is likely to be further arguments and petitions against the certification of soilless production lead by those heavily involved in soil-based organics.

Organic systems are more complex, challenging and unpredictable than conventional hydroponic systems and require a higher degree of management. Organic nutrient sources are variable, usually not optimally balanced for crop production and often contain unwanted elements. With microbial breakdown required to release nutrients for plant uptake, nutrient timing is critical and mineral toxicities and deficiencies are more common in organic hydroponic crops. Conversely, the use of organic materials as both substrates and nutrient sources in soilless cropping offers a potential solution to the management of some waste management streams, with by-products such as poultry and biogas manure, green waste, crop and food waste, aquaculture effluent and others having been successfully used in organic hydroponic systems. Utilization of waste streams and recycling of nutrients are an area of concern for many consumers and producers and even if organic hydroponic systems cannot be certified, absorbing these by-product nutrients for crop production is worth considering.

While organic hydroponics and its current status of not being organically certifiable may mean larger commercial producers lack interest in the system, smaller growers, home gardeners, urban projects, students and aquaponic operations can still operate using organic materials and nutrient sources. These may be used alone or in combination with conventional hydroponic fertilizers and there is a growing market for 'organic additive' products designed to increase microbial diversity within traditional hydroponic systems. Hydroponic systems incorporating organic nutrients may not be universally recognized as fully organic; however, the possibilities of organic and waste material recycling, improved microbial diversity and other potential benefits of the systems should not be overlooked.

References

Adhikary, S. (2012) Vermicompost, the story of organic gold: a review. *Agricultural Sciences* 3(7), 905–917.

Barrett, G.E., Alexander, P.D., Robinson, J.S. and Bragg, N.C. (2016) Achieving environmentally sustainable growth media for soilless plant cultivation systems – a review. *Scientia Horticulturae* 212, 220–234.

Buchanan, M.A., Russelli, E. and Block, S.D. (1988) Chemical characterization and nitrogen mineralization potentials of vermicomposts derived from differing organic wastes. In: Edwards, C.A., and Neuhauser, E.F. (eds) *Earthworms in Environmental and Waste Management*. SPB Academic Publishers, The Hague, The Netherlands, pp. 231–239.

Chanda, G.K., Bhunia, G. and Chakraborty, S.K. (2011) The effect of vermicompost and other

fertilisers on cultivation of tomato plants. *Journal of Horticulture and Forestry* 3(2), 42–45.

Chinta, Y.D., Eguchi, Y., Widiastuti, A., Shinohara, M. and Sato, T. (2015) Organic hydroponics induces systemic resistance against the air borne pathogen, *Botrytis cinerea* (gray mould). *Journal of Plant Interactions* 10(1), 243–251.

Churilova, E.V. and Midmore, D.J. (2019) Vermiliquer (vermicompost leachate) as a complete liquid fertiliser for hydroponic grown pak choi (*Brassica chinensis* L.) in the tropics. *Horticulturae* 5(1), 26. https://doi.org/10.3390/horticulturae5010026

Fascella, G. (2015) Growing substrates alternatives to peat for ornamental plants. In: Asaduzzaman, M. (ed.) *Soilless Culture – Use of Substrates for the Production of Quality Horticultural Crops*. InTech Open. Available at: https://cdn.intechopen.com/pdfs-wm/47996.pdf (accessed 4 September 2020).

Fujiwara, K., Aoyama, C., Takano, M. and Shinohara, M. (2012) Suppression of *Ralstonia solanacearum* bacterial wilt disease by an organic hydroponic system. *Journal of General Plant Pathology* 78, 217–220.

Fujiwara, K., Iida, Y., Iwai, T., Aoyana, C., Inukai, R., et al. (2013) The rhizosphere microbial community in a multiple parallel mineralisation system suppresses the pathogenic fungus *Fusarium oxysporum*. *MicrobiologyOpen* 2, 997–1009.

Goddek, S., Schmautz, Z., Scott, B., Delaide, B., Keesman, J.K., et al. (2016) The effect of anaerobic and aerobic fish sludge supernatant on hydroponic lettuce. *Agronomy* 6(2), 37.

Greer, L. and Diver, S. (2000) *Organic Greenhouse Vegetable Production*. ATTRA, Fayetteville, Arkansas. Available at: https://nofany.org/wp-content/uploads/2020/04/Organic_Greenhouse_Vegetable_Production_ATRRA.pdf (accessed 4 September 2020).

Grunert, O., Hernandez-Sanabria, E., Vilchez-Vargas, R., Jauregui, R., Pieper, D.H., et al. (2016) Mineral and organic growing media have distinct community structure, stability and functionality in soilless culture systems. *Scientific Reports* 6, 18837.

Hadar, Y. and Gorodecki, B. (1991) Suppression of germination of sclerotia of *Sclerotium rolfsii* in compost. *Soil Biology and Biochemistry* 23(3), 303–306.

Haghighi, M., Barzegar, M.R. and Teixeira da Silva, J.A. (2016) The effect of municipal solid waste compost, peat, perlite and vermicompost on tomato (*Lycopersicum esculentum*) growth and yield in a hydroponic system. *International Journal of Recycling and Organic Waste in Agriculture* 5, 231–242.

Han, Z., Zeng, X. and Wang, F. (1989) Effects of autumn foliar application of ^{15}N-urea on nitrogen storage and reuse in apple. *Journal of Plant Nutrition* 12(6), 675–685.

HortiDaily (2017) NOSB: hydroponic and aquaponics are USDA organic. *HortiDaily*, 2 November 2017. Available at: https://www.hortidaily.com/article/6038850/nosb-hydroponic-and-aquaponics-are-usda-organic/ (accessed 4 September 2020).

Khater, E.S.G. and Ali, S.A. (2015) Effect of flow rate and length of gully on lettuce plants in aquaponic and hydroponic systems. *Journal of Aquaculture Research and Development* 6, 318. https://doi.org/10.4172/2155-9546.1000318

Liedl, B.E., Cummins, M., Young, A. and Williams, M.L. (2004) Hydroponic lettuce production using liquid effluent from poultry waste bioremediation as nutrient source. *Acta Horticulturae* 659, 721–728.

Liu, W.K., Yang, Q.C. and Du, L. (2009) Soilless cultivation for high quality vegetables with biogas manure in China: feasibility and benefit analysis. *Renewable Agriculture and Food Systems* 24(4), 300–307.

Neal, J. and Wilkie, A.C. (2014) Anaerobic digester effluent as fertiliser for hydroponically grown tomatoes. *University of Florida, Journal of Undergraduate Research* 15(3), 1–5.

Organic Trade Association (2020) Hydroponics. Available at: https://ota.com/advocacy/organic-standards/emerging-standards/hydroponics (accessed 4 September 2020).

Pant, A.P., Radovich, T.K.J., Hue, N.V., Talcott, S.T. and Krenek, K.A. (2009) Vermicompost extracts influence growth, mineral nutrients, phytonutrients and antioxidant activity in pak choi (*Brassica rapa* cv. Bonsai, Chinesis group) grown under vermicompost and chemical fertiliser. *Journal of the Science of Food and Agriculture* 89(14), 2383–2392.

Pant, A.P., Radovich, T.J.K., Hue, N.V. and Paull, R.E. (2012) Biochemical properties of compost tea associated with compost quality and effects on pak choi growth. *Scientia Horticulturae* 148, 136–146.

Pieterse, C.M.J., van Pelt, J.A., Verhagen, B.M.W., Ton, J., van Wees, S.C.M., et al. (2003) Induced systemic resistance by plant growth promoting rhizobacteria. *Symbiosis* 35, 39–54.

Priadi, D. and Nuro, F. (2017) Seedling production of pak choy (*Brassica rapa* L.) using organic and inorganic nutrients. *Biosaintifika* 9(2), 217–224.

Quaik, S., Embrandiri, A., Rupani, P.F. and Ibrahim, M.H. (2012) Potential of vermicomposting leachate as organic foliar fertiliser and nutrient solution in hydroponic culture: a review. *2nd International Conference on Environment and Bioscience IPCBEE* 44, 43–47.

Ramamoorthy, V., Viswanathan, R., Raguchander, T., Pradasam, V. and Samiyappan, R. (2001) Induction of systemic resistance by plant growth promoting rhizobacteria in crop plants against pests and diseases. *Crop Protection* 20(1), 1–11.

Raviv, M. (2011) The future of composts in growing media. *Acta Horticulturae* 891, 19–32.

Raviv, M. (2013) SWOT analysis of the use of composts as growing media constituents. *Acta Horticulturae* 1013, 191–202.

Raviv, M., Reuveni, R., Krasnovsky, A., Medina, S., Freiman, L. and Bar, A. (1998) Compost as a controlling agent against *Fusarium* wilt of sweet basil. *Acta Horticulturae* 469, 375–381.

Raviv, M., Wallach, R., Silber, A. and Bar-Tal, A. (2002) Substrates and their analysis. In: Savvas, D. and Passam, H. (eds) *Hydroponic Production of Vegetables and Ornamentals*. Embryo Publications, Athens, pp. 25–89.

Reetz, H.F. Jr (2016) *Fertilizers and Their Efficient Use*, 1st edn. International Fertilizer Industry Association, Paris.

Rippy, J.F.M., Peet, M.M., Louws, F.J., Nelson, P.V., Orr, D.B. and Sorensen, K.A. (2004) Plant development and harvest yields of greenhouse tomatoes in six organic growing systems. *HortScience* 39(2), 223–229.

Rogers, M.A. (2017) Organic vegetable crop production in controlled environments using soilless media. *HortTechnology* 27(2), 166–170.

Saharan, B.S. and Nehra, V. (2011) Plant growth promoting rhizobacteria: a critical review. *Life Science and Medical Research* 21, 1–30.

Saijai, S., Ando, A., Inukai, R., Shinohara, M. and Ogawa, J. (2016) Analysis of microbial community and nitrogen transition with enriched nitrifying soil microbes for organic hydroponics. *Bioscience, Biotechnology and Biochemistry* 80(11), 2247–2254.

Shinohara, M., Aoyama, C., Fujiwara, K., Watanabe, A., Ohmori, H., et al. (2011) Microbial mineralization of organic nitrogen into nitrate to allow the use of organic fertilizer in hydroponics. *Japanese Society of Soil Science and Plant Nutrition* 57(2), 190–203.

Ulusoy, Y., Ulukardesler, A.H., Unal, H. and Alibas, K. (2009) Analysis of biogas production in Turkey utilising three different materials and two scenarios. *African Journal of Agricultural Research* 4, 996–1003.

USDA (US Department of Agriculture) (2017) 2016 Certified organic survey: 2016 summary. Available at: https://downloads.usda.library.cornell.edu/usda-esmis/files/zg64tk92g/70795b52w/4m90dz33q/

OrganicProduction-09-20-2017_correction.pdf (accessed 4 September 2020).

Van der Lans, C.J.M., Meijer, R.J.M. and Blom, M. (2011) A view of organic greenhouse horticulture worldwide. *Acta Horticulturae* 915, 15–21.

Wever, G. and Scholman, R. (2011) RHP requirements for the same use of green waste compost in professional horticulture. *Acta Horticulturae* 891, 281–286.

Williams, K.A. and Nelson, J.S. (2016) Challenges of using organic fertilisers in hydroponic production systems. *Acta Horticulturae* 1112, 365–370.

Zhang, H., Tan, S.N., Wong, W.S., Ng, C.Y.L., Teo, C.H., *et al.* (2014) Mass spectrometric evidence for the occurrence of plant growth promoting cytokinins in vermicompost tea. *Biology and Fertility of Soils* 50, 401–403.

8 Propagation and Transplant Production

8.1 Introduction

Crop establishment is a fundamental process in greenhouse and hydroponic production. Seed and seedling delivery systems vary considerably between different types of crops and growing situations and with level of mechanization available. Large-scale commercial crops are largely reliant on fully automated methods of both seeding and transplanting; however, smaller-scale operations may use only manual labour or in combination with low-technology automation.

8.2 Propagation from Seed

Crop establishment from seed is the most common method of propagating self- and cross-pollinated plants and is used for a wide range of different species. Use of seed has the advantage in that it is typically the least expensive method of propagation, allows for convenient and long-term storage, ease of transportation and shipping large distances, and, because of the hard protective seedcoat, can be handled and sown using mechanized methods on an industrial scale. Seeds also provide a method for starting a crop with disease-free planting material where seed health and freedom from pests and pathogens have been maintained (Fig. 8.1). Seed treatments also help reduce the risk of disease infection during the early stages of crop establishment and are now extensively used in horticulture.

8.2.1 Hybrid seed versus open-pollinated seed

Hybrid seed is widely used in greenhouse production and results in significant improvement in yields, improved uniformity of crop growth, more rapid plant development and earlier yields, and disease resistance, as well as a number of other beneficial inbred traits. The process of producing hybrid seed results in heterosis or hybrid vigour and F1 hybrid seed is heterozygous in many genes. To produce hybrid seed, inbred lines are carefully selected based on superior and consistent traits and go through a process of repeated self-pollination. These inbred parental strains are then cross-pollinated with each other to produce first-generation (F1) hybrid seed. Hybrid seed is sown to produce a crop only, as successive seed collected from that crop will not have the same characteristics as the parental hybrid, thus growers must buy hybrids from seed companies each season. Hybrid seed is generally considerably more expensive than open-pollinated seed as the process of hybridization is labour intensive in many species, requiring specialized operations to ensure correct crossing of parental lines. Inbred parental lines of plants which have been selected for desirable characteristics often go through a process of self-pollination for many generations; for the production of tomato hybrids, this can be as many as six generations of inbreeding. Either line can be the female or male parent, but normally the highest seed yielder is selected as the female parent (Opena *et al.*, 2001). For plants which are self-pollinating, such as tomato and many others, the stamens are removed from the flower buds of the female line before they can shed their pollen and cause self-pollination. The female flower must be pollinated by the pollen from the selected male line only and this process of stamen removal is termed 'emasculation'. Emasculation of the female-line flower buds is carried out by hand using sharp pointed forceps and scissors, the flowers from the male parent plants are collected and used to

© L. Morgan 2021. *Hydroponics and Protected Cultivation* (L. Morgan)
DOI: 10.1079/9781789244830.0008

Fig. 8.1. Most hydroponic crops are propagated from seed.

extract pollen. This pollen is then collected, dried, prepared and used to pollinate the female-line flowers. Once pollination and fertilization have occurred, fruit is grown through to maturity when the hybrid seed is extracted using fermentation or mechanical extraction methods.

F2, or second-generation, hybrid seed is produced by self- or cross-pollination of F1 hybrid plants. F2 hybrids often retain some of the advantageous characteristics of the F1 hybrid parents but may also have lower yields and less uniformity. The main advantage of F2 hybrids is that they typically have improved performance over non-hybrid seed and can be produced by open pollination in the field without specific requirements for inbred lines, emasculation or other methods of pollination used in F1 seed production. Several ornamental and some vegetable cultivars are sold as F2 hybrids, largely to reduce the cost of seed over F1 hybrids: retail seed cost of F2 hybrids is less than one-half that of a comparative F1 hybrid cultivar (Bosland, 2005). F2 hybrids are commonly supplied for cultivars of some bedding plants including geranium, pansy and portulaca, and for vegetables such as asparagus and capsicum.

Unlike the highly uniform growth and yields of hybrid seed, open-pollinated varieties will produce plants which are roughly identical to their parents or referred to as 'breeding true', but can be somewhat less stable in their characteristics and prone to a degree of genetic diversity. Open-pollinated seed crops, termed 'standard varieties', may be either self-pollinated or cross-pollinated in the field by insects or wind and are less labour and technology intensive than hybrid seed production.

8.2.2 Seed treatments – pelleting, coating and priming

Applying layers of material to the seedcoat serves a number of purposes. For very small, fine or irregularly shaped seed, applying a thick layer of pelleting material facilitates the ease of separation and handling, uniformity of seed placement and precision sowing for both manual and mechanical sowing operations (Fig. 8.2). Pelleting material may be composed of clay-based mixtures or inert polymers with various chemical compounds incorporated. These may include fertilizers, fungicides, microbial inoculants and other crop promotants which assist seed germination and seedling establishment. Seed pelleting materials must have properties that allow rapid water absorption and dissolution of the pellet after sowing, enabling the radicle and young shoot to emerge without hindrance. Pelletization is commonly used for crop species such as lettuce, herbs, onion and carrot and for a range of small-seeded ornamental crops.

Seed coatings may simply involve the application of a fungicide powder to provide initial protection against fungal pathogens which cause seedling rot or germination failure or seed may be 'film coated' with a thin polymer material applied directly over the

Fig. 8.2. Pelleted seed facilitates ease of handling and precision sowing.

seedcoat. Polymer film coats enclose any other chemicals or compounds such as fungicides or insecticides which have been applied to the seed, holding them in place. The polymer overcoat is typically highly coloured which allows more visibility during sowing seed and protects the handler from seed treatment chemicals. This provides a safe, clean and dust-free product which flows easily through sowing equipment for precision placement. Polymer seed coatings generally break down rapidly once sown and in contact with moisture.

Certain seeds may be 'primed' before sowing to increase the rate and percentage of germination. Many vegetable seeds may be primed; however, lettuce is the most commonly grown species to benefit from this pretreatment. Priming is the process of allowing controlled hydration of the seed so that much of the metabolic processes required for germination can occur while emergence of the radicle is prevented. Germination rates are often at least 40–50% more rapid than non-primed seed in many crop species. This gives faster germination and more uniform seedlings, as well as reducing the amount of time the germinating seed may be exposed to adverse environmental

conditions, pathogens or other factors which might otherwise limit germination.

8.2.3 Seed storage

The viable storage life of a seed varies with species and is highly dependent on the initial moisture content of the seed and the conditions under which it is stored. For commercial storage of most seed, temperature and humidity are typically maintained in the range of 1–10°C and 50–65%, respectively, giving a shelf-life of at least 1 year (Janick, 1986).

Seed which has been harvested, cleaned and graded is labelled with information of the seed lot, including species and cultivar, date, germination percentage and purity. Trace-back seed lot numbers are an essential component of the seed processing industry as they allow information regarding an individual seed line to be retrievable should any issue occur. Since seed is a biologically active product that still respires and ages, the objective of providing the correct storage conditions is to slow this ageing process and maintain a high rate of viability through the shelf-life period. The usable life of seed varies considerably from species to species and with specific seed treatments, thus expiry dates are provided on commercial and domestic seed packages.

Under storage, seed moisture levels and storage temperatures are the most important factors determining usable shelf-life; however, other factors such as whether seed has been primed may reduce storable life and change the optimal temperature conditions for storage. Conditioned storage aims to hold seed under controlled temperature and humidity to maintain viability; this involves a cost and is not generally used for bulk seed of commodity crops. Hermetic storage is commonly utilized for small to medium quantities of valuable seed such as fruit, flower and vegetable hybrids and involves the use of moisture-resistant or hermetically sealed containers or packages for storage, marketing and distribution. For this method, laminate and polyethylene materials are moderately effective in preventing moisture uptake by the stored seed, while hermetically sealed metal cans are highly efficient at maintaining seed at initial moisture levels (Desai, 2004). For some types of packaging, the ambient air can be removed from the container and replaced with gases such as CO_2 and nitrogen (N_2), thus lowering the O_2 levels surrounding the stored seed. In containerized storage, seeds are sealed in a closed container and desiccant products such as silica gel used to control the moisture levels. The type of packaging also influences seed storage life, with the modern technology used to dehumidify and seal seed hermetically into foil packages, metal tins or larger drums giving a considerably longer shelf-life and retention of full viability than seed which was stored in the open or using older methods of packaging.

8.2.4 Production of transplants from seed

The two main methods of crop establishment are: direct seeding, where seeds are sown either manually or mechanically directly where they are to be grown through to maturity, and the use of transplants. Seedling transplants may be either produced onsite by the grower of the mature crop or purchased from a supplier who solely produces these and ships out at the time they are required. By buying in high-health seedlings or transplants from a commercial supplier at the correct stage for planting out, the grower frees up greenhouse space, labour and time to focus on production of the crop to harvest maturity. Use of seedling transplants, as compared with direct seeding, also reduces the time to harvest, particularly as some crop species are slow to develop through the germination and initial seedling stages. Furthermore, for crops which undergo grafting of one variety on to a suitable rootstock before establishment into the greenhouse, this may be done in the seedling stage while still under protection in a nursery area.

Direct seeding, rather than the use of transplants, is typically carried out for short-term crops which are grown at a high density such as baby leaf salad greens, micro greens and herbs where the use of transplants is not practical. Where root crops are to be grown in substrate beds, direct seeding is common as this prevents damage to the tap root. Root crops which are grown hydroponically on a small scale and require direct seeding include carrots, parsnip, radish, turnip, beets and similar vegetables.

8.3 Seedling Delivery Systems

Greenhouses and hydroponic methods may be used to produce transplants for further growth under protected cultivation, or for seedlings which will be established out in the field. Seedlings may be raised in greenhouses, shade houses, cold frames or enclosed plant factories. Seedling transplant production consists of two main delivery methods: containerized and non-containerized. Containerized seedlings, also termed 'cell' or 'plug' transplants, are grown in multi-celled plastic or styrofoam trays, small plastic pots or preformed cubes of inert growing substrates such as stone wool (Fig. 8.3). This method minimizes root disturbance and the potential

Fig. 8.3. Automated cell tray filler and seeder.

for damage during the transplanting and handling process. Non-containerized transplants are also termed 'bare root transplants' and may be grown under protected cultivation in greenhouses or shade houses. The use of bare root transplants (field seedling transplants) is a lower-cost option which can be used for vegetable crops such as onions, pepper, tomato, broccoli, cabbage, cauliflower and a range of ornamental, annual and cut flower crops. Bare root transplants are lifted or pulled from the growing substrate and generally have lower rates of establishment than plug-grown seedlings due to increased root damage.

To produce high-quality seedlings that will withstand transplanting and establish rapidly, a number of criteria must be reached. These include producing strong, compact seedlings with well-formed root systems which are less likely to become damaged through handling and by automated transplanters. Seedling transplants need to be sufficiently 'hardened off' from the protected environment they were raised in so as to avoid transplant shock, wilting and plant losses. The hardening off process of raising

resilient seedlings may involve techniques such as irrigation restriction, high levels of fertilization, exposure to full sunlight and reductions in temperature, as well as the use of growth regulator compounds in some circumstances.

The majority of greenhouse seedling production systems for field-grown crops utilize cell or plug trays which are direct seeded, grow the seedling through to transplant size and are then used to transport these to the field for planting (Fig. 8.4). Cell trays are typically reused for successive seedling crops and require sterilization to prevent disease carryover before refilling and sowing. With some crops, larger seedlings may be removed from cells at the nursery in preparation for feeding into mechanized transplant equipment. While plug plants generally require planting out fairly quickly once a suitable stage of development has been reached, some crop species may be stored or held if planting is delayed due to field or weather conditions.

Cell transplant trays come in a wide range of cell numbers, sizes and configurations

Fig. 8.4. Filled and sown cell trays.

and may be either plastic or polystyrene. The width and depth of the cell tray determine the volume of the root zone as well as the spacing available for each seedling to expand into. The rooting volume also determines factors such as nutrients and water-holding capacity, frequency of irrigation required, and length of time seedlings can be grown or held before seedling quality is compromised. For this reason, different sized cell trays are used depending on the crop transplants being grown. Trays of 200–338 cells per tray are typically used for smaller vegetable field transplants such as cabbage, broccoli, cauliflower, collard, kale and lettuce; these have 2.5–3.5 cm diameter cells. Larger cells, 3.5–6 cm in diameter, are used for the production of field tomato, capsicum, melons, cucumber and squash transplants (Kelley and Boyhan, 2017). Most cell trays contain rows of square cells, however others may have round cells offset in rows, to give additional space for each seedling to develop. Well-grown cell transplants are characterized by a dense root system which rapidly fills the growing medium. This holds the cell plug together

and facilitates removal of the transplant from the tray during the planting-out process. Cell trays with transplants are watered well before transportation to assist with seedling removal. Seedlings which have been held too long in cell trays may be difficult to extract if roots have grown through the base of the tray or if cells of insufficient size have been used.

While multi-celled seedling/plug trays and bare root transplants are most commonly used for large-scale production, other seedling delivery methods may be used for different aspects of crop establishment. These include the use of biodegradable materials to construct various seedling containment systems. Peat pots and compressed discs (Jiffy 7s), coconut fibre pots, cubes and discs, paper pots, compost plugs and other bonded organic materials are available which do not require removal once the seedling is ready for transplanting. These types of materials break down once planted out, allowing roots to penetrate directly into the soil or growing medium with reduced risk of root damage and disturbance. Compressed discs of peat or coconut fibre encased in a fine netting are soaked in water to allow expansion of the substrate before sowing seed. Paper pots, also termed 'paper chain pots', are a low-cost seedling delivery system consisting of compressed sets of pots which expand to form a larger number of paper tubes, each 3–4 cm in diameter. The tubes are attached in rows by perforated paper, the rows fold and are held together in zigzag fashion by water-soluble glue (Scholz and Boe, 1984). Once transplants are mature, the paper pots are separated and planted directly into the substrate where the paper rapidly decomposes as the seedlings establish. Paper pots have the advantage of being relatively inexpensive, while sowing-line equipment and automated field transplanters are available for use with paper pot systems (Scholz and Boe, 1984). Smaller growers may manually plant paper pot transplants or use walk behind hand tractors modified for this purpose (Kumar and Raheman, 2011). Initially paper pots were mostly utilized for the production of sugarbeet transplants; however, they have been adapted for use in a range of vegetable and forestry applications (Robb *et al.*, 1994). Paper pots of different sizes and grades of paper have been developed depending on the type of transplant being grown and the length of time the pot must stay intact before degrading.

Other inorganic forms of seedling delivery systems make use of synthetic or mineral-based materials; these include mineral wool (heated and spun mineral fibres) (Fig. 8.5), oasis and other various expanded foams which are manufactured into propagation cubes of various sizes for a range of mostly hydroponic or high-value nursery crops. Other seedling systems for greenhouse and hydroponic crops may use small plastic pots or tubes (Fig. 8.6) filled with inert growing medium such as perlite, vermiculite, bark fines, expanded clay, sand, scoria, horticultural-grade coconut fibre, peat or untreated sawdust depending on local availability, cost, irrigation method and suitability for the species being grown. These containers are larger than cell transplant trays and allow seedlings of crops such as tomato, cucumber, melon and capsicum to be grown

Fig. 8.5. Small mineral wool propagation cubes are used for a wide range of hydroponic crops.

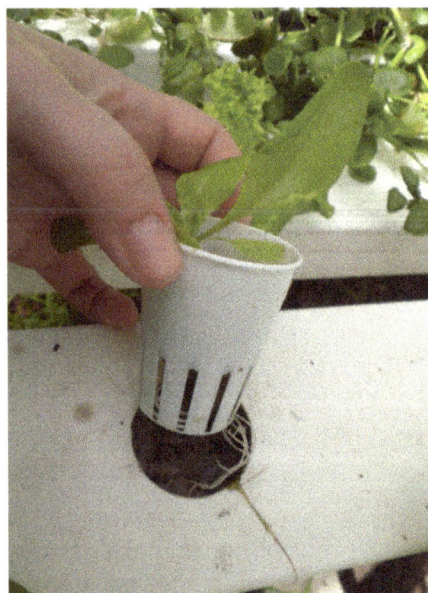

Fig. 8.6. Plastic pots or tubes may be used for propagation of seedlings destined for NFT systems.

on to a later stage of development before planting into hydroponic systems than would typically be transplanted out into the field. Some types of growing medium require a wetting agent to be incorporated to assist with moisture distribution. The physical characteristics required of the propagation medium are that it is lightweight, well drained and well aerated to optimize germination and seedling growth.

8.4 Seeding Methods

Seeding of cell flats, trays and other transplant media is a labour-intensive process which is typically fully mechanized in large commercial greenhouse nursery operations. Smaller growers may hand seed, particularly where pelleted seed is available to facilitate handling and singulation during the sowing process. Automated tray-filling machinery may be used pre-seeding, with trays passed down a conveyor to mechanized seeder units. Seeder equipment varies with speed of operation, their suitability for different types of seed and degree of skill required for operation (Blanchard, 2013). Assisted hand sowing can be carried out with use of automatic vibration wand seeders. These consist of a power source which provides sufficient vibration to move seed down a long chute and drop from the tip on to the substrate surface. Vibration speed can be adjusted to suit the type of seed being sown. Vibration seeders are relatively inexpensive; however, this form of manual seeding is slow and requires labour to operate.

Various types of mechanized tray seeders are used for transplant production. Most of these rely on vacuum seeding processes, with common types including plate seeders and needle seeders. Seeds are delivered on to the surface of the growing-medium-filled cell trays with a high degree of precision and at a rapid rate during automated seeding operations. Plate seeders consist of a flat plate of small holes which match both the size of the seed being sown and the spacing of cells in the tray or seedling flat. A vacuum holds the seeds against the holes in the plate until

the plate is in position and the seeds released. Plate seeders are most suited to species with rounded seeds and are commonly used for brassicas and similar vegetable crops. Seeding rates for plate seeders range from 120 to 300 flats per hour (Blanchard, 2013). Needle seeders use a similar principle to plate seeders in that a vacuum is used through a thin needle to pick up individual seeds which are then transferred into plastic tubes before being released into the tray. Needle seeders have the advantage of being able to individually handle uneven or odd-shaped seed, such as onion seed, with a high degree of precision.

Once seeding has occurred, cells, flats or trays are watered with an overhead fine spray or mist and moved on to germination areas. Some seeding systems have designated areas where germination will occur before trays are transported out to a main greenhouse or other nursery area. This allows the optimal temperatures for germination to be provided during the early stages. Germination chambers may be used which consist of closely spaced vertical shelves where temperature and humidity are controlled to maximize the rate of germination. Since most crop species do not require light to germinate, chambers are usually maintained in darkness until seedling emergence is seen, then transferred out to greenhouse benches. Germination chambers that are cooled to reduce ambient temperatures are used in tropical climates to germinate heat-sensitive seeds such as lettuce. Other systems of seed germination include the use of heated mats, pads, benches, floors or beds which provide gentle heat to the base of the transplant trays. Electric heating cables or hot water pipes are typically used to provide heat during the germination process.

8.5 Germination Problems

Low germination percentages or germination failures can have a number of causes; however, the most common issue, apart from poor quality or incorrectly stored seed, is the environment. Oversaturation of the propagation substrate is a common cause of germination failures as this predisposes the germinating seed to a number of pathogens. Saturation and lack of aeration that causes the seed to rot or temperatures that are either too high or too low are also common causes of failure of otherwise viable seed. If seed has failed to germinate but has not rotted and has not absorbed water, then a hard seedcoat may be preventing germination.

The most damaging disease pathogens during seed germination and transplant production are those that result in 'damping off', which can cause serious loss of seeds, seedlings and young plants. Damping off results in the rot and collapse of young seedlings and is primarily caused by fungi such as *Pythium* and *Rhizoctonia*; however, other species such as *Phytophthora* and *Botrytis* may also be involved. Damping off occurs where substrates become oversaturated, excluding oxygenation, weakening seedlings and where conditions such as high salinity may also occur. Sources of damping-off pathogens include contaminated water or propagation substrates, diseased plant material and debris, contaminated trays and surfaces, and wind-blown spores. Other conditions associated with damping off include inconsistent watering, poor drainage, lack of ventilation, high-density seedlings, sowing seed too deep, and damage to the stem or roots through handling or pests such as fungus gnat larvae. Prevention of seedling disease requires the use of clean water and sterile germination substrates, clean seed (some diseases may be carried on the seedcoat), providing correct temperatures and good rates of air flow through the propagation area, sowing at the correct depth and adjusting seedling density as required during development.

8.6 Transplant Production Systems

Greenhouse seedling transplants may be grown on conventional bench or rack systems (Fig. 8.5). Large-scale commercial transplant producers may also utilize conveyor or moveable bench systems to make the most efficient use of greenhouse space. In these systems benches or tables can be moved to create only one access pathway as required.

Conveyor systems provide efficient movement of trays through the filling and seeding stations out to germination and finishing areas and finally back to packaging and loading bays with minimal requirements for handling by workers. Rail or rack systems for seedling cell transplants use rows of supports which hold the trays (flats) at each end and this is often referred to as the 'speedling system' (Boyhan and Granberry, 2017). Most bench and rail or rack systems for cell transplant production are overhead watered using automated and computer-controlled irrigation systems and sensors. Smaller growers or those producing limited lines of speciality plants may hand water cell trays and this requires a reasonable degree of care and attention to ensure all cells in each tray receive sufficient water without oversaturation. Producers of cell tray transplants may also place flats directly on concrete floors and these often incorporate in-floor heating systems to promote root growth and development.

Another method of seedling transplant production is utilizing hydroponic 'float' or 'raft' systems. In this system, polystyrene multi-celled trays are filled with a suitable growing medium, seeded, irrigated and placed into a germination area. Once germination has occurred trays are moved to a large, shallow reservoir or pond filled with hydroponic nutrient solution. The polystyrene trays are placed on the surface of the solution where they float until seedlings are large enough to be transplanted into the field. Water and nutrients are taken up by the seedlings through the base of the trays from the reservoir in this 'continuous' float system. While the float system has been investigated for use with a wide range of vegetable transplants, its main commercial use is with the production of tobacco transplants.

Ebb and flow (also termed 'flood and drain') systems are another method of irrigation for seedling transplant and cell/plug tray production. Ebb and flow systems operate by providing watering from the base (sub-irrigation), whether this be for cell trays, hydroponic transplants in rockwool propagation cubes or other types of seedling containers. Water or nutrient solution is flooded into a reservoir in which the trays of

Fig. 8.7. Benches of greenhouse-raised cell transplants.

seedlings are grown, providing irrigation; after a specified amount of time the water is drained back into a central holding tank. Water is absorbed by the growing substrate in the trays and moved upwards via capillary action. As the water or nutrient solution drains away, air is drawn from the surface of the trays down into the root zone. While ebb and flow systems can provide efficient and even watering of multi-celled trays, the disadvantage can be the build-up of salts on the surface of the growing medium under certain growing conditions.

8.6.1 Transplant production environment

Greenhouse transplants are grown under conditions which favour not only a good rate of growth development, but also maintain the quality of the seedlings. This involves control over light, temperature and humidity levels. Most seedling transplants are raised under natural light conditions, apart from those produced in plant factories and those being produced under low-winter-light conditions where artificial and supplemental lighting may be used. Low light levels produce poorer-quality transplants, promote elongation of the stem (etiolated seedlings), reduce ability of the seedlings to withstand transplant stress and result in delayed first harvest and yield reductions. Greenhouse transplant producers may use HID lamps to provide supplemental lighting in winter; however, newer, more energy-efficient LED technology is now also becoming more common. LEDs allow more precise control over light spectra, with red, blue, white and combinations of these able to be used for different purposes. It has been found that the beneficial effects of exposure of plants during seedling development to light of different spectra extend beyond the transplanting stage (Javanmardi and Emami, 2013). For both tomato and pepper transplants, exposure to blue and red LED lights during seedling production has been found to increase the rate of first cluster formation and early yields (Javanmardi and Emami, 2013).

Control of seedling transplant height is an important aspect of production. Due to the high density of the cell trays most transplants are grown in, competition for light, high moisture levels and protected growing environments can all result in issues with seedling quality and ability to withstand the transplanting process. High-quality vegetable transplants such as tomato and capsicum are characterized by thick stems, short internodes and well sized, dark green leaves, all features which improve root development after transplanting and affect early yield quality and quantity (Brazaityte et al., 2009). Many methods may be used to control transplant height and 'pre-condition' seedlings before planting out. These include plant growth regulators, restricting water and/or nutrients, temperature control and trimming of shoots (Garner and Bjorkman, 1996). Careful management of these techniques is required as these have the potential to weaken transplants and have longer-term effects on plant growth such as delaying early yields.

Use of chemical growth regulators to restrict height of transplants used for food-producing crops is restricted or prohibited in many countries (Javanmardi and Emami, 2013); however, these are still used during ornamental and bedding plug-plant production. To produce short, compact and high-quality transplants a number of biological, physical and environmental methods are used. These include application of controlled stress such as rates of air movement across the tops of transplants during development and physical brushing of the foliage or vibration techniques which reduce seedling height and keep plants compact. In smaller systems, mechanical conditioning via brushing may be carried out manually by brushing across seedling tops with PVC pipe, cardboard or a wooden dowel. In larger transplant operations, automated brushing equipment provides regular application at the required height. For tomato transplants, brushing treatments begin at the first or second true leaf stage, with ten strokes per day sufficient to reduce seedlings' ultimate height by 20% (Garner and Bjorkman, 1996). Curcubit, aubergine

and cole crops also respond well to brushing treatments; however, it results in significant damage in other transplants such as capsicum, as well as in many bedding plants such as geranium, impatiens and petunia. Use of day/night temperature differentials (DIFs) may be applied to control transplant height in some species. A large difference between day and night temperatures promotes internode elongation, while maintaining similar day and night temperatures restricts height increases (Myster and Moe, 1995).

Hardening off or 'finishing' transplants before planting out is particularly important where field conditions may not be optimal for rapid establishment. Many vegetable transplants such as early-season tomatoes, capsicum and cucurbit crops may be hardened off by reducing temperatures and irrigation in the transplant production area or by holding seedlings outdoors for 5–7 days prior to planting out. Pre-planting conditioning treatments can improve field performance of plug transplants, particularly if transplants have been raised under heated greenhouse conditions.

8.6.2 Seedling nutrition

At the time of expansion of the first two seedling leaves, the young plant requires nutrients. The amount and type of nutrition provided are dependent on the propagation substrate. Inert media such as perlite, vermiculite and rockwool require a complete, but diluted nutrient solution, specific to the crop being grown. Transplants that are being raised in organic composts or potting mixes may already have nutrients blended into the substrate which provide initial nutrition for seedling growth. Where fertilizers have been added to a propagation medium before seed sowing, care needs to be taken to ensure EC levels are not overly high around the developing seedling root system. Nutrient solutions should not be applied before or during the germination process as this can retard development and seeds have sufficient reserves to fuel these early stages of development. Once the first two seedling leaves have expanded the plant is able to carry out photosynthesis and the young root system can start to absorb the nutrients required for growth. Standard, vegetative hydroponic nutrient formulations can be used during the seedling stage, diluted to one-quarter and one-half strength as the transplant develops. Nutrient EC levels can be increased to full strength in the final week of transplant growth to harden seedlings off for transplanting.

8.7 Use of Plant Factories for Seedling Transplant Production

Totally enclosed plant factories which utilize only artificial lighting in stacked or tiered vertical shelf systems are becoming more widely used for seedling transplant production. These allow very high numbers of transplants to be raised under ideal conditions, resulting in a rapid crop turnaround and high-quality seedlings. Precise control over the growing environment, including light, CO_2 enrichment, moisture levels and nutrition via hydroponic production methods, is the main advantage of using indoor plant factories for transplant production. Transplants may be raised for the production of indoor plants, ornamental crops, grafted vegetable transplants, herbs, potted plants, pharmaceutical crops, hydroponic or greenhouse crops as well as those to be established in the field. Use of indoor transplant production with only artificial light is more common in countries which already have a well-established plant factory industry. These include Japan, where small plant factories with a floor area of 15–100 m^2 have been widely used for the commercial production of seedlings over a short time frame and at high planting densities (Kozai and Niu, 2016). The major disadvantage is the increased cost of running such facilities and these are largely used for higher-value transplants.

8.8 Organic Transplant Production

Organic crop production systems must only use approved inputs, and this applies to both seed and seedling transplants. Organically certified seed is produced under organic standards and must not be treated with any chemicals, fungicides, synthetic coatings or other non-organic additives. Greenhouse-raised seedling transplants for organic systems must start with organically approved seed sources and may be produced in either containerized or non-containerized (bare root) systems. Containerized organic transplants require the use of organic substrates which may incorporate materials such as peat, coconut fibre (coir), pine bark, sawdust, coarse sand, composts and vermicast (worm castings). Compost materials used in organic plug-plant production must be fully decomposed before use to prevent nitrogen depletion during the seedling growth stage. While peat moss has long been a component of transplant and germination substrates, it can be conducive to the development of some plant pathogens such as *Pythium* and *Rhizoctonia* (Hoitink and Kuter, 1986) and coir has been identified as a suitable alterative (Colla *et al.*, 2007). Organic plug-plant producers may use commercially available pre-mixed growing substrates of known fertility or blend their own onsite depending on cost and availability of raw materials. Under organic standards guidelines, synthetic fertilizers and wetting agents cannot be used, thus seedling nutrition is dependent on naturally occurring minerals and allowable fertilizer additives. Composts and vermicasts provide much of the initial nutrients for early seedling growth and development as they contain mineralized nutrients readily available for plant uptake. Vermicast (worm castings) has been tested as a component of media for organic production of tomato transplants and it was found seedling development improved as the percentage of worm castings in the medium increased (Ozores-Hampton and Vavrina, 2002). Liquid fertilizers such as fish emulsion may be applied to supplement plant nutrition as required (Brust *et al.*, 2003). Other fertigation materials include compost teas and seaweed extracts which may also be applied as a foliar spray. Additives to the seedling growing mix may include lime (calcium carbonate), rock phosphate, blood and bone meal, organically allowable trace elements and similar naturally occurring nutrient sources. One common issue with the use of organic growing substrates for plug transplant production is that they are more prone to high salinity levels (measured as EC), which may restrict seedling growth and development (Russo, 2005).

8.9 Transplant Establishment

Most commercial hydroponic greenhouse crops such as tomatoes, capsicum, cucumber, melon, aubergine, lettuce, herbs and strawberries are manually transplanted into the soilless production system. These transplants are usually high-value, often hybrid plants, grown through to an exact stage of development before planting out and require careful handling. Manual transplanting systems have the advantage of potentially less damage to seedlings during the handling process; however, on a large scale, this is labour intensive, expensive, repetitive and highly physical work (Orzolek, 1996). Current trends in seed and seedling delivery systems focus on collaboration between horticulturalists and engineers to develop improved systems of mechanization and automation for these labour-intensive operations. Many of these are focusing on increased automation of seedling transplant delivery systems, reducing labour requirements and expenses involved in crop establishment. One such area of development has been into automation of rapid and efficient seedling removal from cell and plug trays during the transplanting process both in the field and with greenhouse crops. Robotic mechanisms have been developed, consisting of a manipulator, an end-effector and two conveyors which efficiently extracts seedlings from cell trays with a pincette-type mechanism and conveys these to their planting site (Mao et al., 2014). New advances focus on the continued development of agricultural robots for both protected cropping and field operations.

Transplanting should occur at the correct stage of seedling development, which is dependent on species. For tomato, cucumber, capsicum and similar crops which are raised in larger blocks or pots of substrate, this can be delayed as compared with those grown in the smaller root volume of cell trays. This limits the amount of time plants are in the vegetative-only state in the production system and decreases the time from transplanting to first harvest. Transplant establishment will occur rapidly under optimal conditions with new roots visible extending from the transplant within a few days. Reasons for slow establishment include root damage during handling and transplanting, pathogen pressure in the new system, waterlogging, extremes in temperature, incorrect EC levels or salinity, and pests such as fungus gnat larvae.

8.10 Grafting

Grafting for improved plant vigour, disease resistance and higher yields is becoming increasingly common under greenhouse production for a number of high-value crops. Most greenhouse-grown cucurbit crops grown in China, Japan, Korea, Turkey and Israel are grafted, while in the Netherlands nearly all the tomatoes produced in soilless culture are grafted on to vigorous rootstocks to increase, or at least secure, yields (Bie et al., 2018). Grafting is the process of joining a rootstock of one plant to the scion (shoot section) of a separate plant by means of tissue regeneration at the graft union. The most commonly grown greenhouse crops which may use grafted planting stock are melons, cucumber, tomato, aubergine and capsicum. Grafting may be utilized with certain crops such as tomatoes for the control of potentially serious diseases such as bacterial wilt (Ralstonia solanacearum) and certain species of nematodes. However, other advantages can include an enhancement of plant vigour, extension of the harvesting period, increases in yield and fruit quality, prolonging postharvest shelf-life, increasing nutrient uptake, and increased tolerance to low and high temperatures and other stresses such as salinity, heavy metal stress, drought and waterlogging (Bie et al., 2018). Suchoff et al. (2018) found that cold tolerance in the tomato cultivar 'Moneymaker' could be increased by grafting on to 'Multi-fort' rootstocks. Colla *et al.* (2008) found that grafted pepper plants produced 22–46% more marketable fruit yield than un-grafted control plants, however this was somewhat dependent on cultivar. Cucumbers grafted on to vigorous rootstocks have been reported to have significantly improved yields as a result of a root system able to enhance photosynthetic rates as well as water and nutrient (particularly nitrogen, phosphorus, calcium and magnesium) uptake efficiencies (Rouphael *et al.*, 2010).

Selection of a suitable rootstock variety is based on a number of factors; these include not only resistance to disease, suitability for the growing environment and purity of the seed, but also compatibility with the scion cultivar. Scion varieties which will bear the fruit are selected for yields, fruit quality and marketable traits including shelf-life as well as being the correct match for the rootstock cultivar available. Commercially available rootstock varieties for commonly grafted species such as tomato have increased in recent years with not only rootstocks from *Solanum lycopersicum* being available, but also rootstocks from other species such as *Solanum torvum* and *Solanum melongena* (Kyriacou *et al.*, 2017).

The procedure of grafting requires both the rootstock and scion to be grown as seedlings. These may be sown at the same time if rates of growth are similar or may be sown on different dates to ensure that at the time of grafting, plants are of a similar size and stem diameters are compatible. The next step involves the creation of the graft union by slicing the stem of both the rootstock and scion and binding these together, followed by heating of the graft union under conditions that help prevent moisture loss from the plant and, finally, acclimatization of the grafted plant (Lee *et al.*, 2010). There are a number of different grafting methods which can be used for hydroponic transplants and this procedure is often carried out by specialist propagation nurseries. Splice grafting and cleft grafting are the two main methods used in the production of grafted greenhouse transplants. Slice grafting involves removing the growing point from the

rootstock using an angled cut and preparing the scion portion to match. The graft union is held in place with binding, clips or a grafting tube. Cleft grafting (also termed 'wedge grafting') requires a vertical cut to be made in the centre of the stem of the rootstock seedling. The scion is prepared by cutting the lower stem from both sides to form a wedge which is inserted into the slit made in the rootstock. Clips, wax or tape may be used to hold the rootstock and scion together until a graft union is fully formed.

Once grafting has been carried out, plants must be maintained in high humidity conditions to prevent excessive moisture loss from the scion section. Humidity levels of 95% are required during the first 48 h when temperatures should be maintained at 27–28°C (Suchoff *et al.*, 2018). Partial shading during the 5–7 days post grafting also assists with transplant survival. Once a graft union has formed and the rootstock is maintaining scion turgor, transplants can be hardened off in preparation for planting out. While grafting represents a significant additional transplant production cost in terms of additional seed, seedlings, increased propagation areas, greater time to transplant maturity and the requirement for skilled labour, increases in plant vigour, disease resistance, yields and potentially in produce quality often more than compensate for these with grafted plants.

8.11 Vegetative Propagation

While most greenhouse vegetable crops are grown from seed, strawberries, some perennial herbs, a number of cut flowers and ornamentals may be grown vegetatively via other methods of propagation. Strawberries are propagated from runners, small plant lets which form on stolons produced by mature plants (Fig 8.8). Runner plantlets may be pinned into a rooting substrate while still attached via the stolon to the parent plant and will rapidly form a new root system. Alternatively, runner-tip cuttings can be severed from the plant and rooted under mist in a separate propagation area. Under frequent misting rooting will occur within

Fig. 8.8. Strawberry propagation from runners.

7–10 days. Detached runners or rooted tip cuttings are then raised as plug or cell transplants to the required size and may be provided with a chilling treatment before planting out into a greenhouse production system.

Propagation from cuttings may be carried out for a number of reasons: these include maintenance of the desirable genetics of a plant since all new plants propagated via cuttings will be 'clones' of the parent, or where growing from seed is slow and germination erratic. Cuttings usually take the form of a section of stem with some foliage still attached; however, in some ornamental species, root cuttings, leaf cuttings and other forms of vegetative propagation such as division may be carried out. Plant material for propagation may be purchased as unrooted cuttings or stock plants can be maintained onsite for the specific purpose of providing cutting material as required. When growing stock plants for propagation it is important that only non-protected varieties are used or that royalties are paid to allow the legal propagation of those which are protected.

Propagation from stem cuttings for most species is relatively straightforward and requires material to be taken from healthy and disease-free parent plants at the correct time and stage of development (Fig. 8.9). This is while the parent plant is vegetative and not in a flowering stage as the presence of flower buds limits new root production in most species. Cuttings are prepared by stripping the lower leaves thus reducing the leaf area from which moisture can be lost, cutting with a sharp knife directly below a node and applying rooting hormone compounds if appropriate. Use of hormone dips, powders and liquids can speed up the rate of root formation and the number of roots produced in many species. Cuttings are then placed in a propagation area, under regular misting with sufficient heating to provide warmth around the base of the cut stem. Rooting substrates range from stone wool and other inert media such as perlite and vermiculite to composts amended with sharp sand, while some cuttings may be rooted in aeroponics or water. A highly aerated substrate is required for callus formation and subsequent root growth as overly saturated substrates can cause the cut stem to rot. Clean water and good hygiene practices are also required in the propagation area to prevent issues with fungal and bacterial pathogens which can cause rapid failures with cuttings. As young roots develop, the heat provided in the rooting substrate can be reduced, misting discontinued and the cuttings permitted to harden off for a few days before transplanting and growing on.

8.12   Tissue Culture

Plant tissue culture or micropropagation is the aseptic culture of cells, tissues, organs and their components under controlled conditions *in vitro* where the environment and nutrition are rigidly controlled. While 'tissue culture' is the most commonly applied and widely recognized term for this process, 'micropropagation', '*in vitro* culture', 'sterile culture' and 'axenic culture' may also be used to describe tissue culture methods (Smith, 2013). Tissue culture has developed to the

Fig. 8.9. Carnation cutting showing root formation.

point where it has become an important tool in both basic and applied studies as well as in commercial application and large-scale plant production (Thorpe, 2007). Tissue culture has now became an important tool with many types of crops including cereals and grasses, legumes, vegetable crops, potatoes and other root and tuber crops, oilseeds, temperate and tropical fruits, plantation crops, forest trees and ornamental species (Thorpe, 2007).

Plant tissue culture techniques have multiple uses: the main commercial application is in the rapid production of large numbers of clones (identical individuals) of plants which have desirable characteristics. These may be newly bred or pre-existing varieties of crops with desirable traits in fruiting, flowering, ornamental value, disease resistance and other factors. Production of large numbers of clones in a rapid time frame avoids the long vegetative propagation period many species would otherwise have using traditional methods. It also avoids problems associated with seed propagation of some species such as the requirement for a suitable pollinator or seed that is difficult to germinate as occurs with certain orchids.

One of the main advantages of tissue culture is that whole plants can be regenerated from plant cells, making this a valuable method during the process of genetic modification to create new varieties with improved cropping characteristics. Another important aspect of tissue culture is the ability to produce specific pathogen-free clones and to clear certain plants of viral and other infections which may be present. Under the sterile conditions inside tissue culture vials, plant material can be transported and shipped long distances without the risk of transferring pests and pathogens; it also allows long-term storage of clonal material (Hartmann and Kester, 1983).

8.12.1 Tissue culture techniques and methods

Successful tissue culture has a number of basic requirements and procedures which must be followed. Many of these are species specific; however, all *in vitro* micropropagation processes need:

1. Specialized equipment and facilities in which to prepare the cultures, to maintain strictly sterile conditions, as well as the correct growing conditions of temperature and light intensity.
2. Correct selection of explant material from suitable stock plants which have been maintained in good condition.
3. Prevention of contamination. Plant material must be disinfected correctly, the culture media sterilized, and all facilities and equipment kept as free from contaminants as possible.
4. Staging. Once explant material has been selected and disinfected it may pass through a number of developmental stages. These include initial establishment from the original explant (plant material used to start the process), multiplication and pre-transplant stages which may include root initiation and acclimation.
5. The final stage involves taking the newly rooted plantlets from the highly protective *in vitro* environment and transplanting into a suitable, sterile growing mix or other

substrate in small pots under greenhouse conditions. Since the young plantlets are prone to desiccation at this stage, humidity tents and misting are applied to increase humidity and light shading is applied. Once the plants have established, they are gradually exposed to lower humidity and higher light levels. During this stage, the young plants are highly susceptible to pathogens and clean, hygienic conditions, including treated, high-quality water supplies for irrigation, are maintained until they are fully established.

References

Bie, Z., Nawaz, M.A., Huang, Y., Lee, J.M. and Colla, G. (2018) Introduction to vegetable grafting. In: Colla, G., Perez-Alfocea, F. and Schwart, D. (eds) *Vegetable Grafting: Principles and Practices.* CAB International, Wallingford, UK, pp. 1–21.

Blanchard, C. (2013) *Transplant Production Decision Tool. Leopold Centre for Sustainable Agriculture,* Iowa Organic Association, Iowa State University, Ames, Iowa.

Bosland, P.W. (2005) Second generation (F2) hybrid cultivars for jalapeno production. *HortScience* 40(6), 1679–1681.

Boyhan, G.E. and Granberry, D.M. (eds) (2017) Commercial production of vegetable transplants. *UGA Extension Bulletin 1144.* University of Georgia, Atlanta, Georgia. Available at: https://secure.caes.uga.edu/extension/publications/files/pdf/B%201144_5.PDF (accessed 7 September 2020).

Brazaityte, A., Duchovskis, P., Urbonaviciute, A., Samuoliene, G., Jankauskiene, J., *et al.* (2009) The effect of light-emitting diodes lighting on cucumber transplants and after effect on yield. *Zemdirbyste Agriculture* 96, 102–118.

Brust, G., Egel, D.S. and Maynard, E.T. (2003) Organic vegetable production. *Purdue University Extension Publication ID-316.* Purdue University, West Lafayette, Indiana. Available at: https://www.yumpu.com/en/document/view/18754705/organic-vegetable-production-purdue-extension-purdue-university (accessed 7 September 2020).

Colla, G., Rouphael, Y., Cardarelli, M., Temperini, O. and Rea, E. (2007) Optimization of substrate composition for organic lettuce transplant production. *Advances in Horticultural Science* 21(2), 106–110.

Colla, G.Y., Rouphael, M., Cardarelli, O., Temperini, E., Rea, A., *et al.* (2008) Influence of grafting on yield and fruit quality of pepper (*Capsicum annuum* L.) grown under greenhouse conditions. *Acta Horticulturae* 782, 359–363.

Desai, B.B. (2004) *Seeds Handbook: Biology, Production, Processing and Storage,* 2nd edn. Marcel Dekker, New York.

Garner, L.C. and Bjorkman, T. (1996) Mechanical conditioning for controlling excessive elongation in tomato transplants: sensitivity to dose, frequency and timing of brushing. *Journal of the American Society for Horticultural Science* 121(5), 894–900.

Hartmann, H.T. and Kester, D.E. (1983) Techniques of *in vitro* micropropagation. In: *Plant Propagation: Principles and Practices,* 4th edn. Prentice-Hall, Englewood Cliffs, New Jersey, pp. 566–594.

Hoitink, H.A.J. and Kuter, G.A. (1986) Effects of composts in growth media on soilborne pathogens. In: Chen, Y. and Avnimelech, Y. (eds) *The Role of Organic Matter in Modern Agriculture.* Martinus Nijhoff Publishers, Dordrecht, The Netherlands, pp. 289–306.

Janick, J. (1986) *Horticultural Science,* 4th edn. W.H. Freeman and Company, New York.

Javanmardi, J. and Emami, S. (2013) Response of tomato and pepper transplants to light spectra provided by light emitting diodes. *International Journal of Vegetable Science* 19(2), 138–149.

Kelley, W.T. and Boyhan, G.E. (2017) Containers and media. In: Commercial production of vegetable transplants. *University of Georgia Extension Bulletin 1144.* University of Georgia, Atlanta, Georgia. Available at: https://secure.caes.uga.edu/extension/publications/files/pdf/B%201144_5.PDF (accessed 7 September 2020).

Kozal, T. and Niu, G. (2016) Introduction. In: Kozai, T., Niu, G. and Takagaki, M. (eds) *Plant Factory: An Indoor Vertical Farming System for Efficient Quality Food Production.* Elsevier, London, pp. 7–33.

Kumar, P.G.V. and Raheman, H. (2011) Development of a walk-behind type hand tractor powered vegetable transplanter for paper pot seedlings. *Biosystems Engineering* 110(2), 189–197.

Kyriacou, M.C., Rouphael, Y., Colla, G., Zrenner, R. and Schwarz, D. (2017) Vegetable grafting: the implications of a growing agronomic imperative for vegetable fruit quality and nutritive value. *Frontiers in Plant Science* 8, 741.

Lee, J.M., Kubota, C., Tsao, S.J., Bie, Z., Echevarria, P.H., *et al.* (2010) Current status of vegetable grafting: diffusion, grafting techniques, automation. *Scientia Horticulturae* 127(2), 92–105.

Mao, H., Han, L., Hu, J. and Kumi, F. (2014) Development of a pincette type pick-up device for automatic transplanting of greenhouse seedlings. *Applied Engineering in Agriculture* 30(4), 547–556.

Myster, J. and Moe, R. (1995) Effect of diurnal temperature alternations on plant morphology in some greenhouse crops: a mini-review. *Scientia Horticulturae* 62(4), 205–215.

Opena, R.T., Chen, J.T., Kalb, T. and Hanson, P. (2001) Hybrid seed production in tomato. *AVRDC Publication #01-527*. Asian Vegetable Research and Development Centre, Tainan, Taiwan. Available at: http://www.sandros.gr/download/texnika-nea/Tomato_Seed_Hybrid_Production.pdf (accessed 7 September 2020).

Orzolek, M.D. (1996) Stand establishment in plasticulture systems. *HortTechnology* 6(3), 181–185.

Ozores-Hampton, M.P. and Vavrina, C.S. (2002) Worm castings: an alternative to sphagnum peat moss in organic tomato (*Lycopersicon esculentum* Mill.) transplant production. In: Michel, F.C., Rynk, R.F. and Hoitink, H.A.J. (eds) *Proceedings of the International Composting and Compost Utilization Science Symposium, Columbus, Ohio, 6–8 June 2002*. The JG Press, Emmaus, Pennsylvania, pp. 105–113.

Robb, J.G., Smith, J.A., Wilson, R.G. and Yonts, C.D. (1994) Paper pot transplanting systems – overview and potential for vegetable production. *HortTechnology* 4(2), 166–171.

Rouphael, Y., Scharz, D., Krumbein, A. and Colla, G. (2010) Impact of grafting on product quality of fruit vegetables. *Scientia Horticulturae* 127(2), 172–179.

Russo, V.M. (2005) Organic vegetable transplant production. *HortScience* 40(3), 623–628.

Scholz, E.W. and Boe, A.A. (1984) Transplanting sugarbeets using Japanese paper chainpots. *Sugarbeet Research and Extension Reports* 15, 181–185.

Smith, R.H. (2013) *Plant Tissue Culture: Techniques and Experiments*, 3rd edn. Elsevier, San Diego, California.

Suchoff, D.H., Perkins-Veazie, P., Sederoff, H.W., Shultheis, J.R., Kleinhenz, M.D., *et al.* (2018) Grafting the indeterminate tomato cultivar Moneymaker onto Multifort rootsock improves cold tolerance. *HortScience* 53(11), 1610–1617.

Thorpe, T.A. (2007) History of plant tissue culture. *Molecular Biotechnology* 37, 169–180.

9 Plant Nutrition and Nutrient Formulation

9.1 Water Quality and Sources for Hydroponic Production

Water quality for hydroponic production is the first consideration when it comes to working through a plant nutritional plan and formulating nutrient programmes for optimal crop growth. Hydroponic systems are characterized by a highly restrictive root zone and thus have minimal buffering capacity for any issues with water quality. Many hydroponic crops including lettuce, many herbs and strawberries are highly sensitive to water quality, while others such as tomatoes are more tolerant of problems such as salinity. Closed hydroponic systems which recirculate the nutrient solution require more attention to water quality as unwanted minerals can accumulate rapidly over time. Open systems where higher rates of drainage can be run are more tolerant of many issues with water quality, as excess salts can be flushed from the root zone with each irrigation.

Water may be sourced from municipal water suppliers, wells, dams, rivers, streams or as rainwater collected from greenhouse building roofs, with all sources having potential issues which must be evaluated before production begins. Water quality is dependent on the concentration of each dissolved mineral, the presence of biotic organisms such as algae, fungi and bacteria, and particular residues (Schroder and Lieth, 2002). Many water quality issues can be either treated or adjusted for with the nutrient solution formulation; however, it is always advisable to obtain a full water analysis before setting up any hydroponic systems. In addition to the nutrient solution used to grow the crop, additional water supplies are required for washing of harvested product, cleaning systems and general hygiene procedures.

9.1.1 Well water

Many commercial hydroponic growers around the world are reliant on water from wells to supply the large volumes required for greenhouse production. The quality of water from wells in different locations varies considerably depending on the soil and minerals naturally present. Deep wells passing through certain soil layers can give partially clarified water although some minerals are always likely to be present in groundwater. Shallow, older or poorly maintained wells can present problems with contamination from pathogens, nematodes, agrochemicals and fertilizers leached through the upper soil layers into the well water. Well water in coastal areas can have intrusion from saltwater into aquifers, resulting in groundwater which is too high in salt for use in plant production (e.g. EC of 10–15 mS/cm) (van Os et al., 2008). Arid, inland regions can also suffer from high sodium levels in well waters with levels in excess of 2000 ppm sodium, however most well waters do not pose such an extreme problem. Sodium is not taken up by plants to any large extent, hence accumulates in recirculating systems displacing other elements. Trace elements in groundwater such as copper, boron and zinc may sometimes occur in high levels and risk toxicity problems in some systems. Many well-water sources have a degree of 'hardness' and contain dissolved minerals such as calcium and magnesium which occur naturally in the soil surrounding the well. These are not necessarily problematic for hydroponic production as their presence can be adjusted for with the nutrient formulation unless levels are excessively high. Well water which has a high alkalinity (more than 300 mg calcium carbonate/l) and pH requires significantly more acid to maintain pH control

DOI: 10.1079/9781789244830.0009

in hydroponic systems. Low-alkalinity well water will have an alkalinity below 100 mg calcium carbonate/l.

The presence of iron in the form of iron hydroxide is often present in well water, particularly those surrounded by large deposits of iron sand or iron ores. Iron hydroxide in water supplies is not necessarily harmful to hydroponic plants, however it can create issues with blockages and deposits inside components of the irrigation system. Deposits of iron also create a medium for the growth of iron bacteria which consume elements that are required for plant growth.

Well water drawn from large aquifers is generally stable and well buffered with composition changing little over time; however, a water analysis should be carried out every 6–12 months to monitor mineral levels and ensure the supply remains safe for hydroponic production. In some areas of intensive horticultural and agricultural production, water drawn from aquifers has been reported to be depleting these natural reserves, raising environmental concerns (van Os et al., 2008). Hydroponic growers should be aware of these issues and consider rainwater collection and alternative water sources where such problems arise.

9.1.2 Surface water

Groundwater sourced from rivers, streams and lakes, or stored in dams or reservoirs, can be variable in quality depending on a number of factors. Water which is continually exposed to air and soil has potential for contamination and can become a source of organic matter, minerals leached from the surrounding soil, and pathogen spore loading can be high (Hong and Moorman, 2005). Other risk factors are agrochemicals, particularly hormone herbicides which can contaminate large volumes of water if sprayed around the perimeter of dams and reservoirs; these can cause considerable crop damage if carried through to crop irrigation. Pathogens of concern to human health can also contaminate groundwater supplies due to the open nature of the water source and

the potential for runoff from farming, sewage outlets and storm water. Many greenhouse operations use open-air storage dams as an economical method of holding large volumes of water collected from greenhouse roofs and other surfaces; however, this water is typically filtered and treated before use. River, stream and lake waters may have inconsistent water quality as operations upstream affect composition of the water and rainfall and flow rates also fluctuate throughout the year.

9.1.3 Rainwater

Rainwater represents an economical and often high-quality water source for many hydroponic operations; however, limitations may be present in some regions where insufficient rainfall does not supply the year-long needs of the hydroponic system. The main advantage of rainwater is that the mineral content is generally very low; however, acid rain from industrial areas, sodium from coastal sites and high pathogen spore loads from agricultural areas can still occur. Much of this contamination has been found to occur when rainwater falls on roof surfaces and collects organic matter, dust and pollutants which naturally collect there. Rainwater should be collected from clean surfaces with a first-flush device installed that allows the first few minutes of rainfall to be discharged from the roof before any is collected for use. Rainwater may also contain traces of zinc and lead from galvanized roof surfaces or where lead flashings and paint may have been used, and this is a greater problem when the pH of the rainwater is low. Collected rainwater for a hydroponic greenhouse operation requires a large storage capacity which may be in the form of lined dams or reservoirs or large tanks; in some areas, rainwater is stored underground in aquifers (van Os et al., 2008). In regions where collection and stored rainwater are insufficient to completely supply the hydroponic operation, another source of water must be secured and water supplies may be blended to provide a uniform year-round supply. In the Netherlands, where rain throughout the year is plentiful, growers are required to have a storage capacity of at least 500 m³ of rainwater per hectare of production facility (van Os *et al.*, 2008), which provides approximately 60% of irrigation water needs. Rainwater collection systems require some additional expense to maintain and regularly clean roofs, gutters and collection tanks; however, rainwater may be treated in much the same way as other water sources to remove pathogens before use in a greenhouse system.

9.1.4   City or municipal water supplies

Municipal water supplies are often used by hydroponic operations sited close to or within city limits and can be an economical choice in many areas. City water supplies are treated to ensure that the water meets the World Health Organization standards for mineral, chemical and biological contamination. This means there is a wide range of water treatment chemicals which may be dosed into a city water supply as well as processes for water softening, pH control, removal of organic matter and addition of fluoride. These aim to produce water that is safe for drinking, does not corrode pipes, leave limescale deposits, have an offensive smell or stain, and is generally acceptable for human use. However, while city water supplies are safe for people, they may not be suitable for hydroponic plants, particularly those in solution culture and recirculating systems with minimal growing medium to act as a buffer.

One of the most common issues with domestic and municipal water supplies can be the use of water softening chemicals, often sodium salts, which result in problematic sodium levels in hydroponic nutrient solutions. Water softeners are not required in hydroponic systems as the elements, such as calcium, which make water 'hard' are usable by plants and can be adjusted for with the nutrient formulation. Sodium levels that are too high, either through the use of water softener chemicals or naturally occurring in the water supply, require RO for removal or

dilution with a low-sodium water source to acceptable levels for the crop being grown. Like well water or groundwater, municipal water supplies often have a degree of hardness with a high mineral content, usually originating from magnesium, calcium carbonate, bicarbonate or calcium sulfate, which causes white limescale to form on surfaces. Hard water may also have a high alkalinity and high pH, meaning that considerably more acid is required to lower the pH in a hydroponic system to ideal levels. While hard water sources can contain useful minerals such as calcium and magnesium, they can create an imbalance in the nutrient solution and make other ions less available for plant uptake. Soft water, in comparison, is a low-mineral water source, and municipal water sources can range from very hard to soft depending on where the city water supply is initially taken from.

Treatment of city water varies depending on the characteristics of the water source and often changes over time depending on incoming water quality. In the past, chlorides were routinely added to city water supplies during processing and for microbial control. High levels of chlorine can be toxic to sensitive plants and young seedlings; however, chlorine is a chemical which rapidly dissipates into the air and can be removed by aeration of the water before use. An increasing number of municipal water suppliers are now using ozone, UV, chloramines, chlorine dioxide and other methods to create safe drinking-water suppliers. While many of these methods are not an issue for hydroponic production, the use of chloramines and other chemicals by city water treatment plants can pose toxicity issues for some plants where high levels are regularly dosed into water supplies. Water treatment chloramines are more persistent than chlorine and slow to dissipate from water supplies, hence they can accumulate in hydroponic systems with the potential for damage of sensitive plants. Diagnosis of chloramine plant damage can be difficult as symptoms appear similar to many root rot pathogens and growers are not usually aware of the origin of the problem. Some hydroponic crops are more sensitive to chloramines than others and determining levels of toxicity can be difficult. If chloramines are known to be present in a city water supply, treatment for removal includes the use of specific activated carbon filters, dechloraminating chemicals or water conditioners which are sold in the aquarium/aquaculture trade to treat water used for fish tanks. The chloramine filters must be of the correct type that has a high-quality granular activated carbon which allows for the long contact time required for chloramines removal. Substrate-based hydroponic systems such as coconut fibre or peat are a safer option where chloramines may be a risk as they provide buffering capacity in a similar way to soil and can deactivate some of the treatment chemicals contained in the water supply. Water culture and recirculating systems are at greater risk from the accumulation of water treatment chemicals such as chloramines and have minimal buffering capacity to such chemicals.

9.1.5 Reclaimed water sources

With growing worldwide concern over the supply and availability of high-quality water sources for crop production, the use of reclaimed water supplies for hydroponics has become of interest over recent years. The most common type of reclaimed water is that discharged from sewage treatment plants, however desalinated water is another source. Sewage and other waste waters which are treated via anaerobic reactor and polishing pond systems have been used successfully for hydroponic production of lettuce (Keller *et al.*, 2008) with no significant microbial contamination of the harvested product. Lopez-Galvez *et al.* (2014) reported that greenhouse hydroponic tomatoes grown with reclaimed and surface water were found to be negative for bacterial pathogens and microbiologically safe for consumption, indicating that good agricultural practices (GAP) were in place which avoided contamination of the fruit. GAP guidelines include avoiding irrigation water and hydroponic nutrient solutions coming into contact with the edible portion of the crop, thus avoiding microbial contamination, while potable or treated water

sources are used for crop sprays and washing of harvested produce. Wastewater may also be used on ornamental hydroponic crops and for the production of fresh animal fodder, where potential contamination with foodborne microorganisms is less of a concern. Reclaimed wastewater such as tertiary treated sewage water has been successfully used for the production of green forage using hydroponic techniques. Treated wastewater was found to produce high yields with less water use than tap water in hydroponically grown barley and proved to be a useful alternative disposal method without the risk of accumulation of heavy metals (Al-Karaki, 2011).

9.2 Water Testing

Water quality may be evaluated by a full analysis from an agricultural laboratory and this should be carried out on water supplies during the planning stages of a hydroponic installation. Ongoing and regular water analysis may be required for some water supplies to ensure water quality is not changing over time or to allow for adjustments to the nutrient formulation if required. A complete water analysis should be carried out for anions, cations, salinity, alkalinity, pH, and specific ion toxicity due to excessive concentrations of sodium, chloride and sulfate (Schroder and Lieth, 2002).

To collect a representative sample for laboratory testing, water should be collected into a clean bottle directly from the water source (well, tap, dam or tank) and not from the irrigation system. Water should be left to run for a few minutes before sample collection, capped tightly, clearly labelled and sent directly to the testing laboratory along with a completed submission form containing all required details such as address of the origin of the sample, date taken, name and contact details of the operation, type of water source and all parameters to be tested for. If water samples need to be held before submission, they can be stored under refrigeration.

A basic water analysis for a hydroponic operation is typically carried out to determine levels of pH, EC, total dissolved solids (TDS), carbonate, bicarbonate, hardness, nitrogen, phosphorus, potassium, magnesium, calcium, sulfur, iron, manganese, boron, zinc, copper, sodium and chloride; however, other parameters may be reported depending on the laboratory. These give a good overall indication of the mineral content of the water supply and can be used to determine if the water supply is suitable for the crop and system of production or whether further treatment is required to remove high levels of unwanted elements. The information provided in a water report also provides valuable data to be used in the nutrient formulation process so that plant-usable minerals present can be adjusted for to prevent accumulation and potential toxicity problems. While this basic water test for minerals is commonly carried out, it does not provide any information on other issues which may occur in some water supplies such as the presence of plant and human pathogenic microorganisms, iron-fixing bacteria, algae and other organic matter. Biological water testing can be carried out; this tends to be costly, however the expense may be warranted for evaluation of the new water source or to assist diagnosis in the case of crop abnormalities (Ingram, 2014).

9.3   Water Analysis Reports

9.3.1   pH and alkalinity

pH should always be considered in conjunction with alkalinity values in a water analysis report. While pH is a measure of the acidity or alkalinity of a water supply, alkalinity is a measure of the strength of that pH, or the ability of the water to neutralize acids. A water supply with a high pH and a high alkalinity will cause the pH around the root zone in substrates to rise quickly over time and require a greater volume of acid for pH control than a water supply with the same pH but low alkalinity. Alkalinity is composed of dissolved calcium, magnesium or sodium bicarbonates and calcium and magnesium carbonates. Laboratory water analysis reports commonly report alkalinity in parts per million (ppm), micrograms per litre (mg/l) or milliequivalents per litre (meq/l). Water is considered

being of low alkalinity if calcium carbonate is below 100 ppm, while levels of 150–300 ppm if combined with a high pH (greater than 7.5) can cause rapid increases in pH in the growing substrate and require the addition of large volumes of acid. High alkalinity (greater than 300 ppm) with a water pH of 7.5 or above will cause rapid increases in pH and should be pretreated with acid to a given pH level (6–6.5), to minimize acid use during production and the nutrient imbalances this can cause (Ingram, 2014).

9.3.2   Electrical conductivity

EC or TDS levels on water analysis reports are a measure of the total amount of dissolved salts in the water. EC may be given in units of decisiemens per metre (dS/m), millisiemens per centimetre (mS/cm) or millimhos per centimetre (mmho/cm). EC levels which are acceptable for hydroponic source water depend largely on whether the dissolved salts are plant usable (such as calcium and magnesium) and thus can be adjusted for, or are likely to accumulate (sodium and chloride), as well as the type of system (closed or open) and crop. Some crops such as tomatoes, grown at higher EC levels and in an open system, can tolerate a water source with a considerably higher level of dissolved salts than more sensitive, lower-EC crops such as lettuce in closed systems. The general recommendation is that water EC levels should be below 1.0 mS/cm for open systems and 0.4 mS/cm for closed (recirculating) systems (Schroder and Lieth, 2002; Ingram, 2014). Water sources with a high EC can lead to a number of issues including salt build-up in growing substrates, displacement of other essential elements, general reduction in growth and plant injury.

9.3.3   Mineral elements in water supplies

Nitrate

The presence of nitrate-N in water supplies does not pose a problem for hydroponic plant growth as this is a plant-usable element.

However, the presence of more than 5 ppm nitrate is an indication that the water source may be polluted (Ingram, 2014).

Phosphorus, Potassium, Calcium, Magnesium and Sulfur

The presence of these minerals in water supplies is rarely a problem as these are plant-usable elements and can be adjusted for with the nutrient formulation. Occasionally water supplies may contain more calcium (greater than 180 ppm) and magnesium (greater than 60 ppm) or sulfur (greater than 70 ppm) than is required for hydroponic crop production, in which case accumulation is likely to occur over time. A suitable solution is often obtained by blending with a lower-mineral water source such as rainwater or by treatment with RO to remove excess salts.

Sodium

All water analysis reports should include sodium as this is a mineral commonly found in many water supplies in varying levels. For many hydroponic systems, sodium and chloride are limiting factors since these minerals are absorbed in low concentration in relation to the concentrations present in many water sources (Voogt and Sonneveld, 1997). The level of sodium which can be tolerated varies between crops; some such as tomatoes can grow and produce well with high sodium levels, others such as lettuce and strawberries are more sensitive. Sodium levels less than 50 ppm are recommended for most hydroponic systems; however, tolerant crops in open systems can withstand higher sodium levels with minimal losses in production.

Chloride

As with sodium, chloride levels are usually associated with sodium chloride contamination of water supplies including well water, surface water and rainwater where salt spray in coastal areas can contaminate supplies. Some crops such as cucumbers in closed systems and strawberries are more sensitive to chloride than others.

Iron

Iron contained in water supplies is present as iron hydroxide and not available for plant uptake; however, iron in this form can cause irrigation blockages and levels greater than 5 ppm can cause toxicity symptoms in some sensitive plants, particularly if the pH in the root zone is below 5.5 (Ingram, 2014).

Manganese

Levels up to 3 ppm in water supplies are acceptable provided this is adjusted for in the nutrient formulation to prevent accumulation over time.

Boron

Levels up to 1 ppm are acceptable, provided this is adjusted for in the nutrient formulation.

Copper

Levels up of 0.3 ppm are acceptable, provided this is adjusted for in the nutrient formulation.

Zinc

Levels up to 0.3 ppm are acceptable, provided this is adjusted for in the nutrient formulation.

9.4   Water Quality and Plant Growth

If water quality issues are suspected, a full water analysis can often determine the cause of some problems such as salinity or excessive levels of trace elements, however many other factors can also play a role. It can be difficult to determine if a water quality issue is responsible for plant growth problems and symptoms that might be occurring. Many diseases and errors with nutrient management or incorrect environmental conditions produce symptoms similar to common water quality problems and detecting issues such as chemical or microbial contamination is more complex. Water testing and laboratory analysis for plant pathogens and other microbial contamination can be carried out and are essential for those producing salad greens, lettuce, micro greens and sprouts to ensure water is of acceptable quality that meets food safety standards. For other water quality issues, the simplest method to determine if water quality is the cause of growth problems is to run a seedling trial and grow sensitive seedlings such as lettuce using RO or distilled water as a control or comparison. Keeping all other growth factors such as nutrients, light and temperature the same between the two sets of seedlings in a solution culture system can give an accurate test of water quality. Comparing growth of the seedlings between the two samples, including appearance of the root systems, can reveal any problems. Water quality issues often show as stunted roots which do not expand downwards, short, brown roots, yellowing of new leaves, stunted foliage growth, sunken brown spots on the foliage, leaf burn and even seedling death.

9.5 Water Treatment Options

There is a wide range of treatment options for hydroponic water supplies, depending on whether the issues are chemical or biological in nature. For excessive levels of salts such as sodium or an overall high EC and alkalinity, some water sources can be blended with lower mineral supplies to reach acceptable levels. Where this is not possible, RO can be used to create relatively pure water for hydroponic production; however, this can be costly to install and run. For biological water quality issues such as the presence of plant or human pathogen microorganisms, more commonly found in well/ground or surface waters, treatment options for water supplies include UV radiation, ozone and slow sand filtration which do not leave the risk of chemical residues that may harm young, sensitive root systems. Correctly designed and run systems have a high kill rate for pathogens and algae. While it is possible to treat water sources with chemical disinfection agents such as chlorine or hydrogen peroxide, this can be a risk to sensitive crops.

Levels of these sanitizer compounds high enough to kill plant pathogens can damage young plants unless they are deactivated or removed before being incorporated into the hydroponic nutrient solution. Hydrogen peroxide is deactivated when it reacts with organic matter in the water and chlorine will dissipate in time, particularly if the water is aerated.

9.6 Water Usage and Supply Requirements

While water requirements of hydroponic crops are less than those of crops grown in soil, an appreciable amount of water can still be required for large plants under conditions of high evaporative demand. Closed systems require lower water inputs than open systems; however, open systems can be managed to minimize leachate/drainage levels and conserve water in many circumstances. Studies have shown that the average water consumption of plants grown in open hydroponic systems was 15–17% higher as compared with closed systems (Tuzel *et al.*, 1999; van Os, 1999). Environmental conditions such as temperature, humidity, air movement and solar radiation and plant factors such as maturity, crop and canopy size all determine water usage and irrigation systems should be designed to supply sufficient water at the greatest rate of usage. Annual water usage rates have been reported to be as low as 8600 m^3/ha for cool-climate cropping systems and as high as 11,400 m^3/ha for a crop with a fully developed canopy under warm conditions in drip-irrigated systems with 30% drainage rates (Schroder and Lieth, 2002). When compared with traditional soil cropping, hydroponic production can result in significant reductions in irrigation water requirements with hydroponics on average using 5–20 times less water than soil agriculture (Al-Shrouf, 2017). Barbosa *et al.* (2015) reported that hydroponic lettuce grown in Arizona, USA required a total of 20 litres of water per kilogram per year, while conventional soil production required 250 l/kg per year.

9.7 Plant Nutrition in Hydroponic Systems

In soilless systems, the substrate itself typically provides no or very little nutrients for plant growth; this must be supplied by the nutrient solution applied to the roots in both substrate and solution culture systems. The first step in this process involves dissolving of mineral ions into water, which in hydroponics is delivered directly to the root system on an intermittent or continual basis so the plant has a constant supply for uptake and growth. In this way the plant does not have to expend a great deal of energy growing roots into considerable depths of soil in search of both water and nutrients, and hence can grow and develop at a more rapid rate than soil crops might. Nutrient ions supplied via the hydroponic nutrient solution are absorbed mostly by the regions of roots that contain high numbers of fine root hairs, they are then transported upward into the aerial regions of the plant. Plant roots are ion selective and certain elements are required in larger quantities by some species than others. The plant must expend some energy during this mineral uptake process.

Hydroponic nutrient formulae aim to supply the plant with all of the essential elements required in ratios close to what the plant will remove from the nutrient solution. This means that a nutrient formula not only supplies all of the plant's requirements, so that no toxicity or deficiency in any one element occurs, but also that ions are removed from the solution in the ratios in which they are continually supplied. The hydroponic nutrient solution must supply the macro- and microelements required for plant growth, these being: nitrogen, potassium, phosphorus, calcium, sulfur, magnesium, iron, manganese, boron, copper, zinc and molybdenum. These essential elements are supplied by dissolving high-quality, greenhouse-grade fertilizers into water, based on the nutrient formulation or recipe, to make concentrated stock solutions. These concentrates are then further diluted with water to create working-strength solutions of the correct EC and pH for application to the crop. Fertilizer salts commonly used to supply all of the essential elements of a nutrient solution include: calcium nitrate (supplies N and Ca), potassium nitrate (supplies K and N), monopotassium phosphate (supplies P and K), magnesium sulfate (supplies Mg and S), potassium sulfate (supplies K and S), iron chelate (supplies Fe), manganese sulfate (supplies Mn), zinc sulfate (supplies Zn), boric acid/Solubor®/borax (supplies B), copper sulfate (supplies Cu) and sodium/ammonium molybdate (supplies Mo). For nutrient formulations that also provide a small amount of nitrogen in the ammonium form (typically less than 15% of N is supplied as ammonium), ammonium nitrate, ammonium phosphate or ammonium sulfate may also be incorporated under certain conditions. Trace element chelates are sometimes used in some nutrient formulations and addition of nitric and/or phosphoric acid (supplying some N and P) may be part of nutrient formulations where pH needs to be continually lowered. While the essential elements for plant growth must be supplied via the hydroponic nutrient solution, there are a number of potentially 'beneficial' elements which may play a role in plant health and productivity that some growers choose to include as part of their nutritional programme. These include silicon, which has been shown to strengthen plant growth, improve resistance to certain diseases and moderate salinity, lower the occurrence of physiological disorders such as blossom end rot, improve compositional quality and increase yields and shelf-life in some plant species under hydroponic cultivation (Schuerger and Hammer, 2003; Stamatakis et al., 2003).

9.8 Essential Elements – Functions in Plants and Deficiency Symptoms

9.8.1 Nitrogen

Nitrogen (N) is an essential component of amino acids in proteins and chlorophyll, and along with potassium is a macroelement required in large quantities for uptake. Nitrogen is mobile within plants and can be redistributed from older foliage to new growth under deficiency conditions. For this reason, deficiencies show first on older, lower leaves as a yellowing or purple coloration depending on plant species. Typically, foliar levels of nitrogen are within the range of 3–6% depending on species and stage of growth (Kay and Hill, 1998). Without nitrogen growth ceases and deficiency symptoms will occur rapidly. An excess of nitrogen, or specifically a high nitrogen to carbon ratio within the plant, causes lush, soft tissue growth which is undesirable in many commercial crops; this can also increase the severity of calcium disorders such as tipburn. Under ideal growing conditions, plants can take up luxury amounts of nitrogen beyond their requirements.

In hydroponics, nitrogen when absorbed as nitrate becomes mobilized first to ammonia and then incorporated into the amino acid glutamine. In hydroponic formulations, nitrogen is commonly supplied as nitrate from calcium or potassium nitrate fertilizers; a small amount of nitrogen may be supplied via the use of nitric acid for water treatment and nutrient solution pH control. For some crops, under low light conditions of winter, a small percentage of total nitrogen may be supplied in the ammonium form from compounds such as ammonium nitrate or ammonium phosphate; however, this should be limited to less than 15% of total nitrogen to maintain balanced vegetative growth and avoid physiological disorders relating to ammonia toxicity. Providing high levels of nitrogen as ammonia-N in hydroponic systems can accelerate the uptake and utilization of nitrogen under certain conditions and can also reduce the uptake of calcium, resulting in a higher occurrence of calcium-related disorders such as tipburn and blossom end rot. Uptake of ammonia can result in a lowering of the pH in the nutrient solution as plants release positive hydrogen ions to balance the charge in the root zone.

9.8.2 Potassium

Potassium (K) is required for the formation of proteins, carbohydrates and fats, and for the function of chlorophyll and several enzymes, particularly those involved with carbohydrate metabolism. Potassium is also responsible for the control of ion movement through membranes and the water status of stomatal apertures and therefore plays a role in controlling plant transpiration and turgor. Potassium, like nitrogen, is highly mobile in the plant and moves freely to new developing leaves as needed, with deficiency symptoms first appearing on older, lower leaves. Foliar symptoms may appear somewhat different between species, but often present as scorched areas towards the margins of older leaves along with reduced growth and susceptibility to fungal diseases (Fig. 9.1). Potassium levels in healthy foliage are typically in the range of 4–7% depending on species and stage of growth (Kay and Hill, 1998). In hydroponic formulations, potassium is supplied as potassium nitrate and monopotassium phosphate; however, it may be further supplied as potassium sulfate. Levels of potassium supplied by hydroponic nutrient formulae vary considerably depending on the crop and stage of growth. For vegetative growth most species have a requirement for potassium which is almost equal to that of nitrogen; with fruiting crops, however, the levels of potassium are increased once fruit set has occurred and have to be adjusted based on fruit loading. Potassium is the predominant cation in fruits such as tomato, cucumber, capsicum, melon and aubergine, and can have major effects on fruit quality. The majority of the potassium absorbed by these crops during the

Fig. 9.1. Potassium deficiency on hydroponic lettuce.

active fruiting stage is incorporated into fruit tissue, requiring high levels to be maintained during the fruiting stages. As fruit load increases on the plant, so too does the requirement for and absorption of potassium and these requirements must be met by the nutrient solution. Potassium is also directly related to fruit quality via the acidity and flavour of the fruit, firmness, ripening disorders, colour and shelf-life.

9.8.3 Phosphorus

Phosphorus (P) is required for cell division and growth and is used in photosynthesis, sugar and starch formation; it is also used in the energy utilization process within plants. Large amounts of phosphorus are required for seed formation and fruiting plants absorb proportionally more phosphorus than non-fruiting or vegetative plants. A deficiency in phosphorus will first appear on the older, lower foliage as a dull green coloration followed by purple and brown coloration as the foliage dies. Root development becomes restricted as phosphorus deficiency progresses due to sugar production and translocation being impeded. Shoot growth is also restricted and leaves are often undersized. Typical foliar levels of phosphorus in healthy plants are between 0.3 and 0.8% (Kay and Hill, 1998). In hydroponic formulations phosphorus is primarily supplied as monopotassium phosphate; however, small amounts may originate from the use of phosphoric acid where this is used for water treatment or nutrient solution pH control.

9.8.4 Calcium

After nitrogen and potassium, calcium (Ca) is the element absorbed in the next greatest quantity. It is a vital component of cell walls and membranes, and is deposited in plants during cell wall formation. Calcium is required for the stability and function of cell membranes and is implicated in a number of physiological disorders. Deficiency of calcium causes cell membranes to become leaky and cell division is disrupted causing abnormal growth, often with twisting or cupping of the newer foliage. Calcium deficiencies show first on young growth as this element is almost totally immobile in the plant and, once deposited in cell walls, cannot be mobilized to transport to other tissues. Calcium moves within the transpiration stream of the plant from roots to leaf tips and developing fruits, thus disorders may develop from both a deficiency of calcium in the root zone and issues with transpiration within the plant. Calcium transport within the xylem is driven by water loss from the foliage, thus under warm, overcast or humid conditions when transpiration is limited (termed 'calcium stress periods'), calcium-related disorders become more common. These include tipburn of sensitive crops such as lettuce, strawberries, celery and other greens, and blossom end rot of fruit such as tomatoes and capsicum. Other symptoms of calcium deficiency include convex or concave cupping of new leaves, or a pale marginal band on new foliage. Water-soaked areas can occur on leaves or stems and the root tips can become soft and break down.

Typical foliar levels of calcium range from 0.6 to 3% in most commonly grown hydroponic crops (Kay and Hill, 1998). In hydroponic nutrient solutions calcium is primarily supplied via calcium nitrate fertilizers, however some available calcium is often present in many water supplies.

9.8.5 Magnesium

Magnesium (Mg) has a primary role in the light collection mechanism of the plant and production of assimilate via photosynthesis as Mg^{2+} is the central ion of the chlorophyll molecule. Magnesium is also involved as a cofactor in the energy utilization process of respiration. Under deficiency conditions, magnesium is mobile within the plant and can move from older to new foliage. Deficiency symptoms therefore show first on older leaves often as yellowing between the veins; if severe these can move up the plant towards the newer growth. A magnesium deficiency can result from a lack of magnesium in the root zone for uptake or can be induced due to periods of low light or if excessive levels of potassium are supplied in the nutrient solution. Typical foliar levels of magnesium are between 0.2 and 0.9% (Kay and Hill, 1998). In hydroponic formulae, magnesium is usually supplied via magnesium sulfate; however, magnesium nitrate may be used where levels of sulfate are not required. Some magnesium is often present in some water supplies and may be adjusted for with the nutrient formulation.

9.8.6 Sulfur

Sulfur (S) is involved in many plant processes including sulfur-containing proteins, amino acids and coenzymes. The vitamins thiamine and biotin use sulfur and so this element plays a key role in plant metabolism. Sulfur has limited mobility within the plant and deficiency symptoms first develop on the youngest leaves; however, sulfur deficiency is rarely seen as plant requirements for this element are reasonably flexible within a wide range. Sulfur deficiency can show up as a general yellowing of the entire foliage, starting with the youngest growth. Unlike deficiencies of iron and manganese where foliage yellowing with green veins occurs, a lack of sulfur will cause leaf blades to take on a dull but uniform yellow colouration. Foliar levels of sulfur range from 0.2 to 0.8% depending on species and growing conditions (Kay and Hill, 1998). In hydroponics sulfur is usually present in adequate amounts from the use of sulfate salts such as magnesium and potassium sulfate; it may also occur in high levels in some water supplies.

9.8.7 Iron

Iron (Fe) is the micronutrient required in the largest quantities in hydroponic nutrient formulations. It is an essential component of proteins contained in plant chloroplasts as well as of electron-transfer proteins in the photosynthetic and respiration chains. Iron deficiency symptoms occur first on the youngest leaves with interveinal chlorosis,

however in the early stages yellowing may be uniform on the leaves. Under severe deficiency conditions, leaves may take on a very pale or even white appearance with necrotic spots and distorted leaf margins on some plant species. Iron deficiency may occur in hydroponic crops for a number of reasons including low uptake due to excessively cool temperatures, waterlogged conditions with low root-zone oxygenation which reduce iron uptake, incorrect pH levels and unsuitable forms of iron such as iron hydroxide. Typical foliar levels of iron are within the 70–350 ppm range depending on species and growing conditions (Kay and Hill, 1998). In hydroponics iron is typically supplied as an iron chelate (Fe-EDTA, Fe-EDDHA or Fe-DPTA) rather than as iron sulfate, which is unstable in solution and tends to form iron hydroxides that are insoluble. Chelation allows the iron to remain available for plant uptake at a wide range of pH values, however maintaining correct pH levels assists with iron remaining available for use. In organic systems, a chelate can be formed from a mixture of iron sulfate and citric acid.

9.8.8 Manganese

Manganese (Mn) is required for chlorophyll development, for photosynthesis, respiration, nitrate assimilation and for the action of several enzymes. Under deficiency conditions symptoms first occur on newer foliage as a dull grey appearance, followed by yellowing between the veins which can be difficult to distinguish from iron deficiency as these often occur at the same time. Under severe deficiency spots of dead tissue may develop on affected leaves. Typical foliar levels of manganese can vary widely from 50 to 400 ppm (Kay and Hill, 1998). In hydroponics, manganese is typically supplied as manganese sulfate, however manganese chelate is also available.

9.8.9 Boron

Boron (B) is required for cell division in plants as new cell walls are produced. Boron also affects sugar transport and is associated with some of the functions of calcium. Being largely immobile in the plants, boron deficiencies show first in the young tissues, growing points and root tips; symptoms include stem cracking, yellow leaf margins, tipburn and death of the shoot apex. Typical foliar levels of boron are between 20 and 90 ppm (Kay and Hill, 1998). In hydroponics boron can be supplied as borax, boric acid or Solubor® with varying percentages of boron.

9.8.10 Zinc

Zinc (Zn) is an integral part of many plant enzymes and contributes to the formation of chlorophyll and the production of the plant hormone auxin. Deficiency first occurs in older leaves, with distortion and interveinal chlorosis followed by retardation in stem development as a consequence of low auxin levels in tissue. Other deficiency symptoms are often crop dependent and include an inward rolling of leaves with restricted leaf expansion common in some vegetative crops. Growing points may eventually die back under extreme deficiency conditions. Typical foliar levels of zinc are within the range of 20–180 ppm (Kay and Hill, 1998). In hydroponics zinc is supplied as zinc sulfate or as zinc chelate (Zn-EDTA).

9.8.11 Copper

Copper (Cu) is essential for photosynthesis and in seed development as well as being required in small amounts as a component of several important enzymes, in the formation of lignin and in metabolism in plant roots. Symptoms of copper deficiency are often species dependent and thus difficult to identify correctly. They may appear as wilted and distorted foliage with a greyish green, yellow or even white coloration. Typical foliage levels of copper are within the range of 7–80 ppm (Kay and Hill, 1998). In hydroponics copper is normally supplied as copper sulfate, however copper chelate may also be used.

9.8.12 Chloride

Chloride (Cl) deficiency is rare in hydroponics and plant requirements for this element are low. Chloride tends to be present through impurities in fertilizer salts, with the use of fertilizers such as potassium chloride and in many water supplies. Chloride deficiency causes wilting of the plant and a highly branched but often stunted root system, whereas toxicity causes marginal leaf scorch, abscission and chlorosis (Scaife and Turner, 1983).

9.8.13 Molybdenum

Molybdenum (Mo) is only required in minute quantities by plants and is involved in nitrogen metabolism in the root system, with the enzyme reducing nitrate to ammonium requiring molybdenum as a cofactor. Deficiencies of molybdenum are rare in hydroponic crops and if it occurs, is highly species dependent. Deficiencies occur first on older leaves and progress to the youngest, causing interveinal yellowing which may also result in cupping of the leaves. Typical foliar levels of molybdenum are 1–5 ppm (Kay and Hill, 1998). In hydroponics molybdenum is supplied as either ammonium or sodium molybdate.

9.9 Beneficial Elements

The role of each of the essential elements contained in hydroponic nutrient solutions is well known and has been widely researched; however, there are a number of potentially beneficial elements which can be incorporated into soilless systems, many of which are currently undergoing renewed research interest. Hydroponic systems are not typically devoid of trace or minute amounts of potentially beneficial additional elements. Many water sources, particularly groundwater or well water that has been in contact with soil, typically contain some, although highly varied, trace amounts of certain naturally occurring elements. Some elements end up in hydroponic solutions from contamination with dust, trace amounts of soil, as contaminants in fertilizers and occur naturally in some growing substrates.

There are more than 60 different mineral elements commonly found in plant tissue; however, only 16 are currently considered to be essential and are supplied through fertilizers, water, carbon dioxide and oxygen. The rest are usually present in plant material in only minute amounts; however, some such as silica are commonly found in reasonably high levels in some plant species. A concern over the use of hydroponics to grow a wide range of food crops is that plants raised on formulated fertilizer-based nutrients do not contain the full range of extra elements required for a healthy diet in the same way that soil-grown crops might. This is not entirely correct; most water sources contain small amounts of a large number of the same minerals as are found in soils and these do find their way into hydroponics crops in quantities comparable to well-grown field crops. However, the use in hydroponic nutrient formulations of trace amounts of elements which are known to be required for the health of higher animals and in human diets has been investigated. These include elements such as iodine, cobalt, selenium, silicon, chromium, tin, vanadium and others which may play a role in human health, but also have some beneficial functions within plants. Hydroponic nutrient solutions can have minute quantities of these potentially beneficial elements typically found in soil added via chemical or naturally occurring compounds and in many causes will then produce more minerally-complete fruits and vegetables than those grown in many heavily cropped soils.

Silicon (Si) is one beneficial element that has been in use by hydroponic growers for decades and is well proven to have positive effects in a number of different crops. The use of silicon has often been restricted to crops that absorb significant quantities of silicon, such as cucumbers, melon, courgette, strawberry, bean and rose (Voogt and Sonneveld, 2001). However, other crops previously thought of as non-accumulators of silicon have also shown beneficial results.

Gottardi et al. (2012) found that addition of silicon to the nutrient solution used to grow hydroponic corn salad (Valerianella locusta) increased edible yield and quality and lowered levels of reducing nitrate in the tissue. It was also found the addition of silica in the nutrient solution slowed the rate of postharvest chlorophyll degradation, delaying leaf senescence and thus prolonging shelf-life in the edible tissue (Gottardi et al., 2012).

Silicon can reduce the incidence and severity of powdery mildew and other fungal diseases on crops such as roses and cucumbers and contributes to the strength and thickness of cell walls. In non-accumulates such as tomato, silicon applied as 100 mg potassium silicate/l has been found to give a significant reduction in powdery mildew incidence and severity when grown at EC levels of 3.9–4 mS/cm (Garibaldi et al., 2011). Stamatakis et al. (2003) found that silicon addition to hydroponic tomatoes as potassium silicate significantly increased the concentration of β-carotene, lutein and lycopene of the fruit as well as fruit firmness and vitamin C. Silicon assists in the absorption and translocation of several macro- and micronutrients and plays a role in allowing plants to survive adverse growing conditions such as salinity and toxicity of excess elements. Khoshgoftarmanesh et al. (2012) found the addition of silicon (1.0 mM Si) increased zinc and iron uptake by hydroponic cucumbers as well as the uptake of silicon itself. Silicon is also known to increase the plant's ability to tolerate environmental stress such as salt, drought, frost and metal toxicities as well as providing mechanical strength (Jana and Jeong, 2014).

In many studies, silicon has been added to nutrient solutions at macroelemental levels since silicon is naturally found in many plant tissues at up to 10% or higher of dry weight. Silicon may be added to nutrient solutions in a number of ways, the most common being the use of potassium metasilicate which, due to the high pH of this compound, requires an increased rate of acidification to maintain optimal pH levels. Mineral sources and supplements containing silicon are another option, however these tend to be a slow-release source of silicon. Many groundwater and well-water supplies as well as some growing substrates also contain naturally occurring silicon at varying levels.

Selenium (Se) is a beneficial element which is currently of renewed interest not only because of its effects on hydroponic plant growth, but also due to its implications for human health because selenium is an antioxidant reported to have a number of health benefits (Mozafariyan *et al.*, 2017). The amount of selenium in fruits and vegetables is dependent on the amount of this element in the soil they were grown in. Thus, hydroponics offers an opportunity to increase the selenium content of food crops in a controlled manner as a means of 'biofortification'. Mozafariyan *et al.* (2017) reported that a level of 5 µM Se applied as sodium selenite ($NaSeO_3$) produced improved root volume, leaf number, carotenoids content and catalase activity in hydroponically grown tomato plants, while levels of 7 and 10 µM Se increased chlorophyll content. For biofortification of edible plants with selenium to enrich levels in the human diet, it has been found that the addition of selenium (as sodium selenate) at rates of 4–12 mg Se/l to hydroponic nutrient solutions gave a dose-dependent increase in selenium uptake rates in basil plants (Puccinelli *et al.*, 2017). Selenium is taken up by plant roots and is translocated to the foliage, particularly the younger leaves. Another study reported that levels of 0.5–1.0 mg Se/l added to the nutrient solution enriched leafy vegetables grown in a floating system as well as having a positive effect on plant yield (Malorgio *et al.*, 2009).

Titanium (Ti) is the tenth most common element found in soils and is considered to be a beneficial element for plants which can under certain conditions improve growth and development (Wadas and Kalinowski, 2017). Titanium applied via the nutrient solution at low concentrations has been found to improve crop performance through stimulating the activity of certain enzymes, enhancing chlorophyll content and photosynthesis, promoting nutrient uptake, strengthening stress tolerance and improving crop yield and quality (Lyu *et al.*, 2017). Choi *et al.* (2015), found that foliar application of titanium dioxide (TiO_2) promoted the growth of strawberry plants under low-light, winter greenhouse conditions, while Hruby *et al.* (2002) reported that the effect of titanium on plant growth was dependent on the nitrogen form in the nutrient solution. Nitrate-containing nutrient solutions resulted in beneficial effects of titanium addition, whereas ammonium-containing nutrient solutions resulted in an inhibitory effect on plant growth, suggesting that an increase in nitrate reductase activity was responsible for the beneficial effect on plants (Hruby *et al.*, 2002). Haghighi and Daneshmand (2018) found that addition of 1–2 mg Ti/l to the nutrient solution of hydroponically grown tomato plants increased plant growth and photosynthesis through increased nutrient uptake and nutrient-use efficiency.

9.10 Nutrient Formulation

Most hydroponic nutrient formulations or 'recipes' are designed to be made up into at least two separate concentrated stock solutions – these are then diluted further with water, often at a rate of 1:100, to create a working-strength nutrient solution for crop production. Concentrated stock solutions consist of a given amount of each of the required fertilizer salts dissolved into a given volume of water, taking care not to exceed the saturation value of any given fertilizer so that all elements stay in solution.

Hydroponic stock solutions have the fertilizer salts divided between two or more separate parts to prevent chemical reactions which occur between certain elements while in a concentrated state. These reactions occur between calcium, phosphate and sulfate which, if combined in a concentrated state, form insoluble precipitates (white cloudy matter). For this reason, the calcium nitrate is dissolved into a separate stock solution (typically termed 'stock solution A') from the phosphate- and sulfate-containing fertilizers, which are contained in stock solution B. Once diluted with water to a working-strength EC, these reactions no longer occur, thus only the concentrated stock solutions need to be maintained separately.

Some nutrient formulations are more complex and may have a number of different stock solution concentrates; often the trace elements may be contained in a stock solution C, allowing for separate adjustment of these if necessary. Commercial growers of some crops, such as fruiting tomatoes, melons or cucumbers, may also have a separate stock solution specifically for potassium to boost this element during times of heavy crop loading.

While there are a large number of published hydroponic nutrient formulations for a wide range of crops, commercial growers typically adjust these to their requirements with use of tools such as onsite or laboratory-based nutrient solution analysis, foliar mineral analysis, spreadsheets and nutrient formulation software to correct for plant nutrient uptake over time. Smaller-scale or hobby growers may make use of manufactured and bottled nutrient solution concentrate products rather than obtaining the individual fertilizer salts and weighing these out. Bottled nutrient concentrates available on the hydroponic market come in a range of formulations designed for different production systems, growing substrates and stages of growth (such as 'vegetative' and 'bloom' nutrient formulations).

Ideal nutrient formulations for soilless cropping are dependent on the climate, crop, stage of growth, type of production system, water quality and other factors, and hence large commercial growers operate a practice of continual nutrient solution monitoring, analysis and adjustment to ensure plant growth and yields are optimized.

For soilless fruiting crops such as tomatoes, cucumbers, peppers and many flowers, plants go through an initial vegetative-only phase, followed by flowering and fruit set and then a fruit development phase – each of these has a different rate and ratio of nutrient uptake from the hydroponic solution. With hydroponic tomato nutrition, nitrogen and potassium are usually absorbed in the largest quantities, followed by calcium at lower rates. Phosphorus, sulfur and magnesium are absorbed in relatively lesser amounts. Large amounts of phosphorus are required for seed formation and a fruiting tomato plant absorbs proportionately more

phosphorus than a non-fruiting or vegetative plant. The trace elements or micronutrients (iron, manganese, boron, copper, zinc and molybdenum) are required in very low concentrations and become toxic at high levels. These trace elements are just as essential as the major (macro) nutrients but are taken up in such small amounts that the main focus is on the macroelements which are used at a much greater rate. Nutrient formulations for specific crops and systems may be adjusted for the presence of minerals in the water supply, for water supplies with a high pH and alkalinity, or for any issues with the soilless substrate such as the presence of naturally occurring potassium or nitrogen drawdown in new coconut fibre or other organic media.

9.10.1 The process of nutrient formulation

Working through the process of nutrient formulation for a hydroponic system needs to take account of many factors. There are nutrient formulation software programs to make this process simpler, and some provide sample nutrient formulations for a range of different hydroponic crops (Nutron 2000+) from which growers can modify and develop their own nutritional programme. There are a number of published nutrient formulations which can make a good starting point; however, it is not always applicable to take a nutrient formulation for one type of crop, system or from a very different climate and use it, without adaptation, in another situation. For example, nutrient formulations developed for hydroponic crops in low-light areas of the world are often not suitable for systems in high-light climates which have a significantly greater requirement for iron and other elements due to differences in uptake rates. Using a vegetative formulation for a heavy fruiting tomato or cucumber crop is another common mistake which results in potassium depletion and risks poor-quality fruit. Growers sharing nutrient formulations without adjustment for system, water supply and other specifics often results in a less-than-optimal nutritional programme.

Before starting to formulate a nutritional programme for a hydroponic crop, the optimal levels of each of the elements required should be determined. It is important that commercial growers have an understanding of the acceptable ratios for all of the elements contained in hydroponic formulations, to ensure the nutrient solution is supplying plant requirements. Generally, the range of acceptable concentrations is wider for the macroelements than for the trace elements and a general guide is given below (Table 9.1). Table 9.2 outlines some examples of hydroponic nutrient formulations for lettuce and tomato crops during different developmental stages.

Note that in the formulations given above, the potassium nitrate is split equally between the A and B stock solutions to assist solubility. Weights of the individual fertilizers will vary depending on the elemental percentage of the fertilizer product used. The specifications of the fertilizer used in the above formulations are given below (Table 9.3). These formulations have not been adjusted for the presence of minerals in water supplies different from pure water (RO treated).

9.11 Hydroponic Nutrient Formulation – Nitrogen Sources

Nitrogen can be supplied as nitrate (NO_3^-) and ammonium (NH_4^+) in hydroponic nutrient

formulations as both are available for plant uptake. In most cases the nitrate form of nitrogen is preferable; however, a small percentage of ammonium can, under certain

Table 9.2. Examples of starting formulations for hydroponic lettuce and tomato crops (in grams per 100 litres of water).

	Lettuce[a]	Tomato vegetative[b]	Tomato fruiting[c]
Stock solution A (100 litres)			
Calcium nitrate	8,036	16,844	14,835
Potassium nitrate	1,196	2,897	5,828
Iron chelate (12% Fe)	500	500	500
Stock solution B (100 litres)			
Potassium nitrate	1,196	2,897	5,828
Potassium phosphate	1,337	3,016	4,952
Magnesium sulfate	3,511	6,590	8,962
Manganese sulfate	98.4	105.0	105.0
Boric acid	39	40	45
Zinc sulfate	11.0	12.4	12.4
Copper sulfate	3.0	3.0	3.0
Ammonium molybdate	1.01	1.01	1.30

[a]Lettuce, fancy types (not iceberg), using RO water supply under cool-climate cropping, recirculating system. Final EC after a 1:100 dilution with water = 1.4 mS/cm (TDS = 980 ppm).
[b]Early vegetative stage, RO water supply in drip-irrigated, open, coco fibre-based system, high-light climate. Final EC after a 1:100 dilution with water = 3.0 mS/cm (TDS = 2100 ppm).
[c]Heavy fruit loading stage, RO water supply in drip-irrigated, open, coco fibre-based system, high-light climate. Final EC after a 1:100 dilution with water = 3.8 mS/cm (TDS = 2660 ppm).

Table 9.1. Levels of macro- and microelements in hydroponic nutrient formulae.

Element	Level (ppm)
Nitrogen	70–450
Phosphorus	20–100
Potassium	70–650
Calcium	70–240
Magnesium	20–95
Sulfur	20–100
Iron	1–6
Manganese	0.5–3
Boron	0.1–0.9
Zinc	0.1–0.5
Copper	0.05–0.1
Molybdenum	0.02–0.07

Table 9.3. Details of fertilizers used in the formulations given in Table 9.2 (note that percentages are given as elements, not oxides).

Fertilizer	Specification
Calcium nitrate	15.5% N
Potassium nitrate	13.7% N, 38% K
Monopotassium phosphate	22.7% P, 28.7% K
Magnesium sulfate	9.6% Mg
Iron chelate	12% Fe
Manganese sulfate	30.5% Mn
Boric acid	20% B
Zinc sulfate	21% Zn
Copper sulfate	24% Cu
Ammonium molybdate	54.3% Mo

circumstances, be incorporated. Ammonium ions are rapidly absorbed by plant roots, but they must not be absorbed more rapidly than they can be utilized in the plant tissues or a toxic reaction will occur. Theoretically, ammonium-N should be the superior source of nitrogen because it could be used more efficiently in the plant than nitrate-N. If ammonium-N only were absorbed by root systems, the internal nitrate to ammonium conversion which occurs inside plant cells and requires a supply of energy would not be required. However, in reality, if nitrogen is only supplied in the ammonium form, the toxic reactions of the accumulation of uncomplexed ammonium override the potential for this greater efficiency of assimilation. Ammonium nitrate also competes for the uptake of potassium, magnesium and calcium, thus increasing the occurrence of calcium-related disorders such as tipburn and blossom end rot. Ammonium toxicity often presents as yellowing or chlorosis between the veins of young leaves, followed by scattered brown, necrotic spots; depending on the species and severity, leaf edges may curl upwards or downwards. Root tips may die back and overall root growth is reduced – this tissue damage then becomes an entry point for root disease pathogens such as *Pythium*.

The addition of ammonium-N as a small percentage (no more than 10–15% of total nitrogen) may be beneficial to plant growth, but only under certain circumstances and this beneficial effect varies between crops. Under conditions of high light and rapid growth, the addition of even small amounts of ammonium nitrate can cause growth reductions and physiological disorders. However, under low light and reduced growth rates, a certain percentage of ammonium-N has been shown to be beneficial for the growth of some crops. Ammonium may be incorporated into nutrient formulae specifically to provide a pH-buffering effect. pH in hydroponic systems tends to gradually increase as plants take up nutrient ions, requiring acidification for pH control. Ammonium-N reduces the rate of pH increase, giving a greater buffering capacity to pH change in the nutrient solution.

9.12 Common Hydroponic Fertilizers

The fertilizers used in modern hydroponic systems have evolved over time; older formulations contained soil fertilizers such as superphosphate, iron sulfate and even seawater as a source of trace elements. As soilless methods developed, higher-quality nutrient sources with a greater degree of purity and solubility became available for hydroponic use, these include the use of microelement chelates and more suitable sources of phosphates. Fertilizer salts used in hydroponics must be completely water soluble and should not contain any additives or insoluble fillers, coatings, or compounds such as insoluble sulfates or phosphates which are common in soil fertilizers. Fertilizers used in standard inorganic hydroponic formulae must not contain high levels of undesirable elements such as sodium, chloride and organic nitrogen nor contaminants such as heavy metals. These unwanted elements can accumulate in hydroponic systems to the extent that the EC measured starts to be made up of a high proportion of usable salts, thus creating imbalances with plant nutrition. It is also important to be aware of any chemical reactions with different fertilizer salts, particularly in concentrate form, and prevent precipitation of insoluble compounds. From a practical point of view, fertilizers selected must be fit for purpose, but also be cost-effective and economical for commercial crop production as well as being readily available from local suppliers. There is little point in utilizing an expensive fertilizer when a more cost-effective one will provide the same nutrient ions and still meet all the criteria as a hydroponic nutrient source.

9.12.1   Calcium nitrate

Calcium nitrate ($Ca(NO_3)_2$), also called 'Norwegian saltpetre' due to its first being synthesized in Notodden, Norway in 1905, is produced by reacting calcium carbonate with nitric acid, followed by neutralization with ammonia. Calcium nitrate may also be manufactured as a by-product of the extraction of calcium phosphate, termed the 'Odda process' (nitrophosphate process), or from an aqueous solution of ammonium nitrate and calcium hydroxide. Being a highly water-soluble fertilizer, calcium nitrate is commonly used for greenhouse and hydroponic crops, providing 15.5% N and 19% Ca. Calcium nitrate is typically found as a tetrahydrate which rapidly absorbs moisture from the air if stored incorrectly. While calcium nitrate provides both calcium and nitrate in nutrient formulae, it also forms about 1% ammonium-N in solution. In hydroponic systems calcium is almost exclusively supplied via calcium nitrate, however some calcium may be present in many water sources. Calcium nitrate is an off-white to cream-coloured granule in appearance and is typically available in 25 kg or 50 lb bags for greenhouse use. Greenhouse-grade calcium nitrate may contain some wax which can appear as a whitish scum floating on the surface of nutrient stock solutions after mixing, however this is harmless to plant growth. Calcium ammonium nitrate is another fertilizer which provides nitrogen in both the ammonium and nitrate forms and may be used in small quantities in some hydroponic formulae.

9.12.2   Ammonium nitrate

Half of the nitrogen in ammonium nitrate (NH_4NO_3) is in the ammonium form and half in the nitrate form. Ammonium nitrate typically contains 34% N with a high rate of solubility and may be dissolved for use in hydroponic systems. Ammonium nitrate is manufactured by reacting ammonia gas with nitric acid. The resulting solution (95–99% ammonium nitrate) is dropped from a tower and solidified to form prills which can be used as fertilizer or made into granular ammonium nitrate by spraying concentrated solution on to small granules in a rotating drum (Reetz, 2016). The total amount of ammonium-N percentage in a nutrient formula should be kept below 15% in most cases.

9.12.3   Ammonium phosphate

Ammonium phosphate ($NH_4H_2PO_4$) is manufactured by reaction of ammonia with phosphoric acid and typically contains 10% N and 22% phosphate. This is not used as a main source of phosphate in nutrient solutions as it would add too much ammonium, it can however be used to supplement phosphate or small amounts of nitrogen in the ammonium form.

9.12.4   Urea

Urea (CH_4N_2O) is a common source of nitrogen for soil application and provides 46% N. Use of urea should be avoided in hydroponics as it causes issues with ammonia toxicity in solution and has no EC charge, thus is it difficult to measure its concentration in hydroponics.

9.12.5   Potassium nitrate

Potassium nitrate (KNO_3) or saltpetre typically contains 13% N and 44–46% K_2O and is widely used for greenhouse and hydroponic crops requiring both nitrogen and potassium in a highly soluble form. Potassium nitrate is manufactured by reacting potassium chloride with a nitrate source such as sodium nitrate, nitric acid or ammonium nitrate. The resulting crystalline material has the appearance of white powder and is fully water soluble. Potassium nitrate, along with monopotassium phosphate, are the principal sources of potassium used in most hydroponic systems.

9.12.6   Potassium sulfate

Potassium sulfate (K_2SO_4), commonly called 'sulfate of potash', typically contains 40% K (48% K_2O) and 17% S, and is used to supplement potassium levels, often in fruiting crops with a high potassium requirement. Potassium sulfate may be found in mineral deposits or is produced from potassium chloride by processes which use either magnesium sulfate or sulfuric acid as sources of sulfate (During, 1984).

9.12.7 Monopotassium phosphate

Monopotassium phosphate (MKP, KH_2PO_4) is a commonly used hydroponic fertilizer which supplies 25% K and 21% phosphate. This fertilizer is a white to off-white fine granule in appearance and is produced by the action of phosphoric acid on potassium carbonate.

9.12.8 Calcium superphosphate

Calcium superphosphate provides 10% phosphate which is highly soluble in solutions (as phosphoric acid), but also produces calcium sulfate/calcium phosphate which is slow to dissolve in hydroponics (Lennard, 2001).

9.12.9 Magnesium sulfate

Magnesium sulfate ($MgSO_4$), or Epsom salts, typically contains 10% Mg and approximately 16% kieserite (magnesium sulfate monohydrate); both are highly water soluble and quick-acting sources of magnesium. Both magnesium sulfate and kieserite also provide sulfate as a soluble source for plant nutrition. Kieserite is a naturally occurring mineral mined from deep-underground, geological marine deposits in Germany (Reetz, 2016) which is finely ground or granulated depending on the application method used. Magnesium sulfate is an inexpensive, whitish crystalline salt which is found as a naturally occurring salt in mineral waters; hence the name 'Epsom salts' indicates the area where it was initially sourced.

9.12.10 Magnesium nitrate

Magnesium nitrate ($Mg(NO_3)_2$) provides 9.5% Mg and 10.5% N and is manufactured by the reaction of nitric acid and various magnesium salts. Magnesium nitrate is water soluble; however, it is an expensive fertilizer salt and usually unnecessary unless counteracting a sulfate accumulation, as both magnesium and nitrate can be supplied in more cost-effective forms with the use of magnesium sulfate and calcium nitrate (Lennard, 2001).

9.12.11 Iron chelates

Iron fertilizers for hydroponics are usually applied in the chelated form, which provides a more stable form of iron under neutral and high pH levels. Iron chelates include Fe-EDTA (6–14% Fe), Fe-EDDHA (6% Fe) and Fe-DPTA (6–11% Fe), which may also be applied as foliar sprays to assist in the correction of visible deficiencies in crop plants. In hydroponics, iron chelates are routinely used to provide Fe at levels of 1–6 ppm in the working-strength nutrient solution. The most commonly used iron chelate is Fe-EDTA which is a brown powder in appearance and is readily soluble in nutrient solutions. The choice of which iron chelate to use is dependent on the pH of the nutrient solution. Fe-EDTA and Fe-DPTA are stable up to a pH of 6.5, whereas Fe-EDDHA should be used where pH levels rise above pH 7.0. Fe-EDDHA is commonly used in substrate systems, often where a high-alkalinity water supply is present and where pH is more difficult to directly control and rises above pH 7.0. Fe-EDDHA is also more suitable for use in systems such as aquaponics where pH levels are typically maintained above 6.5.

Iron sulfate ($FeSO_4$, 20% Fe) is no longer widely used in soilless production due to its instability in solution: iron sulfate tends to form insoluble iron hydroxides which are not available for plant uptake. Despite this, a simple iron chelate can be made from iron sulfate for use in hydroponics in regions of the world where it has proven impossible or not cost-effective to obtain a suitable iron chelate product or for use in organic hydroponic systems where synthetic chelation agents are not permitted. This involves mixing iron sulfate and citric acid (an organic chelation agent) to create a more usable form of iron for hydroponic production. Other organically based chelation agents that can be used in hydroponics include humates and fulvic acid complexes.

9.12.12   Manganese sulfate, manganese chelates

Manganese sulfate ($MnSO_4$) and to a lesser extent manganese chelates may be used to supply manganese in hydroponic nutrient formulae. These are both highly water soluble, with manganese sulfate, a pale pink powder in appearance, containing 24–32% Mn. The manganese content of manganese chelates such as Mn-EDTA varies by source.

9.12.13   Copper sulfate, copper chelates

Copper can be supplied with the use of copper sulfate ($CuSO_4$, 25% Cu), which is a blue crystalline powder, or to a lesser extent as copper chelates. Foliar application of copper-containing compounds may also be used where a deficiency has been identified.

9.12.14   Zinc sulfate, zinc chelates

Zinc can be supplied with zinc sulfate ($ZnSO_4$, 23% Zn) or zinc chelate (Zn-EDTA) which has varying levels of Zn depending on source, in hydroponic nutrient formulae and applied as a foliar spray when required to correct deficiencies. Zinc sulfate is a white powder which readily dissolves.

9.12.15   Boric acid, borax

Fertilizers which supply boron include borax (sodium borate, $Na_2[B_4O_5(OH)_4].8H_2O$, 11–14% B) and boric acid (H_3BO_3, 18% B). Boric acid is a granulated salt which is relatively inexpensive and highly soluble.

9.12.16   Sodium molybdate, ammonium molybdate

Molybdenum can be supplied as ammonium molybdate (($NH_4)_6Mo_2O_{24}$, 48% Mo) or sodium molybdate (Na_2MoO_4, 39% Mo); both can be used in hydroponic formulae depending on what is locally available. Molybdenum fertilizers are usually white powders.

9.12.17   Nitric and phosphoric acids

Both nitric acid (HNO_3) and phosphoric acid (H_3PO_4) are used for pH control in hydroponic systems and for the pretreatment of water supplies to counter high alkalinity and pH levels. These should not be considered a main source of nitrogen or phosphorus and are best applied separately rather than mixed with fertilizers in nutrient stock solutions.

9.13   Fertilizer Composition and Grades

The composition or grade of a fertilizer is expressed as a percentage by weight, either as elements or oxides. This is typically given as an N–P–K grade which is a legal guarantee of the amount of nutrients supplied in the fertilizer product. A fertilizer grade of 6–3–4, for example, provides 6% nitrogen, 3% phosphorus as P_2O_5 and 4% potassium as K_2O. This provides the fertilizer compositional grade as the oxides P_2O_5 and K_2O; to obtain the percentage of actual elemental phosphorus and potassium, a conversion factor must be used. In some countries, fertilizer compositional grades are expressed more simply as the elements P and K rather than as the oxides P_2O_5 and K_2O on fertilizer products, allowing easier calculation of fertilizer programmes.

Fertilizers used for hydroponic formulations should be of 'greenhouse grade' which is suitable for dissolving into water. Greenhouse-grade fertilizers are not laboratory grade in purity but are the most cost-effective way to obtain the large quantities of macronutrients used in hydroponic crops. Trace elements or micronutrients are often purchased as a higher grade of fertilizer salt as only small quantities are required.

9.14 Chelation of Trace Elements

Trace elements such as iron, zinc, manganese and copper may, under certain pH conditions, react with hydroxyl ions in solution to form insoluble oxides which are not readily available for plant uptake. This reaction can be prevented by protecting the metal ion inside a chelating agent or ligand. A typical ligand will have chemical groups which share electrons with a metal ion to form a complex which is resistant to these reactions (Lennard, 2001). Chelating agents include EDTA (ethylenediamine-tetraacetic acid), EDDHA (ethylenediamine-di(o-hydroxyphenylacetic acid)) and DTPA (diethylenetriamine-pentaacetic acid). During plant uptake of the chelated ion, the chelating agent is dissociated and remains in the solution. In organic production, organic chelating agents may be used; these include citric acid, humates, amino acids and fulvic acid complexes.

9.15 Foliar Fertilizers

In some circumstances dilute fertilizer solutions may be applied to crop foliage in order to correct mineral deficiencies or as a supplement to nutritional programmes. While plants can absorb some nutrients through foliar application, this is often limited to those required in small amounts, i.e. micronutrients (trace elements). Foliar-applied nutrients are most commonly used on high-value crops due to the expense of application. Chelated micronutrients are the fertilizer types most suited to foliar application as they allow rapid correction of plant deficiencies with low application rates (Mikkelsen and Bruulsema, 2005). Nitrogen, often in the form of low-biuret urea, is another element which may be supplemented with foliar fertilizer spray application to assist with nitrogen nutrition of crops where required.

9.16 Electrical Conductivity

The EC of a nutrient solution is a measure of the total concentration of dissolved salts and is often referred to as 'salinity' (Adams, 2002). When fertilizer salts are dissolved into water, they dissociate into ions, both cations with a positive charge and anions with a negative charge. These ions in water conduct an electrical charge which is measured with an EC meter. While EC gives an indication of the strength of a nutrient solution, it does not differentiate between the individual nutrient elements; thus, a solution of only sodium chloride can have the same EC as one of potassium nitrate, but contain very different ions. While EC is a reliable measure of the concentration of a conventional hydroponic nutrient solution, some fertilizers do not conduct an electrical charge. These include urea and many organic fertilizer compounds which may be used in hydro-organic systems.

EC meters for hydroponic use come in a range of options and models (Fig. 9.2), from

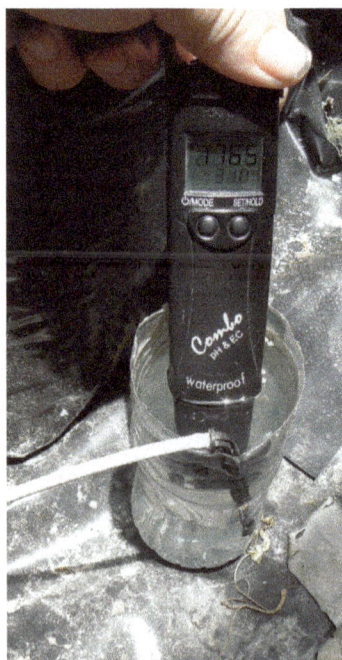

Fig. 9.2. EC meter for measuring the concentration of nutrient solutions.

laboratory-grade, bench-top, highly accurate equipment to the more common 'pen' or hand-held type of portable meter used by growers worldwide. Most EC meters are temperature compensating, have an electronic display, and can be calibrated and checked against a standard solution to ensure ongoing accuracy. Some meters also measure temperature and pH as well as EC, and such 'combo' meters are popular with smaller or hobbyist growers. Worldwide there are a number of different units in which EC can be displayed, the most common being mS/cm or microsiemens per centimetre (µS/cm); however, conductivity factor (CF) is still used in some countries. The conversion between these units is 1 mS/cm = 1000 µS/cm = 10 CF. While EC in mS/cm or µS/cm is the universally correct measurement of EC, some meters may display solution concentration in TDS (ppm as total dissolved solids). TDS meters do not actually measure TDS, but measure EC and then use an inbuilt conversion factor to convert the EC into TDS which is displayed. The issue with using TDS meters is that there is no one accurate conversion from EC to TDS as this is dependent on the combination of salts in the nutrient solution, which varies between different nutrient solution formulations. Thus, different TDS meters often use different conversion factors based on their estimate of the composition of a standard nutrient solution, meaning that TDS is largely an approximate estimate of solution strength and not as accurate as measuring EC directly. Most TDS meters use a conversion factor of 0.5 to 0.7 depending on model, with units expressed in ppm. While a grower using the same TDS meter to continually monitor solution concentration can make comparisons and adjustments based on these readings, comparison between meters with different conversion rates can be an issue. It is more accurate to follow EC guidelines in mS/cm or µS/cm by measuring and adjusting these with an EC meter than to use TDS guidelines not knowing what conversion factor has been used.

Most soilless crops are grown with an EC in the range of 0.8–4.5 mS/cm, with young seedlings requiring a lower EC than more mature plants. Each crop has an optimal EC value based on stage of growth, climate and system of production. Low-EC crops include lettuce, herbs, micro greens and many salad vegetables; higher-EC crops include fruiting plants such as tomatoes where a greater level of nutrition improves fruit flavour quality (Tables 9.4 and 9.5). EC levels, even within crops and systems, may change with growing conditions. Under warmer, high-light conditions, EC is often reduced in many crops, particularly with leafy greens to allow more rapid uptake of water for transpirational demands; this can also assist with reducing the severity

Table 9.4. Recommended EC levels for different crops (in mS/cm).

	Summer[a]	Winter[b]
Lettuce seedlings	0.5–0.8	0.8–1.0
Lettuce (fancy)	0.8–1.2	1.5–2.0
Lettuce (iceberg)	1.8–2.3	2.5–3.0
Basil	2.0–2.5	2.3–2.8
Spinach	1.4–1.8	2.0–2.3
Arugula	0.8–1.2	1.2–1.5
Kale	1.8–2.2	2.4–3.2
Watercress	0.8–1.0	1.2–1.4
Broccoli	1.8–2.4	2.3–2.8
Parsley	0.8–1.3	1.4–1.8

[a]'Summer' refers to higher temperature and light levels.
[b]'Winter' refers to lower light levels.

Table 9.5. Recommended EC levels for different fruiting crops (in mS/cm).

	Vegetative stage	Fruiting stage
Tomato, summer[a]	1.8–2.5	2.3–3.5
Tomato, winter[b]	2.3–2.7	3.0–4.0
Capsicum, summer	1.6–1.8	2.3–2.7
Capsicum, winter	2.0–2.5	2.5–3.0
Chilli (hot)	2.6–2.8	3.0–3.5
Strawberry, summer	1.6–2.0	1.8–2.5
Strawberry, winter	2.0–2.5	2.0–2.5
Cucumber, summer	1.6–1.8	2.0–2.2
Cucumber, winter	1.8–2.0	2.2–2.5
Aubergine	1.8–2.4	2.6–3.0
Melon	1.8–2.0	2.2–2.6
Roses	2.0–2.2	2.4–2.8

[a]'Summer' refers to higher temperature and light levels.
[b]'Winter' refers to lower light levels.

of calcium disorders such as tipburn and blossom end rot. Under lower light and temperatures of winter, EC is often increased in many crops to assist with maintaining produce quality and for plant strengthening.

Under hydroponic production, EC has a significant effect on plant growth regardless of the nutrient content of the solution due to the osmotic potential in the root zone. A higher EC creates a higher osmotic potential which in turn influences the rate of water and nutrient uptake. Increasing the EC will reduce water uptake by the crop and can cause some plants to concentrate organic compounds in fruit and foliage as well as slowing vegetative growth. This is often beneficial for crops such as tomatoes where higher EC has been found to increase the compositional quality of the fruit. A lower EC will promote water uptake and produce softer foliage growth. EC levels may vary between different types of hydroponic systems. EC can be maintained higher in solution culture systems such as NFT, DFT, float or raft systems as the EC can be maintained at a stable level around the root zone. In substrate systems, EC can increase in the root zone after solution application due to evaporative losses from the surface of the substrate and plant uptake of water.

9.17   pH

Along with EC, pH of the nutrient solution and root zone is controlled in soilless systems within the 5.6–6.4 range for most crops. Many commercial systems aim for a tighter pH range of 5.8–6.0. Nutrient formulations will have different starting pH values as individual salts become more or less acidic when dissolved into water. Salts such as monopotassium phosphate lower the pH more than salts such as calcium nitrate. Most formulations made up with pure water (RO) will result in an initial pH of around 5.5–6.0, which is ideal for the growth of hydroponic crops. This pH results from only the commonly used salts being dissolved into stock solutions, and so addition of acid or alkali to stock solutions is usually unnecessary. It is best to correct the pH of

unsuitable water supplies such as those with a high pH and alkalinity before making up the nutrient solutions. In hydroponic solutions, some salts can be used to influence the pH control of the nutrient solution, reducing the requirement for acids during growth development phases of the crop.

pH in nutrient solutions can be adjusted with the use of specifically designed 'pH up and down' products which are diluted and ready to use. Commercial growers make up their own pH adjustment solutions of 10% nitric or phosphoric acid or potassium hydroxide for increasing pH. In some circumstances sulfuric and hydrochloric acid may also be used where only limited volumes are required for pH control. By using a ratio of 50% nitric and 50% phosphoric acid, the solution can be kept in better balance as both nitrogen and phosphorus are being added (however, where large volumes of acid are being used, the nutrient formulation should be adjusted for these additional nitrogen and phosphorus sources). There are 'organic acids' such as citric acid and acetic acid which are occasionally used in hydro-organic or small systems to bring the pH down; however, these are very weak acids and only give a short-term pH reduction.

There is a wide range of pH testing equipment, from the inexpensive liquid test kits and strips sold by aquarium/swimming pool suppliers to high-technology, electronic meters at a wide range of prices (Fig. 9.3). Test kits and strips should be for the lower pH range: many of these measure pH values

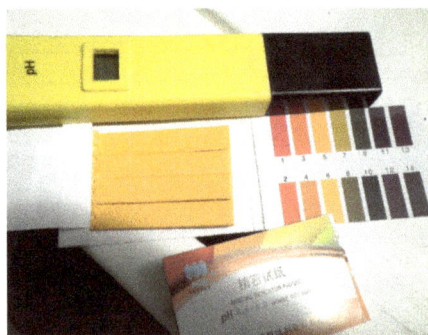

Fig. 9.3. pH meter and test strips.

of 5.0–7.0. Growers with electronic meters may also invest in a liquid pH test kit or strips, to periodically check the accuracy of their electronic meter. pH meter probes tend to 'drift' over time and need to be regularly (weekly) calibrated, cleaned and replaced annually. Calibration of pH meters is essential and often overlooked by many growers; electronic pH meters usually come with 'buffers' – solutions of pH 4 and pH 7 – which can be used to check and adjust the meter's readout, or buffer solutions for this purpose are sold by hydroponics retailers and laboratory supply outlets. Some pH meters require the probe to be stored wet and cleaning the sensitive probe tip is also important for accuracy.

With a recirculating system such as NFT, measurement of pH is simple as the solution contained at the reservoir is what is in direct contact with the roots. For media-based systems which are often drip irrigated and where the nutrient solution may not be recirculated, pH control is slightly more complex. For these types of system pH (and EC/TDS) needs to be measured both in the nutrient supply reservoir and in the solution that drains (i.e. the 'leachate') from the base of the growing beds/bags/slabs containing the plants. With large, rapidly growing plants, there will usually be slight differences in pH between the irrigated nutrient solution and the leachate/drainage solution due to pH changes as the solution flows past the roots and has ions absorbed/released. The growing medium may also influence pH. Adjustment of pH at the nutrient supply reservoir should be based on the pH measured in the leachate draining from the plants' root system. For example, if the pH is 6.9 in the leachate and 6.2 in the feed solution, the pH of the feed solution should be acidified to the point where the drainage solution starts to come down to 5.8–6.0 as this is the pH which the root system will then be experiencing.

9.18   Automation and Testing Equipment

Nutrient solutions in soilless systems may be adjusted and monitored manually or completely automated and computer controlled, depending on the size and complexity of the system. Manual adjustment requires the grower to use hand-held EC and pH meters to monitor and adjust the strength and composition of the nutrient solution on a regular, often daily, basis. Automated systems have inbuilt sensors which feed data back to controller systems, while electronic doser units control the addition of nutrient concentrates and acids on a frequent, almost continual basis. Water and nutrient usage are often recorded via computer programs which also control the environmental conditions in the growing area. In substrate systems, the volume of nutrient leachate, substrate moisture levels, nutrient treatment and sterilization in closed systems and the results of solution analysis are also recorded and used to control the irrigation system.

9.19   Conditions Which Affect Nutrient Uptake Rates

9.19.1   Temperature and humidity

The effect of season on nutrient uptake for most crops is a combination of changes in light levels and temperature. When these two factors decrease so too does crop growth, and with less nutrients required for growth, nutrient absorption from the solution slows down. Not only does nutrient uptake in general fall, but the proportions in which the major elements (nitrogen, phosphorus and potassium) are absorbed also change. For this reason, many growers have standard summer and winter nutrient formulae to compensate for these changes in uptake and provide the plants with the correct ratio of elements during a particular season. Changing nutrient ratios for summer and winter can have other beneficial effects on hydroponic crops, particularly with nitrogen and potassium levels. Potassium plays a role in fruit quality, shelf-life, firmness and many other factors and hence it is essential to maintain good levels of this element in winter. High levels of potassium in the nutrient solution may compensate for low light levels in crops such as heated green-house tomatoes. Similar results have been found with cut flowers such as chrysanthemums, where increasing the level of potassium as the crop matures under low-light conditions helps to improve stem strength (Chu and Toop, 1975).

High air temperatures will increase the nitrogen uptake of crops such as tomatoes; however, at warmer temperatures, if the plants begin to experience low nitrogen levels in solution, then this can induce flower drop and the rapid development of nitrogen deficiency symptoms. At lower temperatures with similar low nitrogen levels, flower drop may not occur and nitrogen deficiency symptoms would take much longer to develop. Phosphorus uptake is also influenced by air and root temperatures. Tomato plants take up substantially more phosphorus at 21°C than at 10–13°C, and this is also true for potassium and a number of other elements. In cucumber plants at temperatures above 20°C, there is little difference in the uptake of potassium and nitrogen; however, if temperatures drop below 12°C then nutrient uptake is sharply reduced.

9.19.2 Time of day

While nutrient uptake ratios at different times of the day do not necessarily affect nutrient formulation, in general both nitrogen and potassium uptake rates are lowest at night and reach a maximum during the brightest part of the day. Rates of uptake then decline towards evening. Therefore, the rates of nitrogen and potassium uptake are highly correlated with light intensity, air temperature and water uptake, all of which are highest during the day.

9.19.3 Light levels

Light levels are an important factor influencing plant growth and development and these also have a major effect on nutrient uptake. In tomato crops, plants growing in the shaded area of a greenhouse were found to accumulate 22% less nitrogen and 19% less potassium than fully exposed plants, although there was little effect on phosphorus uptake (Winsor et al., 1958). If the crop is covered with heavy shading which cuts out 50–67% of the light, then there is likely to be decreased uptake of nitrogen, potassium, phosphorus, calcium and magnesium. The use of supplemental lighting during low light levels will increase the uptake of nitrogen and potassium and result in a greater yield. Under poor light conditions, the uptake of potassium can be reduced to the point where the incidence of 'blotchy ripening' in tomatoes becomes severe, particularly where calcium content of the nutrient solution is high. This emphasizes the importance of maintaining the correct nutrient ratios in solution, which are customized for not only the general season, but current light levels as well.

9.19.4 Root health and size

It has been found that as the amount of root (measured as root dry weight) increases, so too does the uptake of both nitrogen and potassium, with there being little effect on phosphorus uptake (Alwan and Newton, 1984). This is an effect of a greater surface area of root being available to take up the nutrients from solution. Root diseases that adversely affect the health of the root system and cause root death will also result in a reduction in nutrient uptake, even to the point where the plant will show deficiency symptoms.

9.19.5 Aeration and oxygenation

Aeration and the oxygen content of the nutrient solution and/or rooting medium can have an effect on the amount of nutrients taken up by the plant and hence the formulation of the nutrient solution. The growth of tomato plants, for example, increases with the oxygen content of the nutrient solution, thus making some form of aeration essential in NFT and many other systems.

Increases in the oxygen level also mean that more phosphorus, potassium and nitrogen are absorbed by the root system. At low oxygen levels, potassium can actually leak from the roots back into the nutrient solution, stressing the plant and causing imbalances with the nutrient ratios in the formulation. With tomato and other crops, it has been found that there is a direct relationship between the removal of oxygen by the root system and the amount of both nutrients and water taken up. Thus, a highly aerated solution with maximum oxygen levels will allow the plants to absorb the maximum quantities of minerals required for growth, provided all other conditions are maintained at optimal levels.

9.20 Plant Tissue Analysis

Plant mineral tissue analysis is a technique used by growers to monitor plant progress, diagnose nutrient disorders, and assist in planning nutrient programmes and corrections. Samples for plant tissue analysis may be collected by the grower and sent to an agricultural laboratory, or in larger operations may be analysed onsite using a range of portable equipment. Tissue analysis results report the levels of all macro- and micronutrients in the sample tested – as a percentage for macronutrients and as parts per million for micronutrients. Growers can then compare the reported tissue data with optimal levels for the crop being grown to determine the nutritional status of the crop and identify any deficiencies or toxicities (Table 9.6).

9.21 Fertilizer and Environmental Concerns

Under protected cultivation and in hydroponic systems, nursery container culture and other horticultural operations not directly using soil, nitrogen and other nutrient losses can be minimized. These methods include precise control over fertigation, nutrient solution and water application to minimize nutrient losses from containerized plants. The increasing use of 'closed hydroponic systems' which recirculate or collect nutrient solution leachate for treatment and reuse rather than draining to waste are seen as efficient methods of minimizing fertilizer use and preventing environmental issues.

9.22 Water and Nutrient Solution Treatment Methods

Water and nutrient solution treatment to control pathogens is becoming an increasingly important aspect of hydroponic production worldwide. Root diseases are a significant risk to greenhouse operations and can cause considerable plant losses and reductions in yield and quality. Treatment of

Table 9.6. Typical foliar mineral levels by crop. (Data taken from Kay and Hill, 1998.)

Element	Lettuce	Tomato	Rose	Strawberry
N (%)	3.1–4.5	4.5–5.5	3.0–5.0	2.6–3.5
P (%)	0.35–0.60	0.40–0.70	0.25–0.50	0.25–0.35
K (%)	4.5–8.0	4.0–6.0	1.5–3.0	1.0–2.0
Ca (%)	0.8–2.0	1.2–2.0	1.0–2.0	0.7–1.5
Mg (%)	0.30–0.70	0.40–0.70	0.25–0.50	0.25–0.40
S (%)	0.20–0.30	0.60–2.00	0.25–0.70	0.15–0.35
Na (%)	0.00–0.30	0.08–0.15	0.00–0.10	0.02–0.10
Fe (ppm)	50–100	80–200	60–200	100–200
Mn (ppm)	50–300	50–250	30–200	200–500
B (ppm)	25–55	30–60	30–60	30–100
Zn (ppm)	25–250	30–60	17–100	30–80
Cu (ppm)	7–80	15–50	7–25	5–12

incoming water supplies is dependent on the water source, with municipal supplies often having sufficient treatment to meet drinking-water standards which also controls most potential plant pathogen problems. Water supplies which come from wells, dams or sources that may have been in contact with soil, vegetation or wind-disseminated spores are common sources of plant pathogens. Apart from water supplies which carry infectious diseases into a crop, nutrient solutions that have been in use may carry pathogens back out into the environment via wastewater systems and many countries now require these to be treated before discharge. In recirculating systems, the nutrient solution may be a vector of pathogen spores, thus one or two infected plants can potentially infect the rest of the crop via the nutrient solution. Nutrient solution treatment to control pathogens can assist with reducing the risk of pathogen dissemination in this way; however, most treatments do not have a residual action and root-to-root transfer of diseases can still occur. There is a range of different treatment options for pathogen control of source water, nutrient solution and wastewater; these include UV radiation, ozone, heat, surfactants, filtration and chemical treatments.

9.22.1 Ultraviolet disinfection

UV radiation may be used for treating water supplies, recirculating nutrient solutions and waste solutions before discharge. Antimicrobial activity largely occurs within the UVC range of 200–280 nm, with plant pathogens becoming inactivated when UVC irradiation affects the nucleic acid which strongly absorbs at or close to 260 nm (Wohanka, 2002). UV treatment of water or nutrient solutions is carried out with specialized equipment which passes light from low- or high-pressure UV lamps through a thin film of solution. For this process to work effectively the solution must be clear and often prefiltration is required to remove excess organic matter or other material. The recommended UV dose rates for disinfection

of recirculating systems are $100 \ mJ/cm^2$ for control of pathogenic fungi and $250 \ mJ/cm^2$ for more complete treatment and control of all pathogens including viruses (Runia, 1994); however, these rates are only effective when the transmission of water is sufficient and the quartz tube does not contain appreciable precipitation of salts (van Os, 2001). Most UV systems use either high- or low-pressure gas inside the bulbs and while the performance of both types of pressure lamps is similar, high-pressure lamps are less energy efficient (Ehret et al., 2001; van Os 2001; Scarlett et al., 2016).

In recirculating systems such as NFT, UV disinfection efficacy is a function of the dose of UV irradiation in terms of light intensity, but also transmittance and the flow rate of the circulating solution (Zhang and Tu, 2000). For these reasons, filtration methods such as slow sand filtration or treatment with membrane filters may be used to remove organic material so that transmittance of the solution is maintained above 50% (Zhang and Tu, 2000). While UV may effectively control pathogens within the nutrient solution, it does not have a residual effect and thus diseases already present within the root system are not controlled via UV treatment. Gharbi and Verhoyen (1993) found that UV treatment of a recirculating nutrient solution in NFT confined the spread of infection to only plants downstream of the infection source and prevented inoculation of plants between different channels.

While UV is an effective disinfection method for water supplies and can assist pathogen control in recirculating systems, it does not discriminate between pathogens and non-target or beneficial bacterial populations. Maintaining levels of healthy microbial species in hydroponic systems has become of increased interest in recent years as they may play an important role in suppressing root disease (van Os, 2001). Studies have found that the control of Pythium root rot with UV treatment in recirculating systems also affects non-target bacterial populations and requires further investigation (Zhang and Tu, 2000). Another potential issue with the use of UV treatment on nutrient solutions is the effect on iron chelates, which may reduce the availability of iron for plant uptake and has been reported to cause iron deficiencies in some circumstances. It is possible to prevent this by increasing levels of iron chelate and using the most stable chelate forms, which are Fe-EDDHA, followed by Fe-DPTA and Fe-Na-EDTA (Acher *et al.*, 1997).

9.22.2 Ozone

Ozone has been used for drinking-water treatment since the early 1900s and may be used for water supplies and recirculating solutions. Ozone (O_3) is a strong oxidizing agent with no residual activity against microorganisms in the water and therefore leaves no residues that might be toxic to plants (Hong and Moorman, 2005). As no residue remains after ozone treatment, the development of pathogen resistance to this treatment method is unlikely (Guzel-Seydim *et al.*, 2004). Ozone treatment of nutrient solutions requires an ozone generator onsite which discharges ozone-enriched air as small bubbles into the solution flowing through a venturi. Ozone dissolves from these bubbles into the solution over a certain contact time. The ozone reacts with organic matter including pathogens and the efficiency of disinfection can be increased by lowering the pH of the solution to 4.0 with acid (Runia, 1994). One of the major benefits of the use of ozone for water or nutrient solution disinfection is that any ozone which has not reacted with chemical or biological compounds reverts to oxygen; this increases the DO content of the solution which has the potential to improve crop growth. Ozone-generation systems must be installed correctly as off-gassing into the air surrounding the crop can cause considerable plant damage. When applied directly to substrates as drip irrigation, the risk of off-gassing is greatly reduced as the solutions are not exposed to the bulk atmosphere (Graham *et al.*, 2011).

Like UV, ozone not only destroys pathogenic microorganisms but also beneficial bacteria carried within the treatment solution. Despite this, the fact that ozone has no residual activity ensures the populations of beneficial microorganisms may still exist and colonize root systems and system surfaces where ozone does not reach. Ozone treatment can break down iron chelate and may cause precipitation of manganese; both of these compounds may require adjustment or monitoring where ozone is in use. In some systems, ozone may also break down other additives or organic compounds applied to the nutrient solution and these breakdown products may become phytotoxic under certain conditions.

9.22.3 Filtration

A number of filtration methods may be incorporated into hydroponic systems for different purposes. Coarse or mechanical filters on recirculating systems are used to catch debris in the nutrient solution as it returns to the reservoir, these may include particles of growing substrate, pieces of root systems and undissolved nutrient scale and precipitates. Other types of filtration aim to remove pathogens from the recirculating nutrient solution but may also be used to treat water supplies before addition to a hydroponic system. These are membrane filtration, which includes microfiltration (pore size 100–1000 nm), ultrafiltration (10–100 nm), nanofiltration (1–10 nm) and RO (<1 nm), and slow sand filtration or biofiltration (Postma *et al.*, 2008). There are a number of membrane-type filtration systems that can be used to remove microbes from the water (Hong and Moorman, 2005). The most efficient membrane systems use a combination of different filters to progressively remove smaller and smaller particles as the solution flows past Many systems have automated back washing of the filters required to prevent frequent clogging and improve efficiency. Membrane filtration has been shown to be effective for a number of different pathogens (Wohanka, 2002); however, a high level of investment and maintenance is required to maintain an effective system.

Activated carbon or charcoal filters may be used on water supplies, particularly where municipal water may have been treated. Carbon filters consist of a canister filled with activated carbon which removes certain chemicals and compounds from water supplies, including chlorine and other water treatment chemicals and organic compounds. While these are effective for some water supplies, the activated carbon requires frequent replacement for the filter to remain effective.

9.22.4 Slow sand filtration

Slow sand filtration or 'biofiltration' is a method of water purification which has been utilized for over a century and has been adapted for use with hydroponic nutrient solutions to control a wide range of pathogens (Calvo-Bado et al., 2003). Slow sand filtration works on a number of different levels. First, the filter material screens out any organic or suspended matter from the water or nutrient solution, this forms a skin or dirt layer on the surface of the filter. Second and more importantly, the filter material provides a large surface area which is colonized by a diverse range of beneficial microorganisms, these provide the biological filtration which assists in the removal of pathogens. The principle behind slow sand filtration is that the nutrient solution applied must flow through the filter at the correct rate. If flow rates are too rapid, the removal of plant pathogens such as *Pythium* is compromised and the filter may not be effective for disease control. Flow rates of 100 l/m^2 per h and selection of the correct grade of fine sand (0.15–0.35 mm) have been found to increase the performance of slow sand filters (van Os et al., 1997); however, Wohanka et al. (1999) also found a grain size of 1–2 mm and flow rates of 300 l/m^2 per h to be effective. Other factors which determine the effectiveness of slow sand filtration include the oxygen level of the flowing solution and temperature, both of which are likely to influence the activity of the microbial communities residing within the filter structure. Apart from removal of organic material and pathogen control obtained via

slow sand filtration, another advantage of this method is that the treated solution leaving the system is enriched with populations of beneficial microorganisms which may assist with further reduction of pathogens in the hydroponic system (Postma et al., 2000).

The construction of a slow sand filter is relatively straightforward and requires the top of the filter to be open to the air as oxygen is a vital component of biological filtration. The base of the filter is filled with coarse drainage sand or gravel (8–16 mm), the middle levels of the filter with finer sand (2–8 mm) and the top layer with the finest grade of sand (1–2 mm). The top layer of sand should be at least 30 cm deep as this is where the majority of the biological filtration will occur. Granulated rockwool has also been used as the filter body material as an alternative to sand and this has proven to be highly effective, particularly in smaller filters (Wohanka et al., 1999). Granulated rockwool provides a high surface area for colonization by microorganisms and may also have the advantage of being cleaner and less likely to 'leak' fine sand into the lower layers of the filter. A newly constructed sand filter requires a few weeks for the microbial communities to establish and populate the filter material. While it is possible to inoculate a new filter with microbial products, these will naturally develop over time and providing warm conditions, sufficient oxygen and moisture will assist with this process.

To operate an effective slow sand filter, the nutrient solution or water to be treated must be dripped or sprayed on to the surface of the filter so as not to dislodge or disrupt the filter surface. This process also assists with oxygenation of the solution which is important as the bacteria in the filter bed require oxygen to function. A shallow layer of water (supernatant water) must remain over the surface of the sand to keep it moist while the slow flow rate is controlled by the outlet in the base of the filter system within an inline tap. Nutrient solution flowing through a slow sand filter will undergo biological filtration; however, this process will not change the physical or chemical nature of the solution, therefore pH, EC and levels of individual ions will not alter during filtration. It has been found the levels of DO in a nutrient solution fall as it flows through the filter material (Wohanka, 2002). This is caused by the high levels of microorganisms contained within the filter materials, which increase the BOD. Aeration of the solution as it returns to the nutrient reservoir will increase DO back to normal levels. While slow sand filters are simple to maintain, they require cleaning or scraping of the top layer of filter bed to remove any organic material build-up to prevent clogging. While this process is required to maintain flow rates, it should only be carried out when necessary as the top layer is heavily colonized by biologically active microorganisms which assist with the breakdown of organic material.

9.22.5 Chlorine

Chlorination is commonly used to disinfect water supplies and in the postharvest treatment of many fruits and vegetables. Chlorination treatment efficacy is dependent on solution pH and is most effective at a pH of between 5 and 6.5. Other factors which influence the disinfection process with chlorination are temperature, organic loading and microbial content of the water or solution being treated. Solutions treated with chlorine require frequent testing of free available chlorine present for effective pathogen control as contact with organic matter deactivates chlorine. Another method utilized to measure the real-time oxidizing potential of a chorine solution is the use of 'oxidation reduction potential' (ORP) (Lang et al., 2008). The pathogen species and life stage of the pathogen being controlled are also factors as chlorine sensitivity can vary with genera, species and even type of propagule of a single pathogen (Hong et al., 2003).

Chlorine may be applied as sodium hypochlorite (NaOCl), calcium hypochlorite (Ca (ClO)$_2$) or chlorine gas (Cl$_2$), with application rates depending on the organic loading of the treatment solution and pathogens being controlled. It has been reported that zoospores of six Phytophthora species and some Pythium

species were controlled by chlorine exposure at 2 mg/l for 0.25 min (Hong et al., 2003). Lang et al. (2008) reported that Pythium aphanidermatum zoospores were killed after 0.5 min exposure where ORP ranged from 748 to 790 mV at a pH of 6.3. Another study (Cayanan et al., 2008) found that a chlorine level of 2.5 mg/l controlled Pythium species only, while a level of free chlorine of 14 mg/l was required to control the species studied (Phytophthora infestans, Phytophthora cactorum, Fusarium oxysporum and Rhizoctonia solani); however, this level was also phytotoxic to many common nursery species. Chlorination toxicity symptoms include necrotic mottling, leaf necrosis, stunting and premature leaf abscission (Cayanan et al., 2008).

9.22.6 Hydrogen peroxide

Hydrogen peroxide (H$_2$O$_2$) is a strong, unstable oxidizing agent which reacts to form water and an oxygen radical that in turn reacts with any type of organic material including pathogens (Postma et al., 2008). The by-product of the use of hydrogen peroxide is the release of oxygen into the water or nutrient solution, thus increasing DO levels. The use of hydrogen peroxide in recirculating nutrient solutions carries a risk of damaging root systems. Nederhoff (2000) reported that 100 ppm of H$_2$O$_2$ controlled pathogens; however, 85–100 ppm nearly killed young lettuce seedlings and 8–12 ppm reduced the growth of hydroponic lettuce. Other studies have found that a concentration of 50 ppm was sufficient to kill Pythium in drain water within 5 min and 100 ppm to control Fusarium; however, viruses required a level of 500 ppm which was also harmful to plant roots (Postma et al., 2008). Hydrogen peroxide is also used as surface disinfectant for greenhouses, irrigation systems after crop removal and to treat discharge waters.

9.22.7 Heat

Heat treatment (pasteurization) is one of the most reliable methods for treating water to eliminate all types of plant pathogens (Hong and Moorman, 2005) and is used commercially by greenhouse growers, particularly in the Netherlands and the UK (Ehret *et al.*, 2001). Systems commonly use heat exchangers to first preheat and then further heat water to the correct temperature (Poncet *et al.*, 2001), with prefiltration often required to remove organic matter and other debris. Most commercial heat treatment systems for recirculating solutions heat to a temperature of 95°C with a holding time of 30 s (Ehret *et al.*, 2001). This is based on studies carried out by McPherson *et al.* (1995), who found that control of *Phytophthora cryptogea* and *P. aphanidermatum* in recirculating nutrient solutions for tomato and cucumber was possible when heated to 95°C for 30 s, while introduced pathogens readily dispersed through the system and infected a large percentage of plants with untreated solutions.

While heat treatment of water is relatively straightforward, heating of nutrient solutions causes a build-up of scale deposits on the heating coils and pipes and acid may be added to the nutrient solution prior to heat treatment to minimize the effect. After heat treatment, water is pumped back through an exchanger to recover some of the heat and to start the cooling process. While the use of heat for disinfection is highly effective, the major drawbacks are the high energy requirements for this treatment, the time taken for treatment and solution cooling, and use of tanks for treating and holding water or nutrient solutions.

9.23 Surfactants

Surfactants are one of the most widely used additives in agriculture – they essentially lower surface tension and allow the spreading and adhesion of liquids as well as enhancing the absorption of compounds and sprays. Surfactants are also used as 'wetting agents' in soil and soilless growing mixes, allowing the substrate to initially saturate up easily as many media such as peat can be water repellent when fully dry. Another

use of surfactants in hydroponic nutrient solutions is based on the ability of these compounds to rapidly lyse mobile zoospores of pathogens such as *Olpidium* (a vector for lettuce big vein virus), *Phytophthora* and *Pythium* (Stanghellini *et al.*, 1996; Stanghellini and Miller, 1997; Nielsen *et al.*, 2006). Many of the common root rot diseases which are problematic in hydroponics are spread via 'zoospores'. What makes these particularly damaging in hydroponics is that diseases producing zoospores release these into the nutrient solution or irrigation water. Zoospores survive in water and are able to swim, locate and infest new root systems. Thus, zoospores in hydroponic systems can rapidly spread an isolated disease outbreak through this highly efficient system of zoospore infection. Zoosporic fungal diseases such as *Pythium* thrive under warm, wet hydroponic conditions which favour the spread via motile spores particularly if plants have been weakened or stressed in any way.

A number of non-ionic surfactants such as 'Agral' have been found to disrupt the plasmalemma of fungal structures such as zoospores which lack a cell wall; however, only zoospores are affected with other structures such as mycelium or encysted spores remaining intact. Studies have found the zoospores of *Pythium* and *Phytophthora* will lyse within 1 min of exposure to a non-ionic surfactant at a concentration of 20 µg/ml (Stanghellini and Tomlinson, 1987). While surfactants can destroy large numbers of zoospores being disseminated by the nutrient solution and hence prevent or slow the spread of these pathogens, they have no effect on plants already infected when the disease was contained inside plant tissue. Thus, the use of non-ionic surfactants as a nutrient solution additive is more of a preventive action rather than curative. For this to be effective the correct rate of non-ionic surfactant needs to be continually maintained in the nutrient solution as the compounds degrade over time. Since surfactants cause foaming at the nutrient tank, the lack of foaming may be used as a guide to the timing of subsequent surfactant doses (Stanghellini *et al.*, 1996).

References

Acher, A., Heuer, B., Rubinskaya, E. and Fischer, E. (1997) Use of ultraviolet-disinfected nutrient solutions in greenhouses. *Journal of Horticultural Science* 72(1), 117–123.

Adams, P. (2002) Nutritional control in hydroponics. In: Savvas, D. and Passam, H. (eds) *Hydroponic Production of Vegetables and Ornamentals*. Embryo Publications, Athens, pp. 211–261.

Al-Karaki, G.N. (2011) Utilisation of treated sewage wastewater for green forage production in hydroponic system. *Emirates Journal of Food and Agriculture* 23(1), 80–94.

Al-Shrouf, A. (2017) Hydroponics, aeroponic and aquaponics as compared with conventional farming. *American Scientific Research Journal for Engineering, Technology and Sciences* 27(1), 247–255.

Alwan, A.H. and Newton, P. (1984) Dissolved oxygen, root growth, nutrient uptake and yields of tomatoes. In: *Proceedings of the 5th International Congress on Soilless Culture, Wageningen, The Netherlands, 18–24 May 1984*. ISOSC Secretariat, Wageningen, The Netherlands, pp. 81–111.

Barbosa, G.L., Gadelha, F.D.A., Kublik, N., Proctor, A., Reichelm, L., *et al.* (2015) Comparison of land, water and energy requirements of lettuce grown using hydroponic vs. conventional agricultural methods. *International Journal of Environmental Research and Public Health* 12(6), 6879–6891.

Calvo-Bado, L.E., Pettitt, T.R., Parson, N., Petch, G.M., Morgan, J.A.W. and Whips, J.M. (2003) Spatial and temporal analysis of the microbial community in slow sand filters used for treating horticultural irrigation water. *Applied Environmental Microbiology* 69(4), 2166–2125.

Cayanan, D.F., Zheng, Y., Zhang, P., Graham, R., Dixon, M., *et al.* (2008) Sensitivity of five container grown nursery species to chlorine in overhead irrigation water. *HortScience* 43(6), 1882–1887.

Choi, H.G., Moon, B.Y., Bekhzod, K., Park, K.S., Kwon, J.K., *et al.* (2015) Effects of foliar fertilization containing titanium dioxide on growth, yield and quality of strawberries during cultivation. *Horticulture, Environment and Biotechnology* 56(5), 575–581.

Chu, C.B. and Toop, E.W. (1975) Effects of substrate potassium, substrate temperature and light intensity on growth and uptake of major cations by greenhouse tomatoes. *Canadian Journal of Plant Science* 55(1), 121–126.

During, C. (1984) *Fertilisers and Soils in New Zealand Farming*. Government Printing Office, Wellington.

Ehret, D., Alsanius, B., Wohanka, W., Menzies, J. and Utkhede, R. (2001) Disinfestation of recirculating nutrient solutions in greenhouse horticulture. *Agronomie, EDP Sciences* 21(4), 323–339.

Garibaldi, A., Gilardi, G. and Gullino, M.L. (2011) Effect of potassium silicate and electrical conductivity in reducing powdery mildew of hydroponically grown tomato. *Phytopathologia Mediterranea* 50(2), 192–202.

Gharbi, S. and Verhoyen, M. (1993) Sterilization by UV irradiation of nutrient solutions with a view to avoiding viral infections transmitted by in hydroponic culture of lettuce. *Mededelingen van de Faculteit Landbouwwetenschappen, Rijksuniversiteit Gent* 58(3a), 113–1124.

Gottardi, S., Lacuzzo, F., Tomasi, N., Cortella, G., Manzocco, L., *et al.* (2012) Beneficial effects of silicon on hydroponically grown corn salad (*Valerianella locasta* (L.) Laterr) plants. *Plant Physiology and Biochemistry* 56, 14–23.

Graham, T., Zhang, P. and Dixon, M. (2011) Aqueous ozone in the root zone: friend or foe? *Journal of Horticulture and Forestry* 3(2), 58–62.

Guzel-Seydim, Z.B., Greene, A.K. and Seydim, A.C. (2004) Use of ozone in the food industry. *LWT Food Science and Technology* 37(4), 453–460.

Haghighi, M. and Daneshmand, B. (2018) Beneficial effect of titanium on plant growth, photosynthesis and nutrient trait of tomato cv. *Foria. Iran Agricultural Research* 37(1), 83–88.

Hong, C.X. and Moorman, G.W. (2005) Plant pathogens in irrigation water: challenges and opportunities. *Critical Reviews in Plant Sciences* 24(3), 189–208. https://doi.org/10.1080/07352680591005838

Hong, C.X., Richardson, P.A., Kong, P. and Bush, E.A. (2003) Efficacy of chlorine on multiple species of *Phytophthora* in recycled nursery irrigation water. *Plant Disease* 87(10), 1183–1189.

Hruby, M., Cigler, P. and Kuzel, S. (2002) Contribution to understanding the mechanism of titanium action in plants. *Journal of Plant Nutrition* 25(3), 577–598.

Ingram, D.L. (2014) Understanding irrigation water test results and their implications on nursery and greenhouse crop management. *Cooperative Extension Service Publication HO-111*. University of Kentucky, College of Agriculture, Food and Environment, Lexington, Kentucky. Available at: http://www2.ca.uky.edu/agcomm/pubs/HO/HO111/HO111.pdf (accessed 8 September 2020).

Jana, S. and Jeong, B.Y. (2014) Silicon: the most under-appreciated element in horticultural crops. *Trends in Horticultural Research* 4(1), 1–19.

Kay, T. and Hill, R. (1998) *Field Consultants Guide to Soil and Plant Analysis – Field Sampling, Laboratory Processing and Interpretation*. Hill Laboratories Limited, Hamilton, New Zealand.

Keller, R., Perin, K., Souza, W.G. and Cruz, L.S. (2008) Use of polishing pond effluents to cultivate lettuce (*Lactuca sativa*) in a hydroponic system. *Water Science and Technology* 58(10), 2051–2057.

Khoshgoftarmanesh, A.H., Mohaghegh, P., Sharifnabi, B., Shirvani, M. and Khalili, B. (2012) Silicon nutrition and *Phytophthora drechesleri* infection effects on growth and mineral nutrients concentration, uptake and relative translocation in hydroponic-grown cucumber. *Journal of Plant Nutrition* 35(8), 1168–1179.

Lang, J.M., Rebits, M., Newman, B. and Tisserat, N. (2008) Monitoring mortality of *Pythium* zoospores in chlorinated water using oxidation reduction potential. *Online, Plant Health Progress*. https://apsjournals.apsnet.org/doi/pdf/10.1094/PHP-2008-0922-01-RS

Lennard, S. (2001) *Nutron 2000+ Edition 3 User Manual and Nutrient Formulation Guide*. Suntec NZ Ltd, Tokomaru, New Zealand, pp. 7–8.

Lopez-Galvex, F., Allende, A., Pedrero-Salcedo, F., Alarcon, J.J. and Gil, M.I. (2014) Safety assessment of greenhouse hydroponic tomatoes irrigated with reclaimed and surface water. *International Journal of Food Microbiology* 191, 97–102.

Lyu, S., Wei, X., Chen, J., Wang, C., Wang, X. and Pan, D. (2017) Titanium as a beneficial element for crop production. *Frontiers in Plant Science* 8, 597.

McPherson, G.M., Harriman, M.R. and Pattisson, D. (1995) The potential for spread of root diseases in recirculating hydroponic systems and their control with disinfection. *Mededelingen van de Faculteit Landbouwwetenschappen, Rijksuniversiteit Gent* 60(2b), 371–379.

Malorgio, F., Diaz, K., Ferrante, A., Mensuali-Sodi, A. and Pezzorossa, B. (2009) Effects of selenium addition on minimally processed leafy vegetables grown in a floating system. *Journal of the Science of Food and Agriculture* 89(13), 2243–2251.

Mikkelsen, R.K. and Bruulsema, T.W. (2005) Fertilizer use for horticultural crops in the US during the 20th century. *HortTechnology* 15(1), 24–30.

Mozafariyan, M., Pessarakli, M. and Saghafi, K. (2017) Effects of selenium on some morphological and physiological traits of tomato plants grown under hydroponic conditions. *Journal of Plant Nutrition* 40(2), 139–144.

Nederhoff, E. (2000) Hydrogen peroxide for cleaning irrigation systems. *Commercial Grower* 55(10), 32–34.

Nielsen, C.J., Ferrin, D.M. and Stanghellini, M.E. (2006) Efficacy of biosurfactants in the management of *Phytophthora capsici* on pepper in recirculating hydroponic systems. *Canadian Journal of Plant Pathology* 28(3), 450–460.

Nutron 2000+ Edition 3 hydroponic nutrient formulation software. Available at: www.suntec.co.nz/nutron.htm (accessed 8 September 2020).

Poncet, C., Offroy, M., Bonnet, G. and Frun, R. (2001) Disinfection of recycling water in rose culture. *Acta Horticulturae* 547, 121–126.

Postma, J., Klein, W. and van Elsas, J.D. (2000) Effect of indigenous microflora on the development of root and crown rot caused by *Pythium aphanidermatum* in cucumber grown on rockwool. *Phytopathology* 90(2), 125–133.

Postma, J., van Os, E. and Bonants, P.J.M. (2008) Pathogen detection and management strategies in soilless plant growing systems. In: Raviv, M. and Lieth, J.H. (eds) *Soilless Culture: Theory and Practice*. Elsevier, London, pp. 425–457.

Puccinelli, M., Malorgio, F., Rosellini, I. and Pezzarossa, B. (2017) Uptake and partitioning of selenium in basil (*Ocimum basilicum* L.) plants grown in hydroponics. *Scientia Horticulturae* 225, 271–276.

Reetz, H.F. Jr (2016) *Fertilizers and Their Efficient Use*, 1st edn. International Fertilizer Industry Association, Paris.

Runia, W.T.H. (1994) Elimination of root-infection pathogens in recirculation water from closed cultivation systems by ultra-violet radiation. *Acta Horticulturae* 361, 361–371.

Scaife, A. and Turner, M. (1983) *Diagnosis of Mineral Disorders in Plants. Volume 2: Vegetables*. MAFF/ARC, London.

Scarlett, K., Collins, D., Tesoriero, L., Jewell, L., van Ogtrop, F. and Daniel, R. (2016) Efficacy of chlorine, chlorine dioxide and ultraviolet radiation as disinfectants against plant pathogens in irrigation water. *European Journal of Plant Pathology* 145(1), 27–38.

Schroder, F.G. and Lieth, J.H. (2002) Irrigation control in hydroponics. In: Savvas, D. and Passam, H. (eds) *Hydroponic Production of Vegetables and Ornamentals*. Embryo Publications, Athens, pp. 263–298.

Schuerger, A.C. and Hammer, W. (2003) Suppression of powdery mildew on greenhouse-grown cucumber by addition of silicon to hydroponic nutrient solution is inhibited at high temperature. *Plant Disease* 87(2), 177–185.

Stamatakis, A., Papadantonakis, N., Savva, D., Lydakis-Simantiris, N. and Kefalas, P. (2003) Effects of silicon and salinity on fruit yield and quality of tomato grown hydroponically. *Acta Horticulturae* 609, 141–147.

Stanghellini, M.E. and Miller, R.M. (1997) Biosurfactants, their identity and potential efficacy in the biological control of zoosporic plant pathogens. *Plant Disease* 81(1), 4–12.

Stanghellini, M.E. and Tomlinson, A. (1987) Inhibitory and lytic effects of a nonionic surfactant on various asexual stages in the life cycle of *Pythium* and *Phytophthora* species. *Phytopathology* 77(1), 112–114.

Stanghellini, M.E., Rasmussen, S.L., Kim, D.H. and Rorabaugh, P.A. (1996) Efficacy of nonionic surfactants in the control of zoospore spread of *Pythium aphanidermatum* in a recirculating hydroponic system. *Plant Disease* 80(4), 422–428.

Tuzel, I.H., Irget, M.E., Gul, A., Tuncay, O. and Eltex, R.Z. (1999) Soilless culture of cucumber in glasshouses: II. A comparison of open and closed systems on water and nutrient consumptions. *Acta Horticulturae* 491, 395–400.

Van Os, E.A. (1999) Closed soilless culture systems: a sustainable solution for Dutch greenhouse horticulture. *Water Science and Technology* 39(5), 105–112.

Van Os, E.A. (2001) Design of sustainable hydroponic systems in relation to environment-friendly disinfection methods. *Acta Horticulturae* 548, 197–206.

Van Os, E.A., Bruins, M.A. and van Buuren, J. (1997) Physical and chemical measurements in slow sand filters to disinfect recirculating nutrient solutions. In: *Proceedings of the 9th International Congress on Soilless Culture, St. Helier, Jersey, 12–19 April 1996*. ISOSC Secretariat, Wageningen, The Netherlands, pp. 313–328.

Van Os, E., Gieling, T.H. and Lieth, J.H. (2008) Technical equipment in soilless production systems. In: Raviv, M. and Lieth, J.H. (eds) *Soilless Culture: Theory and Practice*. Elsevier, London, pp. 157–207.

Voogt, W. and Sonneveld, C. (1997) Nutrient management in closed growing systems for greenhouse production. In: Goto, E., Kurata, K., Hayashi, M. and Sase, S. (eds) *Plant Protection in Closed Ecosystems: The International Symposium on Plant Production in Closed Ecosystems held in Narita, Japan, 26–29 August 1996*. Kluwer Academic Publishers, Dordrecht, The Netherlands, pp. 83–102.

Voogt, W. and Sonneveld, C. (2001) Silicon in the nutrient solution for horticultural crops. In: *Silicon in Agriculture Program Agenda and Abstracts*, p. 5. Available at: http://www.issag.org/1st---usa.html (accessed 8 September 2020).

Wadas, W. and Kalinowski, K. (2017) Effect of titanium on growth of very early maturing potato cultivars. *Acta Scientiarum Polonorum Hortorum Cultus* 16(6), 125–138.

Winsor, G.W., Davies, J.N. and Messing, J.H.L. (1958) Studies on potash/nitrogen ratio in nutrient solutions using trickle irrigation equipment. *Annual Report of the Glasshouse Crops Research Institute*, 1957, 91–98.

Wohanka, W. (2002) Nutrient solution disinfection. In: Savvas, D. and Passam, H. (eds) *Hydroponic Production of Vegetables and Ornamentals*. Embryo Publications, Athens, pp. 345–372.

Wohanka, W., Ludtke, H., Shlers, H. and Lubke, M. (1999) Optimisation of slow filtration as a means for disinfecting nutrient solutions. *Acta Horticulturae* 481, 539–544.

Zhang, W. and Tu, J. (2000) Effects of ultraviolet disinfection of hydroponic solutions on *Pythium* root rot and non-target bacteria. *European Journal of Plant Pathology*, 106(5), 415–421. https://doi.org/10.1023/A:1008798710325

10 Plant Health, Plant Protection and Abiotic Factors

10.1 Introduction

Hydroponic cultivation utilizing clean, sterile substrates largely eliminates the problems of contamination with soilborne pests and disease pathogens. Clean and sterilized water supplies also help prevent many of the root rot diseases which can infect many crops; however, pest and disease control is still required in most greenhouse production systems at some stage. Insect pests such as greenhouse whitefly (*Trialeurodes vaporariorum*), aphids (*Myzus persicae*), mites (*Tetranychus urticae*, *Aculops lycopersici*), caterpillar larvae (*Autographa gamma*, *Chrysodeixis chalcites*, *Helicoverpa armigera*), thrips (*Frankliniella occidentalis*), leaf miner (*Liriomyza bryoniae*), tomato psyllid (*Bactericera cockerelli*) and others can enter greenhouse systems through vents, air intakes, on workers, equipment and with seedling transplants. Disease spores can be blown in by wind and many common outdoor diseases such as powdery mildew (*Oidium lycopersicum*, *Leveillula taurica*, *Erysiphe cichoracearum*), downy mildew (*Bremia lactucae*), grey mould (*Botrytis cinerea*), blights (*Phytophthora infestans*, *Alternuria solani*), sclerotinia drop (*Sclerotiorum minor*), bacterial rots (*Erwinia carotovora*), anthracnose (*Colletotrichum coccodes*), leaf mould (*Cladosporium fulvum*) and others occur seasonally in greenhouse cultivation. Keeping foliage dry helps prevent many disease issues in most crops; however, spray programmes, biological control and/or IPM are used in hydroponic production systems. Virus diseases can also infect a wide range of hydroponic crops and require plant removal to prevent spread as they are largely unable to be controlled once infection has occurred. Common viruses in hydroponic systems include tomato and tobacco mosaic viruses, cucumber mosaic virus, tomato spotted wilt virus, lettuce mosaic virus, big vein virus and a large number of others.

In recirculating systems, the nutrient solution can aid the spread of certain pathogens, most notably those that produce zoospores which are disseminated through water and can swim towards host root systems. *Pythium aphanidermatum*, *Phytophthora* spp., *Fusarium oxysporum*, *Verticillium* spp. and *Olpidium brassicae* are examples of root rot pathogens which occur in hydroponics systems and may be spread via the nutrient solution (Ehret *et al.*, 2001). In recirculating systems, treatment of the nutrient solution with sterilization agents such as UV radiation, ozone, chemical sterilants, membrane filtration, surfactants, heat or slow sand filtration may be used to help control the spread of root disease pathogens (Stanghellini *et al.*, 1996). In many healthy hydroponic systems, a wide and diverse range of beneficial microbial life develops in the root zone and nutrient solution which assists in the suppression and prevention of root pathogen attack. Inoculation of the nutrient solution with commercially prepared microbial control agents is also used in recirculating soilless systems to help control pathogens in the root zone (Paulitz *et al.*, 1992). Root pathogens tend to become more of an issue in soilless systems where conditions have created an environment where opportunist diseases can thrive; these include low levels of root oxygenation, oversaturation of the growing medium, weakened plants, high salinity, excessive temperatures, and overloading of organic matter such as old root systems or organic supplements.

10.2 Major Greenhouse Pests

10.2.1 Whitefly

The two main species of whitefly which may infest hydroponic greenhouse crops are the greenhouse whitefly (*T. vaporariorum*) and the tobacco whitefly (*Bemisia tabaci*) (Perdikis *et al.*, 2008). While both species are pests of

© L. Morgan 2021. *Hydroponics and Protected Cultivation* (L. Morgan)
DOI: 10.1079/9781789244830.0010

a wide range of crops, the greenhouse white-fly is the most common worldwide under protected cultivation. Whitefly adults are small (2 mm), moth-like in appearance and will fly up from the foliage when disturbed (Fig. 10.1). In crops such as tomato, cucumber and capsicum, greenhouse whitefly adults will congregate within the tops of the plants to feed and lay eggs. Whitefly develop through three larval instars and a pupal stage to adult. Eggs are laid on the undersides of the newest leaves in the crop and these take 4–10 days to hatch into the first instar. Larval instars are flat, oval in shape and transparent. Development time from egg to adult is dependent on temperature and host plant with rapid population increases occurring under warmer conditions. The first instar is the only mobile nymphal instar and selects a location on the underside of the leaf to establish, where it remains stationary (Perdikis et al., 2008). Later nymph stages and pupae are typically located on the older leaves. Removal of these lower, older leaves of tall plants such as tomatoes is one means of slowing population

Fig. 10.1. Whitefly adults on the underside of a basil leaf.

increase by removing foliage containing late instar nymphs or pupae.

Whitefly damage crops by both sucking sap from the foliage and depositing honeydew, a sticky substance which contaminates produce and is rapidly colonized by black sooty mould. The presence of honeydew and sooty mould not only reduces photosynthesis and assimilate production by the plant, but also necessitates the washing of fruit after harvest to remove this contamination. Heavily infested plants become weakened by whitefly feeding and may exhibit yellowing leaves, lack of plant vigour, reduced fruit size and yields. Whitefly may also transmit virus diseases in many crops and are an efficient vector of the tomato yellow leaf curl virus (Mehta et al., 1994).

It has become increasingly difficult to control greenhouse whitefly using insecticides as this pest has developed genetic resistance to a variety of pesticide compounds (Kamikawa et al., 2018). Classes of pesticide must be rotated regularly and persistent chemicals avoided to prevent spray resistance. Other spray options such as soaps and oils may be used to a limited extent on pests such as whitefly; however, frequent, long-term use has been known to cause extensive spray damage in many greenhouse crops. Control of greenhouse whitefly often requires an integrated approach. Screening of vents and double-door entries can assist with exclusion of flying adults. Removal and control of host plants such as weed species in the immediate vicinity of the greenhouse can reduce the source of infestations. The use of yellow sticky traps placed close to vents, doors and other entry points can serve as a monitoring aid to determine when the pests may have entered a crop (Lemic et al., 2020). Monitoring pest numbers by counting the number of nymphs found on the undersides of leaves can be used to determine when control measures need to be applied, or the effectiveness of IPM programmes. Biological, botanical and chemical control options are available for whitefly. These include a number of predator and parasites which are commercially available for release into greenhouse crops. Parasitic wasps, *Encarsia formosa* and *Eretmocerus eremicus*, are the most commonly utilized species; however, long-term control requires an environment conducive to survival of the parasite as well as maintaining a balance between pests and control agents. Botanical control agents, such as extract from the seed kernel of the neem tree (*Azadirachta indica*), which contains insect growth regulatory, antifeedant and other activities, have been developed into a number of spray products that are used for greenhouse whitefly control.

10.2.2 Aphids

A number of aphid species may become pests of greenhouse crops; however, these are usually seasonal in nature and more common in spring and early summer as temperatures begin to increase. Aphids are small, approximately 1–3 mm in length, with a soft, oval body (Fig. 10.2). Colour of these pests is dependent on species and may range from light green to yellow, tan or black. Common species infesting greenhouse crops include the green peach aphid (*M. persicae*), the melon/cotton aphid (*Aphis gossypii*), the chrysanthemum aphid (*Macrosiphoniella sanborni*), the potato aphid (*Macrosiphum euphorbiae*), the rose aphid (*Macrosiphum rosae*) and the foxglove aphid (*Aulacorthum solani*) (Parrella and Lewis, 2017). Damage caused by aphids include weakening of the plant through removal of plant sap, secretion of sticky honeydew and the development of sooty mould on foliage

and fruit, transmission of virus diseases, and for some species, severe distortion of the newer leaves as a result of feeding damage (Fig. 10.3). Hydroponic crops such as capsicum are particularly sensitive to developing twisted and deformed foliage in the tops of the plant as a result of aphid feeding.

Early infestations of aphids are often seen in proximity to vents and other entry points where winged adults fly in from outside. Once established, populations of aphids may consist of both winged and wingless forms and the life cycle can be complex. Aphid infestations typically originate from host plants outside the greenhouse, with only a few individuals required to start an infestation because aphids reproduce parthenogenically with young developing within the parent. Under high population pressure, winged forms then develop as a result of overcrowding and spread the population further within the crop. Aphids are attracted to the young growing points of most plants and may be found in the centre of lettuce heads and on the undersides of young leaves.

Control of aphids is relatively straightforward as most remain susceptible to a number

Fig. 10.2. Adult and juvenile aphids.

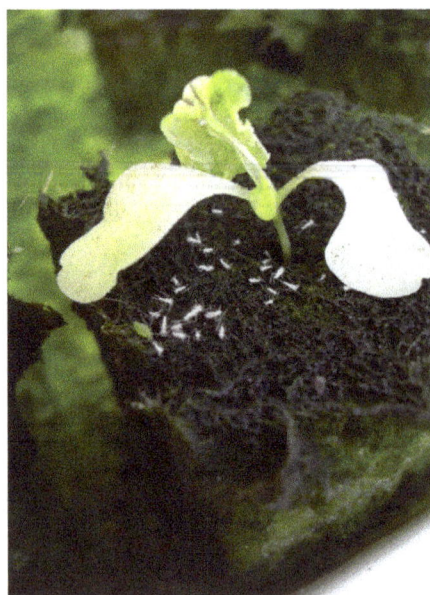

Fig. 10.3. Aphid damage on a young seedling.

of systemic insecticides as well as growth regulators and biological controls including entomopathogenic fungi and insect predators and parasites. The use of yellow sticky traps positioned close to vents and entry ways may be employed to provide an early indication of the arrival of winged adults into a growing area (Fig. 10.4). Important insect predators and parasites include the gall midge (*Aphidoletes aphidimyza*), the ladybird (*Adalia bipunctata*), green and brown lacewings, and parasitic wasps of the genus *Aphidius*, which appear to be the most effective (Parrella and Lewis, 2017). As with all natural controls, a balance needs to be maintained between pest and beneficial insects for effective control.

10.2.3 Thrips

A number of thrips species may become pests within greenhouses (Parrella, 1999); however, the main species of thrips which infest greenhouse crops are onion thrips (*Thrips tabaci*) and western flower thrips (*F. occidentalis*), both species are common worldwide and are a pest of a number of different crops. Thrips are small, elongated, dark-coloured insects which can move rapidly to avoid detection and may be concealed within buds and leaf axils. Identification of thrips is often by the damage this pest creates, which is a characteristic silvering of the leaf surface due to removal of the epidermal cell contents. Western flower thrips are most prevalent in the upper part of the plant, feeding on developing tissue such as growing points, flowers and fruitlets. Apart from causing damage and deformation (Fig. 10.5) to developing plant parts, thrips also transmit viruses such as the tomato spotted wilt virus which can be a significant disease in hydroponic tomato and capsicum crops. Population growth of western flower thrips is optimized at 25°C where the life cycle can be completed within 12–14 days. Thrips are spread via winged adults which can enter via vents, entrance ways and infested plant material such as seedlings and transplants. Blue sticky traps can be used to monitor and detect flying thrips adults to allow for the application of early control options.

Thrips are another insect pest which has become increasingly difficult to control due to widespread genetic resistance to a range of chemical pesticide options. Prevention of infestations is vital for control of this pest and includes the use of fine mesh 'thrips screens' over vents and use of double-door entrances (Fig. 10.6). Control of thrips with insecticide sprays is difficult as complete coverage of spray material is required and thrips inhabit areas where spray penetration is not always achieved. Thrips develop resistance to chemical insecticides relatively quickly and most control compounds are limited to two to four applications per season. Different chemical classes of insecticides should be rotated to assist with the prevention of genetic spray resistance and

Fig. 10.4. Yellow sticky traps can be used close to vents and entrance ways to provide an early indication of the arrival of aphids.

Fig. 10.5. Deformation of cucumber fruit caused by thrips.

Fig. 10.6. Thrips screening over greenhouse vents.

combined with exclusion and other control methods. There are currently a number of predators available for biological control of thrips, these include predatory mites (*Amblyseius cucumeris*) which attack immature thrips and the soil-dwelling mite (*Hypoaspis miles*) (Parrella and Lewis, 2017). Minute pirate bugs (*Orius* spp.) may also be used for the control of adult and immature thrips.

10.2.4 Mites

The main species of mites infesting greenhouse hydroponic crops are the two-spotted mite (*T. urticae*) and, to a lesser extent, the tomato russet mite (*A. lycopersici*). While the tomato russet mite only infests tomato crops, causing significant fruit damage (Fig. 10.7), the two-spotted mite can become a significant pest of a wide range of greenhouse crops, particularly under warm temperatures and low humidity. The tomato russet mite causes bronzing of tomato leaves and deformed fruit with a coarse browned skin; considerable damage can occur at high temperatures where

Fig. 10.7. Damage caused by the tomato russet mite.

population growth is rapid and the life cycle less than 7 days. Adult russet mites are small, cone shaped and orange brown in colour; a hand lens is usually required to see these mites and damage is often misdiagnosed as other conditions. The two-spotted mite, which is common on a wide range of greenhouse fruit, vegetable and ornamental plants, is also an extremely small insect pest which requires use of magnification for identification. Two-spotted mites lay eggs in webbing, these develop into six-legged larvae which then develop through two nymph stages to become adults. The adults have four pairs of legs and vary in colour from different stages of red through to brown and green. Adults feature a pair of large spots on the rear of the body.

Damage caused by the two-spotted mite includes leaf necrosis, interveinal chlorosis and distortion of leaf and petiole. This occurs through feeding on the epidermal cells by puncturing the cell wall and removing the contents. Plants eventually lose vigour, becoming stunted and discoloured, and in severe cases may die back. Fine webbing is produced in established populations which cover leaf surfaces; this provides protection for the mite pests and also hinders control spray penetration. Mites can be dispersed via air currents, being transported by threads of fine webbing; however, in cropping situations, mites are also easily spread on workers' clothing, on equipment and plant debris and along crop wires and other structures.

Mites are often a seasonal pest in temperate-zone climates, with rapidly growing infestations occurring in late spring, through summer and autumn when warmer temperatures accelerate population growth. Under winter conditions, pest numbers are slower to establish and build. As with many insect pests, mites have developed widespread resistance to a number of chemical spray options and require a rotational insecticide programme to achieve control which should be applied early in the infestation stage. Predatory mites may be used either alone or after initial control with spray compounds and include the use of *Phytoseiulus persimilis* and *Amblyseius californicus* (Parrella and Lewis, 2017) (Fig. 10.8). As with all biological control, a balance must be maintained between the released predator and the two-spotted mite population for control to be effective. This includes regular monitoring of the mite

pest and biological control species' populations, continual release of predators and use of other control methods when required.

10.2.5 Caterpillars and leaf miner larvae

Various caterpillar pests may infest a wide range of greenhouse crops and are usually easily distinguished from other insect damage by large holes in foliage and black frass, or insect excrement, over plant surfaces. The larvae of a number of species of moth are significant pests of greenhouse crops although most are seasonal. Most commonly found species are the green looper (*A. gamma* and *C. chalcites*), larvae of the silver Y moth (*A. gamma*), diamondback moth (*Plutella xylostella*) and cabbage white butterfly (*Pieris rapae*), which are pests of brassica crops, and armyworm, leaf roller and leaf miner species. Most species of caterpillar vary in length from 3 to 5 cm with colour dependent on species and including green, grey, brown, banded or striped. Leaf roller larvae species cause damage by both feeding on foliage and also webbing leaves together to provide protection. Leaf roller larvae may damage fruit and are the larvae of a number of different species of moth.

Prevention and control of caterpillar larvae is relatively straightforward but relies on early detection before foliage damage becomes severe. The adult moths which lay eggs on the crop can be excluded through the use of insect screens over vents and control

Fig. 10.8. Mites and biological control predators.

of entrance ways. Where young larvae are seen, these can be controlled with a number of spray options. These include the use of the bacterial control agent Bt (*Bacillus thuringiensis*), which is available in a number of commercial preparations. Bt is applied as a foliar spray, ensuring complete coverage, and when ingested by caterpillars produces a bacterial toxin in the alkaline digestive system which immediately stops the insect feeding, and results in dehydration and death within a few days. There are also a number of pesticides with control activity for caterpillars in a wide range of crops, these need to be applied early before significant damage occurs. For biological control, the parasitic wasp *Trichogramma brassicae* can be used to control the larvae of the green looper, while the predatory bug *Macrolophus caliginosus* consumes moth eggs.

Leaf miners are members of the order Diptera and may be particularly damaging in warm tropical areas. *Liriomyza* leaf miners are a problem on a wide range of ornamental and vegetable crops grown in greenhouses (Perdikis *et al.*, 2008; Parrella and Lewis, 2017). Leaf miners damage foliage by feeding inside the leaf which creates visible pale markings on the leaf surface. Heavy infestations can make produce such as leafy greens unmarketable. Control of leaf miner for limited infestations is often by removal of the infested leaves with the larvae inside. For larger outbreaks, a systemic pesticide is required to penetrate the leaf surface and control the pest which is protected from contact insecticides.

10.2.6 Fungus gnats

Fungus gnat (*Bradysia* spp.) adults are small, delicate, dark-coloured flying insects 2–3 mm in length which are attracted to overly damp substrates where fungi may also be growing. Fungus gnat adults lay eggs on the surface of moist growing media which hatch within 5–6 days. The resulting larvae, which are white and have a black head, feed on organic material and root systems before pupating and emerging as adults. While fungus gnats are mostly attracted to substrates such as peat and compost which have a high percentage of organic matter, they are seen in a wide range of hydroponic systems, including those using perlite and rockwool as a substrate. While the adults do not generally damage plants, the larvae of this pest feed on plant roots as well as decaying organic matter and tunnel into the crown and stems of plants (Smith, 2012). Further damage includes the possible transmission, via gnat larvae feeding, of fungal spores of plant-pathogenic *Pythium* and *Fusarium*. Gnat-damaged root systems are also prone to infection with a wide range of pathogens which enter through the damaged tissue.

Control of fungus gnat larvae requires prevention via environmental modification, particularly maintaining a dry surface to growing media and preventing ponding of the nutrient solution near the surface. General greenhouse satiation and hygiene including prevention of wet floors and stagnant water, removal of plant debris and eliminating algae assist with control of adult fungus gnats. Early detection can be carried out with use of yellow sticky traps. To monitor for the presence of larvae in a growing medium, pieces of raw potato can be placed on the substrate surface which will attract larvae and should be checked regularly for their presence (Smith, 2012). For established populations of fungus gnats, insecticide control options are limited; however, biological control organisms include a type of entomopathogenic nematode, *Steinernema feltiae*, which naturally parasitizes larvae. There are also predatory mites such as *H. miles* which feed on the first instar of fungus gnat larvae. The soil-borne bacterium *Bacillus thuringiensis israelensis* contained in the product Gnatrol® may be applied as a substrate drench or via the irrigation system for gnat larvae control but requires repeated applications and must be ingested by the larvae to be effective.

10.2.7 Nematodes

Despite hydroponic systems being soilless, nematode infestations can still occur and can cause considerable damage. Dutch commercial hydroponic rose growers have reported that nematode infestations can cause production losses of up to 40% (Garcia Victoria and Amsing, 2007). Nematodes, also termed 'eel worms' or 'needle worms', can only be seen under a microscope, which complicates identification by growers. Not all species of nematode are parasitic on plants and a formal identification of the species present is often advisable. Damage from nematode infestations includes root knots, swellings, galls, stunting of growth, root death and overall loss in plant vigour. Some of the most common nematode species which may infest hydroponic crops are the root lesion nematodes (*Pratylenchus penetrans* and *Pratylenchus vulnus*), which cause reddish brown root lesions, stunted plants and leaf yellowing; the root knot nematode (*Meloidogyne hapla*), which causes round to spindle-shaped swellings (galls) on roots as well as plant stunting and wilting; and the needle nematode (*Longidorus elongatus*), which inhibits root elongation and causes swelling of the regions just behind the root tip resulting in a forked and shortened root system. Further to root damage, nematodes can transmit a number of plant viruses via their feeding activity.

Nematodes present in hydroponic systems may initially be carried by soil contamination, in water supplies, particularly those from sources such as wells, streams and dams which are in contact with soil, and on infected planting stock. Seedlings and transplants which have been raised in soil and then planted into a hydroponic system are a common source of nematode infestation and this practice should be avoided. Nematodes can be further spread through a crop by the circulating nutrient solution (Moens and Hendrickx, 1992). Nematodes can also spread from soil to hydroponic systems via equipment, animals and human activity. Fungus gnat adults and larvae also carry nematodes and spread these from infected to healthy crops.

Prevention of nematode infestations includes treatment of at-risk water supplies, general hygiene practices and greenhouse sanitation which includes avoidance of soil contamination of floors, equipment and transplants. Drain water and nutrient solutions may use a series of fine filters composed of polyester-felt filter bags to remove plant-parasitic nematodes (Moens and Hendrickx, 1992). Nematode-resistant varieties are available for some crops where persistent pest problems arise. Once an infestation is correctly identified, severely infected hydroponic systems may require immediate shutdown and removal of all plants and substrate followed by replanting with clean stock.

10.3 Pest Control Options – Integrated Pest Management

While synthetic pesticides are in use in many greenhouse operations, there has been an increasing move towards the use of more widely integrated pest management programmes. These incorporate the use of 'softer' control options such as biological and microbial control combined with physical exclusion, pest trapping, resistant crops and other methods. The greenhouse environment, being highly protected and often with advanced climate control, is particularly suitable for the effective use of commercially available natural pest enemies (Perdikis *et al.*, 2008; van Lenteren, 2012). Biological pest control based on utilizing commercially available and mass-produced arthropod natural enemies such as pest predators and parasites is well developed in greenhouse crops (Pilkington *et al.*, 2010). However, there are cases where such natural predator and parasite introductions are not sufficiently effective, not available or too expensive to be viable (Gonzalez *et al.*, 2016). For some greenhouse pests, natural enemies may be commercially available, but not necessarily effective on all crops. Some predator or parasite biocontrol agents form specific relationships with certain host crops due to their feeding or other life cycle requirements. For example, predatory mirid bugs prefer plants with leaf hairs such as tomato and aubergine, while phytoseiid predatory mites on capsicum may be effective due to the presence of pollen and nectar but not on many ornamental plants which lack these supplemental food sources (Messelink *et al.*, 2014). The most successful of the predator or parasite biological control agents are those which can reach equilibrium with the pest they are controlling and maintain this for a long time. To facilitate this, environmental modification and changes in management practices within greenhouses may be applied to assist the survival of introduced beneficial insects. The use of banker plants to provide alternative hosts or food sources is one such example, combined with restriction of the use of spray compounds which may harm biocontrol species (Weintraub *et al.*, 2017).

Microbial pest controls are an area of research focus and although some commercially available products have been utilized by greenhouse growers for quite some time, new applications are currently under investigation. These include the applied use of combinations of arthropod natural enemies with microbial products to give a more integrated approach to pest management (Gonzalez *et al.*, 2016). Microbial control agents, which are also another form of biological control, may be based on entomopathogenic microorganisms such as viruses, bacteria and fungi. These types of controls may be used as an alternative to pest parasites where these are not available or, increasingly, when they are not sufficiently effective (Chandler *et al.*, 2011). Products based on microorganisms are often termed 'biopesticides' or 'bioinsecticides'. The majority of microbial pesticide products currently in use are based on subspecies of Bt. Bt forms spores which contain crystals that lyse gut cells when consumed by susceptible insect pests (Gill *et al.*, 1992), which then stop feeding, become paralysed and die. Other microbial insecticides have different modes of action and some have the ability to invade the pest directly without requiring ingestion. This is particularly effective on insects such as aphids and whitefly which feed directly from the phloem and do not ingest microbials that are deposited on the leaf surface (Gonzalez *et al.*, 2016). While Bt dominates the bacterial microbial control options for pests such as caterpillars and fungus gnats, there is a wider range of fungal species which have been developed into commercial biocontrol agents. These include *Beauveria bassiana* for whitefly and thrips

control, *Isaria fumosorosea*, *Metarhizium* and *Lecanicillium* (formerly *Verticillium*) spp. (Gwynn, 2014). These have all been reported to be effective against pests such as whitefly, thrips and aphids (Khan *et al.*, 2012).

Commercially available, formulated microbial control agents are often prepared as wettable powders, technical concentrates or oil dispersions (de Faria and Wraight, 2007). These are applied in a similar manner to other pesticides through spraying devices. Other methods of application may include seed treatments for a more targeted approach to certain insects or incorporation into growing substrates for control of soil-dwelling larvae such as the pupae of western flower thrips (Ansari *et al.*, 2008). The survival and effectiveness of applied microbial control agents are also dependent on the greenhouse environment being favourable. For example, increasing greenhouse humidity can help optimize pest control with *B. bassiana*; however, such adjustments may adversely affect crop growth and disease management (Shipp *et al.*, 2003).

10.4 Selected Diseases of Hydroponic Crops

10.4.1 *Botrytis*

B. cinerea or 'grey mould' causes a fungal disease with a worldwide distribution and is one of the most important pathogens of vegetable, fruit and ornamental crops in greenhouses in many countries (Jarvis, 1989; Elad and Shtienberg, 1995). *Botrytis* can be particularly damaging under protected cultivation as high humidity and more fragile plant tissue favour disease establishment. Infection generally first appears as water-soaked areas, which rapidly develop grey-brown mycelium with long branched conidiophores, spore-producing bodies which are able to disperse large quantities of conidia (grey-brown spores) into air currents. On tomato fruit, airborne *Botrytis* spores which infect green fruit cause the occurrence of 'ghost spot' and the development of pale cream-coloured circular rings on the fruit surface. *Botrytis* can infect all above-ground parts of crops including stems, leaves, petioles and flowers, and damaged tissue is often the first site of infection. Being largely a disease that requires a high humidity and cooler conditions to infect plant tissue, *Botrytis* mostly occurs in winter and early spring under greenhouse production. Humidity levels of over 90% combined with a lack of adequate ventilation and the formation of condensation are conditions which favour development.

The main method of *Botrytis* control and prevention is environmental modification, which includes reducing humidity and improving air movement. This may be achieved by venting warm humid air in the late afternoon and allowing cooler air to flow in, which can be heated before dark to reduce the relative humidity. Air-movement fans should be set up over the crop to provide a constant flow of air around the plants so that no humid areas remain within the crop. Density levels and spacing need to be maintained to allow each plant sufficient ventilation. Crop nutrition is important in producing plants which are less susceptible to *Botrytis* infection; careful nitrogen nutrition and maintaining good levels of leaf calcium are important to produce less-susceptible plants. Calcium nutrition has been shown to reduce grey mould of rose flowers, tomatoes, pepper and aubergine through strengthening of cell walls, delayed senescence of plant tissue and delaying softening of fruit flesh (Ferguson, 1984; Volpin and Elad, 1991; Yunis *et al.*, 1991). Greenhouse sanitation and hygiene are also important measures for the prevention of initial *Botrytis* infection and include removal of infected plants and plant debris at crop completion and during the growing season. Sanitation can be carried out by steaming or intensive heating, using chemical agents, or solar heating (Elad and Shtienberg, 1995).

While environmental control, air movement and humidity reduction are vital in the control of *Botrytis*, there are several fungicides which may be used as part of an integrated control programme. As with many diseases, *Botrytis* can develop rapid genetic resistance to individual spray compounds

and products of different chemical classes should be rotated to prevent this from occurring. A number of biofungicides have been tested and developed into commercial products for greenhouse *Botrytis* control. Most of these have been focused on *Trichoderma* spp., *Ulocladium* spp., *Bacillus subtilis* and plant extracts including essential oils (Paulitz and Belanger, 2001; Abbey *et al.*, 2019; Ni and Punja, 2020b).

10.4.2 Mildew diseases

Powdery and downy mildews are caused by a range of different pathogens, many of which are species dependent; *B. lactucae*, for example, causes downy mildew of lettuce while *O. lycopersicum* and *L. taurica* cause powdery mildew of tomato crops. Powdery mildew first appears as diffuse, white powdery patches on the upper surfaces of leaves composed of fungal mycelium (Fig. 10.9). Foliage may then display yellow, chlorotic patches and eventually die back. In fruiting crops,

Fig. 10.9. Powdery mildew causes white patches on the upper leaf surface.

the resulting reduction in photosynthetic activity can lead to premature leaf drop and a reduction in fruit size and yield (Pace *et al.*, 2016). Powdery mildew is spread by conidia which can be dispersed by air movement. Conditions which favour the disease are low light and warm temperatures (18–25°C) such as those prevalent during late summer and autumn in greenhouse crops. Control of powdery mildew involves reducing fluctuations in humidity levels, especially during the night, and increasing plant spacing to allow improved light penetration. Fungicide compounds are often required for control in susceptible cultivars. Other methods include the use of microbial spray products such as those based on *B. subtilis* QST 713 (Ni and Punja, 2020a), potassium bicarbonate products, ultrafine spraying oil and sulfur (Pace *et al.*, 2016) for powdery mildew control on tomato crops.

Downy mildew is often confused with powdery mildew as the spore development can look similar in some situations. Downy mildew is less prevalent than powdery mildew and generally infects fewer species under greenhouse production. Crops such as lettuce and other salad greens are most susceptible particularly under temperate climates and winter conditions. Greenhouse roses are a crop which is highly susceptible to downy mildew (*Peronospora sparsa* (Berk.)) and this pathogen can become a significant problem for growers in many parts of the world (Salgado-Salazar *et al.*, 2018). Downy mildew infections may occur at all stages of crop growth from seedlings through to maturity. Infection first occurs as light green to yellow spots bordered by the leaf veins which eventually become necrotic. Spore-producing bodies develop on the necrotic areas, emerging through the stomata on the lower surface of the leaf tissue as sparse white to grey mildew growth. Sporulation occurs at night during periods of high humidity when temperatures are between 5 and 24°C, followed by release of sporangia into the air in the morning as temperature increases and humidity drops. Control of downy mildew involves using resistant varieties where possible and modification of the growing environment. Fungicides may also be used; however, they are most effective when used as a preventive.

10.4.3 *Pythium* root rot

Pythium is a commonly found pathogen in hydroponic systems; however, infection is most likely when plants have become weakened or damaged and allow entry into root tissue. *Pythium* is the generic name for over 50 species of fungi in the *Oomycete* class with *Pythium ultimum*, *P. aphanidermatum* and *Pythium dissotocum* being among the most common species found in hydroponic cultivation. *Pythium* is easily dispersed through hydroponic systems via the nutrient solution or water supply via mobile zoospores. Optimum conditions for *Pythium* infection are root volumes which are low in oxygen, overwatered and/or stressed by other factors such as unfavourable temperatures and high salinity or EC. Early symptoms of *Pythium* infection include a browning of the root tips followed by dark brown or black dieback of the feeder roots and a brown staining of the central root core (Fig. 10.10). Infected plants remain stunted and wilt during the middle of day under strong heat and light as the infected roots are unavailable to take up sufficient water and nutrients (Figs 10.11 and 10.12). Roots become short and stunted and may die back completely in severe cases. In milder outbreaks some root death may occur, but replacement roots may be formed.

Pythium control in hydroponics requires multiple approaches. This is due to the fact that most control options are usually ineffective against one or more of the life stages of the pathogen. *Pythium* can exist as a number of different spore types in solution and can infect plants via zoospores or hyphae fragments. Once infected, plant tissue can act as a host for the multiplication of hyphae and sporangia which in turn release zoospores into solution. Under adverse conditions (e.g. drying/unsuitable temperatures) zoospores can form a very hardy, long-lived and durable survival stage (oospore), only to germinate to form hyphae containing sporangia which

Fig. 10.10. Comparison of roots browned by *Pythium* root rot (right) with those of healthy plants (left).

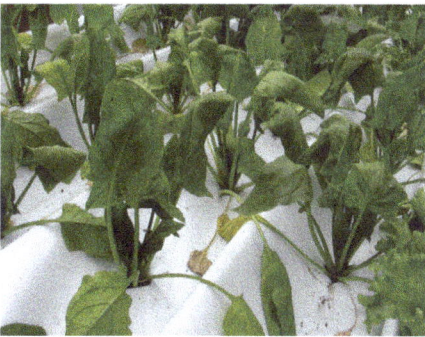

Fig. 10.11. *Pythium* causes wilting and stunting as plants are unable to take up sufficient water and nutrients.

erupt releasing more zoospores into solution once conditions become favourable (Ehret *et al.*, 2001). Zoospores are mobile in solution and will move towards sources of root exudates and coatings on roots especially root tips and junctions. These areas are primary locations for infection.

Control of an established *Pythium* infection is difficult and prevention by promoting a healthy root system and reduction in inoculum are the most effective means of preventing damage. *Pythium* is prevalent where oversaturation of the root zone occurs. If watering is too frequent, the air-filled pores in a substrate remain saturated and the plant has less access to the oxygen contained in air. Overwatering is the most common cause of *Pythium* in hydroponics and an understanding of the need of the root system for oxygenation required for respiration is important.

High levels of oxygen around the roots are proven to be suppressive of *Pythium*. Oxygen in most hydroponic systems comes from pores in the substrate (the air-filled capacity of the growing medium) which hold air. Air has approximately 21% oxygen, while nutrient solution or water can only hold 6–13 ppm depending on temperature. Forced aeration of the nutrient solution is also beneficial in all hydroponic systems. Creating a good level of turbulence in solution culture systems may also be a valuable technique as when the flow of nutrient solution is turbulent, *Pythium* zoospores shed their flagella and encyst, losing the ability to sense and be attracted to infection sites on the roots (Sutton *et al.*, 2006). *Pythium* has been proven to be far more aggressive under high temperatures and less oxygen is also held in the nutrient solution as temperature increases. Chilling the nutrient solution in recirculating systems and maintaining suitable air temperatures can assist with prevention. Healthy plants which are not under any cultural stress are more resistant to opportunist *Pythium* attack. Plants which have been weakened or damaged will release phenolic compounds which signal to *Pythium* spores and promote infection.

Heavy use of sanitation chemicals was once thought to be the answer to preventing *Pythium* infection; however, even with very thorough cleaning and use of disinfectants, *Pythium* is often not completely eradicated, and sources of inoculum are widespread in the environment. Encouraging a healthy root zone with a highly diverse range of microbes is a more effective long-term approach. Recently

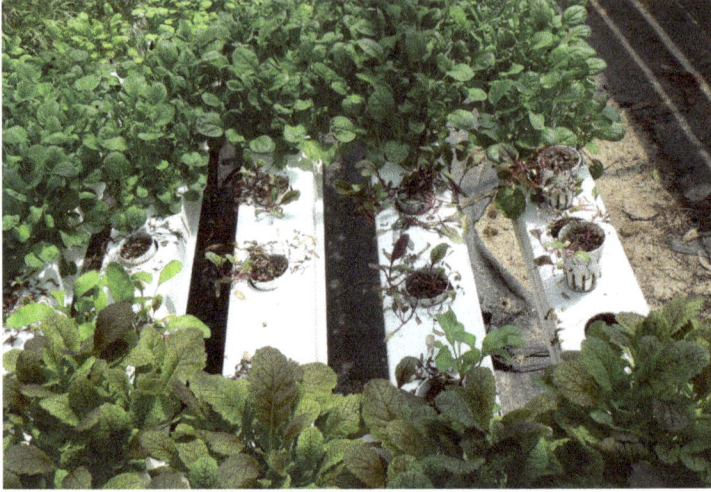

Fig. 10.12. Weakened seedlings caused by *Pythium* infection.

disinfected hydroponic systems are devoid of useful microbes, allowing opportunists such as *Pythium* an open field for attack. A well-seasoned system may be a better approach, as is introductions of beneficial microbes. Hydroponic systems are often low in microbial diversity and density, particularly of those species which may antagonize root rot disease pathogens such as *Pythium*. However, several species of specific microbes have been shown to have potential for controlling root rot pathogens in certain crops. These include *Pseudomonas chlororaphis*, *Pseudomonas fluorescens*, some species of *Bacillus*, *Gliocladium*, *Lysobacter* and *Trichoderma*, indicating that inoculation with beneficial microbes of the correct species could be an effective control option if carried out correctly. The use of slow sand filtration is a method which provides microbial degradation of organic matter in nutrient solutions and destruction of pathogen propagules by beneficial bacteria.

Many researchers have carried out trials into the effect of surfactants for *Pythium* control in hydroponics with promising results, however dose rates and surfactant products differ from study to study. Stanghellini *et al.* (1996) found that non-ionic surfactants added to recirculated nutrient solutions effectively controlled the spread of *Pythium*

in cucumber and *Phytophthora* in pepper with no apparent phytotoxicity. Surfactants are thought to disrupt the plasmalemma of fungal structures such as zoospores, causing them to rapidly lyse within a short exposure time (Stanghellini and Tomlinson, 1987).

While *Pythium* is often the most common root rot pathogen found in hydroponic systems, other disease-causing species include *Phytophthora cryptogea*, which causes *Phytophthora* root rot, and *Rhizoctonia solani*, which causes damping off or bottom rot. Symptoms can be difficult to distinguish between these pathogens, particularly when they infect young seedlings and cause general dieback and decline.

10.4.4 Wilt diseases

The most commonly occurring fungal wilt diseases affecting hydroponic crops include *Verticillium* wilt (*Verticillium dahliae* and *Verticillium albo-atrum*) and *Fusarium* wilt (*F. oxysporum*). Wilt pathogens infect the root system, destroying root tissue which prevents the plant from taking up sufficient water to maintain plant turgor. Wilting is normally most prevalent during the middle of the day under warmer growing conditions.

Verticillium can infect more than 200 plant species including capsicum, aubergine, strawberry and many others, with infection occurring at relatively cool root-zone temperatures of 13–23°C (Sanoubar and Barbanti, 2017). *Verticillium* wilt causes wilting of the older leaves, which turn yellow and become desiccated, combined with a tan to pink discoloration of the vascular tissue at the base of the plant stem. Infected plants often have characteristic V-shaped lesions occurring at the leaf edges.

Fusarium wilt is the most common wilt disease found in greenhouse tomato crops and is prevalent worldwide, even resistant varieties may be affected (Sanoubar and Barbanti, 2017). *Fusarium* wilt can resemble *Verticillium* wilt; however, wilted leaves turn yellow only on one side of the plant and vascular tissue discoloration is a reddish-brown colour that extends further down the stem. Control of wilt diseases involves prevention via use of resistant cultivars and rootstocks where applicable, combined with general greenhouse hygiene and sanitation. Use of sterile growing media, water supply disinfection and control over insect pests which may spread these diseases are the main form of control.

10.4.5 Common bacterial diseases

Bacterial diseases typically require a high humidity and/or the presence of free water on leaf surfaces for infection to occur. These can be persistent on outdoor hydroponic crops which are exposed to warm conditions and frequent rainfall. Commonly occurring bacterial pathogens include bacterial soft rot (*E. carotovora*) and various bacterial leaf spots such as varnish spot (*Pseudomonas cichorii*) and *Xanthomonas campestris*. *Erwinia* is found worldwide and causes damage to a wide host range of plants both during production and in the postharvest stage. Symptoms of infection include sudden wilting as the bacteria invade and multiply within the xylem vessels. Where infection of the foliage occurs, dark brown spots may form which progress to produce a slimy black rot. *Pseudomonas* and *Xanthomonas* leaf spots are common under warm,

wet conditions and prevention is largely through keeping the foliage dry. Both pathogens infect leaf tissue through natural openings in the leaf or via sites which have been damaged.

Bacterial wilt of tomato (*Ralstonia solanacearum*) is prevalent in subtropical and tropical regions and initially begins with wilting of the lower leaves. Vascular tissue becomes stained dark brown and pith becomes pink then hollow as the disease progresses. Cut stems of infected plants exude a grey slime when immersed in water, which is used as a simple diagnostic test of this disease. Bacterial wilt has a high optimum infection temperature range of 28–35°C combined with high moisture levels. Control of bacterial wilt with the use of grafted planting stock which has a high level of resistance to this pathogen is one method of control and can increase yields in areas where this pathogen occurs (McAvoy *et al.*, 2012).

10.4.6 Virus diseases

While there are large number of viruses which may infect greenhouse crops, the most widely occurring are tomato mosaic, cucumber mosaic, tobacco mosaic, lettuce mosaic, tomato spotted wilt, double streak, tobacco etch, curly top, tomato yellow leaf curl and pepino mosaic. Viruses may produce a range of symptoms including leaf mottling, chlorosis and other colour disorders, twisting, curling and deformity such as 'shoestring'-like growth, stunting of growth, dwarfism, fruit disorders and deformities. Virus damage can range from a few mildly infected plants to serious plant and yield losses. Most viruses only survive long-term in the living plant tissue of certain hosts or briefly within the insects which spread them. However, some viruses such as tobacco mosaic virus can survive for many months or even years in soil, plant debris, dried plant material such as tobacco, on tools, in growing substrates or as seed contaminants.

Viruses in some crops can become a serious problem due to the ease with which most are transmitted from plant to plant. In a greenhouse setting, sap-to-sap transmission can rapidly occur during training, pruning and other plant maintenance procedures via equipment, knives and workers' hands. Chewing and sucking insect pests also carry virus both from outside the greenhouse structure and within the crop from plant to plant. The most common insect vectors for tomato viruses are whitefly, aphids, thrips and leaf hoppers. One potential method to assist with prevention of virus introduction is the use of UV-blocking plastic films which disrupt normal insect behaviour so as to discourage pest activity (Diaz et al., 2006). Since pests such as aphids, thrips and whitefly transmit virus both within a crop and from outside, prevention of such pests is one of the main methods of virus prevention and control. Furthermore, UV-absorbing films were effective in reducing the population density of thrips and the spread of tomato spotted wilt virus (Diaz et al., 2006).

Many modern hybrids of crops such as tomatoes, capsicum and cucumbers have inbred resistance to a range of common viruses, and this is the most effective form of prevention of virus infection. Use of commercial seed supplied by a reputable company is also recommended for virus prevention as some can be seedborne and easily transferred to a new crop during propagation. Crops should be regularly scouted for signs of unusual growth such as leaf twisting, mosaic or stunting and culled rapidly to prevent further spread. Where virus-infected plants have been removed, sterilization of floors, surfaces, tools and pruning equipment should be carried out to prevent any carryover infection.

10.5 Abiotic Factors and Physiological Disorders

Abiotic disorders in greenhouse crops are those caused by non-living or non-infectious factors, which may also be referred to as 'abiotic disease' or 'physiological disorders'. 'Abiotic factors' refers to a wide range of plant and crop problems with varied symptoms and differing degrees of damage. Abiotic disorders are often associated with causal factors such as temperature, substrates, water, chemicals, mechanical injuries, cultural practices and in some cases a genetic predisposition within the plant itself (Schutzki and Cregg, 2007). While abiotic factors which have a negative effect on plant growth primarily reduce yields and productivity, they also influence produce quality, postharvest shelf-life, susceptibility to pathogens and marketability of a wide range of food and ornamental plants. There is also a complex relationship between many abiotic factors, with light, temperature, moisture and nutrient levels all interacting to create multiple plant stresses. Plant responses to abiotic stresses are dynamic and complex; they are both elastic (reversible) and plastic (irreversible) (Cramer et al., 2011), and many responses are heavily dependent on genetic susceptibility or tolerance.

With abiotic disorders in plants being very common, it is critical to understand abiotic disorders fully in order to monitor and maintain plant health (Kennelly et al., 2012). Identification of the early symptoms of many abiotic disorders is vital for horticultural producers so that corrections may be applied where possible and yield losses prevented or minimized. Many disorders, such as water stress or nutrient deficiency, are relatively simple to identify by crop observation and basic testing; however, some abiotic issues are more complex, often overlooked or incorrectly diagnosed.

10.5.1 Temperature damage

Incorrect and damaging temperatures are abiotic factors which can severely restrict yields and quality if not managed correctly. The injury caused by temperature extremes is dependent on a number of factors such as the crop species, stage of growth, length of exposure to incorrect temperatures and interaction with other crop stresses such as moisture levels.

Heat injury

Abiotic heat damage is defined as where temperatures are high enough for sufficient time that they cause irreversible damage to plant function or development (Hall, 2001). Cool-season crops, or those adapted to lower temperatures, tend to be susceptible to high-temperature damage which results in a number of physiological disorders depending on species. Many plants are retarded by high temperatures and may stop growing, develop wilt during the warmest part of the day and suffer root dieback if extreme temperature conditions persist for an extended length of time. Plants may also exhibit foliar symptoms such as leaf scorch, tipburn, premature leaf drop, leaf bleaching and abnormal coloration. Fruit may develop disorders such as blossom end rot of tomatoes and capsicum or fruit skin disorders such as blistering, crazing, discoloration, cracks and splits. Fruiting crops such as tomatoes and capsicum are particularly prone to damage to surface and internal tissues by combinations of high temperature and intense solar radiation (Hall, 2001). Some species react with a lack of fruit set caused by poor pollen viability under high temperatures. Heat injury in some species may also damage floral development so that plants do not produce flowers (Hall, 2001). A significant effect of heat stress is the reduction in photosynthesis which may be due to damage to components of photosystem II located in the thylakoid membranes of the chloroplast and membrane properties (Al-Khatib and Paulsen, 1999). High temperatures also damage root zones, with root growth of many species retarded at temperatures greater than 30°C (Mathers, 2003) and root growth in many woody species stopping at temperatures exceeding 39.4°C (Johnson and Ingram, 1984).

Low-temperature and chilling injury

Exposure to temperatures in the 1–10°C range can cause chilling injury in some tropical and subtropical crops (Lukatkin et al., 2012). Exposure of chilling-sensitive plants to low temperatures causes disturbances in a number of physiological processes including water regulation, nutrition, photosynthesis, respiration and metabolism, with the degree of symptoms dependent on both temperature and duration of exposure to chilling

conditions. Chilling injury symptoms include wilting or water-soaked areas, surface lesions, water loss and desiccation or shrivelling, internal discoloration, tissue breakdown, failure of fruit to ripen or uneven or slow ripening, accelerated senescence and ethylene production, shortened storage or shelf-life, compositional changes, loss of growth or sprouting capacity, and increased decay due to leakage of plant metabolites (Skog, 1998). Low-temperature injury is often mistaken for other conditions such as fungal pathogens causing dieback or root rot-induced wilting.

10.5.2 Light

While insufficient or excess light can damage growth, reduce yields and create physiological symptoms, fruiting crops in particular may be prone to cosmetic damage such as scorch, sunscald and other forms of surface damage. Symptoms of insufficient light are typically straightforward to identify on most greenhouse crops and include elongated growth, etiolated seedlings, pale weak stems, pale foliage coloration, reduced growth rates, lower yields, poor quality fruit often with low solids levels, and weakened plants which are more prone to disease infection. Low light may be common under shorter day lengths in winter or where excessive shading is used in nursery situations. Excess light may cause a range of symptoms depending on species and adaptation to radiation levels. Plants which have been grown under shaded or protected conditions, such as transplants raised under cover in a nursery situation, may suffer damage if not gradually hardened off to full light levels in a production system. Such damage often shows as scorched or bleached foliage with a growth check until acclimatization has occurred. Persistent excessive light damage is more common in shade plants and often results in stunted, hard and dark-coloured growth. Some plants, when exposed to excessively high and damaging light levels, will develop strategies to avoid or limit cell damage. This may include leaf rolling, leaf drooping or orienting leaves upwards to restrict the amount of foliage surface area exposed to the damaging radiation levels. High light intensity can also lead to physiological disorders of fruiting crops, these include sunscald injury and uneven ripening of tomatoes and other fruit which are brought on by direct effects of light on fruit and an interaction with high temperature (Adegoroye and Jolliffe, 1987).

10.5.3 Root-zone abiotic factors

Oversaturation of the root zone in soilless substrates can become a problem where irrigation rates are either too frequent or provide excessive volumes of nutrient solution which is held around the root zone. Issues with oversaturation are more common in highly moisture-retentive substrates and under cooler growing conditions where plant requirements for moisture are low. Saturated growing substrates induce a number of symptoms in plants, these include wilting which appears similar to drought stress but is caused by a lack of oxygen restricting root function. Without oxygen, roots become anaerobic and toxic compounds are produced within the plant which destroy plant cell function. Under mild waterlogging conditions, plants may be stunted and fail to thrive, exhibiting mineral deficiency symptoms due to a lack of root activity; however, this abiotic disorder also makes root systems more susceptible to infection by biotic diseases which cause root rot (e.g. *Phytophthora* spp.).

Water stress is a less common occurrence in hydroponic crops due to the frequent application of nutrient solution and management of root-zone moisture levels. However, issues with irrigation blockages, electricity or pump failures cause interruptions to the supply of nutrient solution, while irrigation programmes that are not well matched to plant moisture requirements can lead to water stress and associated physiological disorders. If water stress occurs after crop establishment then symptoms include wilting during the heat of the day, slow and stunted growth, foliage burn or scorch and marginal leaf necrosis, leaf abscission,

and under prolonged water stress, complete, irreversible plant wilting and death. Mild water stress may not cause significant symptoms of an abiotic disorder but will affect fresh weight and water content. If prolonged, a lack of water will result in secondary abiotic disorders such as mineral deficiencies as water stress restricts the movement of nutrients to the root system. Mild water stress also causes a reduction in chlorophyll levels and an acceleration in ethylene biosynthesis which inhibits plant growth through several mechanisms. Photosynthesis is restricted under drought stress and CO_2 assimilation by leaves is reduced mainly by stomatal closure, membrane damage and disturbed activity of various enzymes (Farooq et al., 2009). Leaf expansion is also decreased and premature leaf senescence often occurs, which further add to reductions in the photosynthetic ability of the crop.

10.5.4 Irrigation water quality and salinity

Water quality can contribute either directly or indirectly to a number of abiotic disorders in a range of greenhouse crops. An increasingly common issue with irrigation water in many regions of the world is salinity. The salinity hazard of irrigation water is measured by EC. The primary effect of high EC on crop productivity is the imposition of physiological drought (Bauder et al., 2011). If this condition is prolonged, abiotic drought disorders develop which include wilting or a darker bluish-green colour and sometimes thicker waxier leaves (Ayers and Westcot, 1994) reduced growth and yields, and in severe cases foliar burn and increased occurrence of physiological fruit disorders such as blossom end rot. Plant species vary widely in their response to salinity of irrigation water with crops such as tomatoes being relatively tolerant and others such as many leafy greens being highly sensitive.

Other water quality issues affect plants indirectly via nutrition. High levels of ions such as chlorine, sodium and boron can cause direct toxicity effects on sensitive crops. Disorders due to ion toxicity from irrigation water often first appear as marginal leaf burn and interveinal chlorosis as absorbed ions are transported to the leaves where they accumulate during transpiration. These ions build up to the greatest extent where water loss is greatest at the leaf tips and leaf edges (Ayers and Westcot, 1994). The degree of crop damage varies depending on exposure, concentration of the toxic ion, crop sensitivity and the volume of water taken up by the crop. Chloride is generally safe for most plants below 70 ppm (Bauder *et al.*, 2011); however, sensitive plants such as lettuce, cucumber and strawberry can show leaf burn injury at lower levels. Studies have shown that strawberry crops develop leaf scorch due to an increase in foliar Cl^- concentration (Martinez Barroso and Alvarez, 1997). Boron toxicity due to the presence of high boron levels in water supplies can occur and results in necrotic patches on foliage.

10.5.5 Chemical injury (phytotoxicity)

Chemical injury or phytotoxicity can result in a wide range of damage and is caused by incorrect application of herbicides, pesticides, fungicides, plant growth regulators, sanitation agents and via contamination with chemicals from other industries. 'Phytotoxicity' is the term used to describe such interactions and damage can range from mild, where plants may recover over time, to severe with resultant crop losses. Herbicides can cause particularly severe damage as they are initially designed to kill plant life and can cause symptoms at low levels on sensitive plant species.

In greenhouse systems, issues with herbicide damage are most likely to occur where there has been spray drift from chemicals applied outside the structure or where herbicides contaminate the water supply. Contamination of irrigation water with persistent herbicide compounds is a known cause of abiotic disorders which may affect crops some distance from where the herbicide was initially sprayed for weed control. Distinguishing herbicide injury from other abiotic disorders and biotic factors such as plant

viruses can be difficult if the damage occurs some distance from where the herbicide was initially applied. Some crops such as tomatoes are particularly sensitive to these types of hormone herbicides and can suffer spray drift inside greenhouses where minute traces of the herbicide can be carried on wind currents and through open vents. Mild symptoms on non-target crops such as tomatoes to contamination with synthetic auxin herbicides appear similar to infection by some viruses and can be difficult to distinguish.

Chemical injury may also occur from the use of certain crop sanitation agents such as sodium hypochlorite, hydrogen peroxide and similar compounds. Accidental exposure may occur where sprays of these compounds are used to clean floors and other surfaces adjacent to actively growing crop plants, or where resides remain in irrigation lines and tanks after disinfection procedures. Use of disinfection agents such as chlorine or hydrogen peroxide added to water supplies or nutrient solutions may also damage sensitive or young plants at relatively low levels. Other water or nutrient solution sterilization procedures such as ozone may also damage crops if correct practices are not followed for degassing before application to the crop.

Pesticides, fungicides and other plant protection compounds applied to crops may cause damage under certain circumstances. Damage or phytotoxicity usually occurs when compounds are applied at an incorrect rate or stage of development, applied outside label-recommended usage, or mixed with incompatible products before application. Environmental conditions and the physiological condition of the plants these compounds are applied to can also play a role in spray injury occurrence. Application of some fungicidal sulfur compounds under high temperatures may cause crop damage, while copper materials can cause tissue bronzing on sensitive crops. Repeated applications of spraying oil and soap insecticide sprays have been known to cause crop damage to sensitive crop species, particularly those grown under greenhouse conditions. Spray injury from pesticides and fungicides can be difficult to recognize as it may take a period of time from application until visible symptoms are seen. These often include leaf chlorosis, marginal burn and spotted necrosis on leaf surfaces. Plant growth regulators, which are chemicals used to promote favourable growth characteristics and are used in the production of a number of horticultural crops, may cause damage under certain circumstances. Many plant growth regulators are applied in order to affect flowering, root growth, reduce vegetative growth making plants more compact, control or promote plant height, thin fruit and improve fruit quality. If plant growth regulators are applied at the wrong rate, the wrong time or under certain environmental conditions (e.g. high light or high temperature), the consequences can be severe with long-lasting discoloration, stunted growth and other injury symptoms (Kennelly *et al.*, 2012).

10.5.6 Ethylene

Ethylene is a risk factor under greenhouse production where it is a by-product of incomplete combustion of fossil fuels in faulty heating systems. Injury is typically a downward curling of the leaves and shoots, termed 'epinasty', followed by stunting of growth; however, leaf, petal, flower and fruitlet abscission may also occur (Kennelly *et al.*, 2012). Ethylene causes leaf abnormality in tomato plants, leaf irregularities and poor flower formation in narcissus and tulip, failure of carnation flowers to open and flower drop in snapdragon crops (Middleton *et al.*, 1958). Symptoms of ethylene exposure are often misdiagnosed as herbicide injury and virus infection as these appear very similar. Crops such as tomatoes, capsicum, cucumber, orchids and roses are particularly sensitive to low levels of ethylene contamination: exposure to levels as low as 0.1 ppm for 6 h cause epinasty in tomato and capsicum plants (Sikora and Chappelka, 2004), with levels of 1–10 ppm causing significant plant damage or even plant death (Kennelly *et al.*, 2012).

10.6 Cultural Practices Causing Abiotic Disorders

Incorrectly carried out or timed cultural practices and processes can result in a number of abiotic disorders. These may originate from physical damage during transplanting, pruning, grafting and propagation or a lack of maintaining the correct environment immediately after carrying out cultural operations. One of the most common abiotic disorders of greenhouse crops related to cultural practices is transplant shock, where young seedlings or planting stock fails to thrive after planting out into a production system or repotting during nursery operations. Transplant shock can have a number of causes. Seedlings or sensitive young plants can easily become damaged when handled, with stems, leaves and root tissue becoming bruised. Incorrect planting depth is another cause of poor transplant performance and physiological disorders such as wilting, plant losses and slow establishment. Other cultural practices such as budding and grafting may cause abiotic disorders such as graft failure or graft incompatibility due to a union failure between scion and rootstock.

10.7 Identification of Abiotic Disorders

Accurate diagnosis of abiotic disorders is reliant on observation of symptoms, experience with the crop under review and where necessary further analytical testing. The main difficulty is determining whether symptoms are biotic or abiotic in nature as both may occur concurrently on the same plant and many may appear similar. Biotic or 'living' factors such as plant disease should be first ruled out either by correct identification of symptoms such as fungal spores and the nature of spread of the infection or by laboratory diagnosis of tissue samples. Once biotic factors have been excluded as the cause of injury, abiotic disorders require a systematic review to determine the cause of the damage. This includes observing patterns and progress of the damage through a crop. While biotic issues such as disease tend to spread and progress over time, often transferring from plant to plant from the initial site of outbreak, abiotic disorders tend to occur all at once, after a particular event such as a period of extreme temperatures or incorrect chemical application. If chemical damage is suspected, a review of recent spray application records can be useful in determining if this was the cause of damage. A complete investigation of the entire plant may also provide further information on the causes of abiotic disorders, this includes inspection of the root zone. Plants which have been subjected to waterlogging, low oxygenation, salinity, incorrect irrigation practices or issues with irrigation water quality generally show damage to the root system. This may include stunted, poorly formed root systems, root browning and dieback, and the invasion of secondary pathogens in damaged areas.

More complex abiotic disorders such as those induced by nutrient deficiency or toxicity may be diagnosed by foliar symptoms alone; however, this can be confirmed and quantified by use of laboratory tissue analysis of the damaged leaves. This may be followed up by a nutrient leachate test to determine if the abiotic nutritional disorders are caused by an excess or deficiency in the root zone or an induced uptake problem such as incorrect pH, nutrient interactions or other plant factors. Abiotic disorders which are caused by multiple factors are perhaps the most difficult to correctly identify, these include many nutrient interactions which may occur in soilless production. Examples of these are calcium and iron deficiency symptoms in plants, which often occur in crops despite there being more than sufficient of these elements in the nutrient solution and root zone. Iron uptake in many crop plants can be limited by low temperatures, thus iron chlorosis on the new growth may occur in winter or early spring irrespective of the iron status in the growing medium. While identification of iron chlorosis symptoms is relatively straightforward, understanding the cause of the problem is more complex as increasing iron fertilization around the root system will not remedy the situation.

Understanding nutrient interactions, uptake and how the environmental conditions may influence these is an important process in the diagnosis of abiotic disorders of a nutritional nature.

Abiotic factors may be complex where there is more than one cause or multiple disorders occurring at the same time. This is particularly difficult where tissue damaged by an abiotic factor is then invaded with secondary colonizers such as rot pathogens. These may mask the initial cause of the tissue damage, making correct diagnosis and treatment of the original problem more difficult.

10.8 Crop-Specific Physiological Disorders

10.8.1 Blossom end rot

Physiological disorders in greenhouse crops can be common under certain conditions and are often misdiagnosed as pathogenic or nutritional problems. Blossom end rot (BER) affects fruiting crops such as tomato and capsicum and can range from mild to severe depending on environmental conditions and cultivar susceptibility. Mild symptoms appear as small water-soaked or browned, sunken areas at the blossom end of the fruit, which progress into a dark lesion which later turns black. This is attributed to the disintegration of the cell membranes and an increased ion permeability in the fruit tissue (Saure, 2001). In severe cases as much as 50% of the fruit may become blackened and eventually rot. Apart from the external symptoms of blossom end rot, 'internal BER' may also occur that only affects tissues inside the fruit including some seeds (Adams and Ho, 1992). In greenhouse crops such as tomatoes, it has been reported that the susceptibility of different cultivars to blossom end rot appears to be related to the development of the xylem network and the rate of cell expansion during the early fruit development phase (Ho and White, 2005).

Blossom end rot occurs when calcium is not transported into the fruit at rates sufficient to prevent tissue breakdown and is generally attributed to a deficiency of calcium in the fruits (Shear, 1975). This problem is compounded when warm growing conditions favour high rates of cell division and fruit development, increasing the requirement for calcium during the formation of new tissue. In many fruits such as tomato, calcium levels of 0.2% have been found at the stem end, while only 0.04–0.07% gets transported to the blossom end where this disorder typically occurs (Morgan, 2008). While blossom end rot is a calcium-related disorder, it is rarely caused by a deficiency of calcium in the nutrient solution or root zone but is an uptake and transportation issue within the plant. Calcium moves within the transpiration stream through the xylem tissue, thus any condition which restricts transpiration also lowers the uptake and movement of calcium in the plant and out to the newly developing tissue regions. Other factors which contribute to blossom end rot and directly affect the rate of calcium uptake are water stress, high EC and high salinity, imbalances of elements in the nutrient solution, high temperatures and humidity. In hydroponic production, high levels of potassium and magnesium in the nutrient solution can result in K:Ca and Mg:Ca antagonism, while high nitrogen levels in the ammonium-N form have been shown to produce more blossom end rot than those in only the nitrate-N form (Morgan, 2008).

Blossom end rot prevention includes increasing greenhouse ventilation and air movement to promote transpiration and calcium movement within the plant. Lowering EC to increase water and hence calcium uptake may also be used alongside nutrient formulation balancing. Reducing temperatures and controlling humidity as well as selection of cultivars with good resistance to blossom end rot may also be effective.

10.8.2 Tipburn

Tipburn is another calcium-related disorder which affects crops such as leafy greens, strawberries, celery and herbs. Tipburn is characterized by browning and dieback of the leaf edges, usually on the newer growth in plant centres, and it can be particularly prevalent in tissue with a low rate of transpiration such as the inner leaves of heading lettuce (Fig. 10.13). Cool-season salad crops grown under higher-than-optimal temperatures are most likely to be affected by this disorder. While there are different types of tipburn, the most common form is caused by a lack of calcium deposition in the leaf edges. Calcium-deficient tissue and the subsequent cell-wall breakdown results in necrotic areas which may either desiccate or rot depending on the ambient humidity. The deficiency in calcium at the leaf edges is usually not attributed to a lack of calcium in the nutrient solution or root zone, but a lack of calcium transport within the plant itself. Much like blossom end rot, tipburn occurs when a lack of transpiration from the leaf surfaces results in a reduction in the flow of calcium which moves within the transpirational stream. Leaf tips, being the end of the transpirational stream, may not receive sufficient calcium to prevent cell breakdown under certain environmental and cultural conditions. Warm, still, humid environments favour tipburn which is often a seasonal issue. Prevention of tipburn involves growing less-susceptible cultivars wherever possible, reducing solution EC to assist water and calcium uptake, increasing air flow and ventilation, and decreasing temperatures and humidity levels. Avoidance of using the ammonium form of nitrogen in the nutrient solution for susceptible crops

Fig. 10.13. Tipburn on lettuce.

can also assist calcium uptake and tipburn prevention. Research has also shown that calcium deficiency in chervil and lettuce can be prevented or reduced by growing plants in negative DIF conditions where night temperature is higher than day temperature (Olle, 2015). Negative DIFs increase transpiration during the dark period and can lead to higher tissue calcium content (Olle and Williams, 2017).

10.8.3   Bolting

'Bolting' is the term used to describe premature flowering in vegetable crops such as lettuce, salad greens and herbs. Lettuce in particular may bolt at any stage of development, from the seedling stage through to harvest, if certain adverse conditions occur; however, bolting is more common during the intermediate stages of growth. Bolting in crops such as lettuce first appears as an upward stem elongation which in the early stages can be difficult to detect unless the plant is cut through the centre (Fig. 10.14).

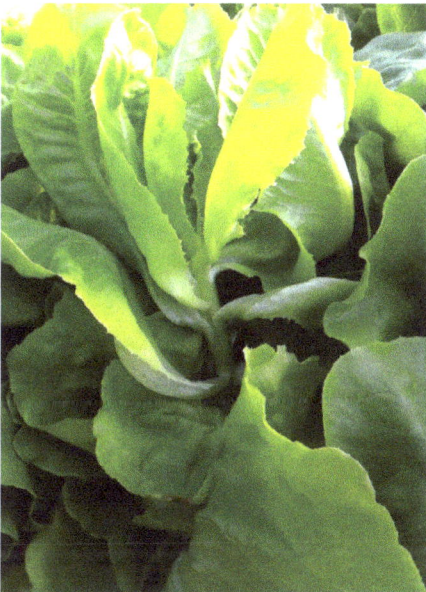

Fig. 10.14. Bolting of lettuce.

In the later stages, the stem will rapidly elongate upwards until flowering occurs at the apex. Bolting makes most produce unmarketable and may not be visible inside the tight heads of crisphead or butterhead lettuce where internal spiralling of leaves inside the heart can only be seen if these are cut open. Premature bolting is largely triggered by plant stress, often high temperatures in cool-season crops. When warm conditions are combined with low light or overcrowding, the severity of bolting is increased and this may occur at an earlier stage of growth. In some lettuce cultivars, long days can initiate bolting, while other cultivars are day neutral and do not have this response. It has been reported that the effect of long days is additive in that as the age of the plant increases, the influence of high temperature on bolting becomes more significant (Waycott, 1995).

Prevention of bolting requires an integrated approach as cultivar selection is important where conditions favour this disorder. Greenhouse environments may require modification to reduce temperatures and increase air flow and the use of nutrient solution chilling can also assist with bolting prevention in tropical climates. High EC and water stress are also known plant stress triggers which promote bolting, as is any condition which reduces the health of the plant or affects the ability to take up water.

10.8.4   Fruit shape and splitting/cracking disorders

Greenhouse crops such as tomatoes, cucumbers and capsicum may experience a number of issues with fruit shape and deformities (Fig. 10.15). Misshapen fruit often result from issues during pollination and fruit set; however, some cultivars are more susceptible than others to these disorders. Seedless cucumber types which set fruit without the need for pollination are the exception, as these may become deformed if unwanted pollination actually occurs resulting in seed development in the flesh. Tomato fruit can develop a condition termed 'catface' which describes deformity of fruits which typically occurs due to fruit set under cooler-than-optimal conditions.

Fig. 10.15. Fruit shape deformities can have a number of causes.

In capsicum, fruit with uneven locule development and malformed fruit are also due to lower temperatures during flowering and fruit set. These problems are most common in unheated crops in temperate regions.

Russeting or cracking and splitting of greenhouse-grown fruit such as tomatoes, melons and capsicum can have a number of causes. Fruit cracking is the splitting of the epidermis around the calyx and mainly occurs when there is a rapid net influx of solutes and water into the fruit (Olle and Williams, 2017). At the same time, genetic susceptibly, ripening or other factors reduce the strength and elasticity of the fruit skin (Leonardi et al., 2000). Cuticle cracking can occur at all stages of fruit growth; however, it is more common as fruits mature, particularly as colour develops (Olson, 2004). These physiological disorders are often linked to irrigation irregularities and are more prevalent during the later stages of fruit development or as fruit are starting to ripen. Irrigation of fruiting crops should be monitored and managed to prevent any drying within the root zone. Crops which have suffered some

drying and then given a large irrigation volume will rapidly take up a large influx of water which can cause the fruit to swell and split. Rapid changes in nutrient solution EC can have a similar effect, with a sudden intake of water causing cracking and splitting of the fruit cuticle. Another cause of fruit splitting is heavy pruning of plants because sudden removal of large amounts of foliage may cause fruit to split as moisture is diverted into the fruit. High temperature and light levels are also associated with increased rates of fruit cracking (Peet, 1992). In some tomato types such as the small cocktail varieties, fruit may split postharvest in response to rapid changes in temperature; this can be avoided by careful regulation of media moisture and EC levels in the later stages of fruit development. Harvesting cherry tomatoes in the morning has been found to give the highest rates of postharvest fruit cracking, while the lowest rates were found in fruit harvested in the evening (Lichter et al., 2002). Selection of cultivars of fruiting crops that have resistance to russeting and cracking can assist with prevention of these disorders. Crack-resistant cultivars of tomato are generally associated with a thicker cuticle (Sadhankumar et al., 2001) and an epidermis that stretches well during fruit expansion.

References

Abbey, J.A., Percival, D., Abbey, L., Asiedu, S.K., Prithiviraj, B. and Schilder, A. (2019) Biofungicides as alternative to synthetic fungicide control of grey mould (*Botrytis cinerea*) – prospects and challenges. *Biocontrol Science and Technology* 29(3), 207–228. https://doi.org/10.1080/09583157.2018.1548574

Adams, P. and Ho, L.C. (1992) The susceptibility of modern tomato cultivars to blossom end rot in relation to salinity. *Journal of Horticultural Science* 67(6), 827–839.

Adegoroye, A.S. and Jolliffe, P.A. (1987) Some inhibitory effects of radiation stress on tomato fruit ripening. *Journal of the Science of Food and Agriculture* 39(4), 297–302.

Al-Khatib, K. and Paulsen, G.M. (1999) High-temperature effects on photosynthetic process in temperate and tropical cereals. *Crop Science* 39(1), 119125.

Ansari, M.N., Brownbridge, M., Shah, F.A. and Butt, T.M. (2008) Efficacy of entomopathogenic fungi against soil-dwelling life stages of western flower thips, *Frankliniella occidentalis*, in plant growing media. *Entomologia Experimentalis et Applicata* 127(2), 80–87.

Ayers, R.S. and Westcot, D.W. (1994) Water quality for agriculture. *FAO Irrigation and Drainage Paper 29 Rev. 1*. Food and Agriculture Organization of the United Nations, Rome.

Bauder, T.A., Waskom, R.M., Sutherland, P.K. and Davis, J.G. (2011) Irrigation water quality criteria. *Fact Sheet No. 0.506*. Colorado State University Extension, Fort Collins, Colorado. Available at: https://extension.colostate.edu/docs/pubs/crops/00506.pdf (accessed 9 September 2020).

Chandler, D., Bailey, A.S., Thatchell, G.M., Davidson, G., Greaves, J. and Grant, W.P. (2011) The development, regulation and use of biopesticides for integrated pest management. *Philosophical Transactions of the Royal Society B: Biological Science* 366(1573), 1987–1998.

Cramer, G.R., Urano, K., Delrot, S., Pezzotti, M. and Shinozaki, K. (2011) Effects of abiotic stress on plants: a systems biology perspective. *BMC Plant Biology* 11, 163.

De Faria, M.R. and Wraight, S.P. (2007) Mycoinsecticides and mycoacaricides: a comprehensive list with worldwide coverage and international classification of formulation types. *Biological Control* 43(3), 237–256.

Diaz, B.M., Biurrun, R., Moreno, A., Nebreda, M. and Fereres, A. (2006) Impact of ultraviolet-blocking plastic films on insect vectors of virus diseases infesting crisp lettuce. *HortScience* 41(3), 711–716.

Ehret, D., Alsanius, B., Wohanka, W., Menzies, J. and Utkhede, R. (2001) Disinfection of recirculating nutrient solutions in greenhouse horticulture. *Agronomie* 21(4), 323–339.

Elad, Y. and Shtienberg, D. (1995) *Botrytis cinera* in greenhouse vegetables: chemical, cultural, physiological and biological controls and their integration. *Integrated Pest Management Reviews* 1, 15–29.

Farooq, M., Wahid, A., Kobayashi, N., Fujita, D. and Basra, S.M.A. (2009) Plant drought stress: effects, mechanisms and management. *Agronomy for Sustainable Development* 29, 185–212.

Ferguson, L.B. (1984) Calcium in plant senescence and fruit ripening. *Plant, Cell & Environment* 7(6), 477–489.

Garcia Victoria, N.G. and Amsing, J.J. (2007) A search for the sources of root knot nematodes in commercial rose nurseries. *Acta Horticulturae* 751, 229–235.

Gill, S.S., Cowles, E.A. and Pietrantonio, P.V. (1992) The mode of action of *Bacillus thuringiensis* endotoxins. *Annual Review of Entomology* 37, 615–636.

Gonzalez, F., Tkaczuk, C., Dinu, M.M., Fiedler, Z., Vidal, S., *et al.* (2016) New opportunities for the integration of microorganisms into biological pest control systems in greenhouse crops. *Journal of Pest Science* 89, 295–311.

Gwynn, R.L. (2014) *The Manual of Biocontrol Agents: A World Compendium*, 5th edn. BCPC, Alton, UK.

Hall, A.E. (2001) *Crop Responses to Environment*. CRC Press, Boca Raton, Florida.

Ho, L.C. and White, P.J. (2005) A cellular hypothesis for the induction of blossom-end rot in tomato fruit. *Annals of Botany* 95(4), 571–581.

Jarvis, W.R. (1989) Managing diseases in greenhouse crops. *Plant Disease* 73(3), 190–194.

Johnson, C.R. and Ingram, D.L. (1984) *Pittosporum tobira* response to container medium temperature. *HortScience* 19, 524–525.

Kamikawa, S., Imura, T. and Sato, H. (2018) Reduction of sooty mold damage through biocontrol of the greenhouse whitefly *Trialeurodes vaporariorum* (Hemiptera: Aleyrodidae) using selective insecticides in tomato cultivation greenhouses. *Applied Entomology and Zoology* 53, 395–402.

Kennelly, M., O'Mara, J., Rivard, D., Millar, G.L. and Smith, D. (2012) Introduction to abiotic disorders in plants. *The Plant Health Instructor.* https://doi.org/10.1094/PHI-I-2012-10-29-01

Khan, S., Guo, L., Maimaiti, Y., Jijit, M. and Qiu, D. (2012) Entomopathogenic fungi as microbial biocontrol agent. *Molecular Plant Breeding* 3, 63–79.

Lemic, D., Dvecko, M., Drmic, Z., Viric Gasparic, H., Cacija, M. and Bazok, R. (2020) The impact of visual cards on pest populations in greenhouse tomato production. *European Journal of Horticultural Science* 85(1), 22–29.

Leonardi, C., Guichard, S. and Bertin, N. (2000) High vapour pressure deficit influence growth, transpiration and quality of tomato fruits. *Scientia Horticulturae* 84(3–4), 285–296.

Lichter, A., Dvir, O., Falllk, E., Cohen, S., Golan, R., *et al.* (2002) Cracking of cherry tomatoes in solution. *Postharvest Biology and Technology* 26(3), 305–312.

Lukatkin, A.S., Brazaityte, A., Bobinas, C. and Duchovskis, P. (2012) Chilling injury in chilling sensitive plants: a review. *Zemdirbyste Agriculture* 99(2), 111–124.

McAvoy, T., Freeman, J.H., Rideout, S.L., Olson, S.M. and Paret, M.L. (2012) Evaluation of grafting using hybrid rootstocks of management of bacterial wilt in field tomato production. *HortScience* 47(5), 621–625.

Martinez Barroso, M.C. and Alvarez, C.E. (1997) Toxicity symptoms and tolerance of strawberry to salinity in the irrigation water. *Scientia Horticulturae* 71(3–4), 177–188.

Mathers, M.H. (2003) Summary of temperature stress issues in nursery containers and current methods of production. *HortTechnology* 13(4), 617–624.

Messelink, G.J., Bennison, J., Alomar, O., Ingegno, B.L., Tavella, L., *et al.* (2014) Approaches to conserving natural enemy populations in greenhouse crops: current methods and future prospects. *BioControl* 59, 377–393.

Mehta, P., Wyman, J.A., Nakhla, M.K. and Maxwel, D.P. (1994) Transmission of tomato yellow leaf curl geminivirus by *Bemisia tabaci* (Homoptera: Aleyrodidae). *Journal of Economic Entomology* 87(5), 1291–1297.

Middleton, J.T., Darley, E.F. and Brewer, R.F. (1958) Damage to vegetation from polluted atmospheres. *Journal of the Air Pollution Control Association* 8(1), 9–15.

Moens, M. and Hendrickx, G. (1992) Drainwater filtration for the control of nematodes in hydroponic type systems. *Crop Protection* 11(1), 69–73.

Morgan, L.S. (2008) *Hydroponic Tomato Crop Production*. Suntec NZ Ltd, Tokomaru, New Zealand.

Ni, L. and Punja, Z.K. (2020a) Management of powdery mildew on greenhouse cucumber (*Cucumis sativus* L.) plants using biological and chemical approaches. *Canadian Journal of Plant Pathology*. https://doi.org/10.1080/070606 61.2020.1746694

Ni, L. and Punja, Z.K. (2020b) Management of fungal diseases on cucumber (*Cucumis sativus* L.) and tomato (*Solanum lycopersicum* L.) crops in greenhouses using *Bacillus subtilis*. In: Islam, M., Rahman, M., Pandey, P., Boehme, M. and Haesaert, G. (eds) *Bacilli and Agrobiotechnology: Phytostimulation and Biocontrol*. Springer, Cham, Switzerland. https://doi.org/10.1007/978-3-030-15175-1_1

Olle, M. (2015) *Methods to Avoid Calcium Deficiency on Greenhouse Grown Leafy Crops*. LAP Lambert Academic Publishing, Saarbrucken, Germany.

Olle, M. and Williams, I.H. (2017) Physiological disorders in tomato and some methods to avoid them. *Journal of Horticultural Science and Biotechnology* 92(3), 223–230.

Olson, S.M. (2004) Physiological, nutritional and other disorders of tomato fruit. *Publication HS-954*. Florida Cooperative Extension Service,

Institute of Food and Agricultural Sciences, University of Florida, Gainesville, Florida.

Pace, H., Vrapi, H. and Gixhari, B. (2016) Evaluation of some reduced-risk products for management of powdery mildew in greenhouse tomatoes. *International Journal of Ecosystems and Ecology Sciences* 6(4), 505–508.

Parrella, M.P. (1999) Arthropod fauna. In: Stanhill, G. and Enoch, H.Z. (eds) *Ecosystems of the World 20: Greenhouse Ecosystems*. Elsevier, Amsterdam, pp. 213–250.

Parrella, M.P. and Lewis, E. (2017) Biological control in greenhouse and nursery production: present status and future developments. *American Entomologist* 63(4), 237–250.

Paulitz, T. and Belanger, R.R. (2001) Biological control in greenhouse systems. *Annual Review of Phytopathology* 39, 103–133.

Paulitz, T.C., Zhou, I.T. and Rankin, L. (1992) Selection of rhizosphere bacteria for biological control of *Pythium aphanidermatum* on hydroponic grown cucumber. *Biological Control* 2(3), 226–237.

Peet, M.M. (1992) Fruit cracking in tomato. *HortTechnology* 2(2), 216–223.

Perdikis, D., Kapaxidi, E. and Papadoulis, G.T. (2008) Biological control of insect and mite pests in greenhouse solanaceous crops. *European Journal of Plant Science and Biotechnology* 2(1), 125–144.

Pilkington, L.K., Messelink, G., van Lenteren, J.C. and Le Mottee, K. (2010) 'Protected biological control' – biological pest management in the greenhouse industry. *Biological Control* 52(3), 216–220.

Sadhankumar, P.G., Rajan, S. and Peter, K.V. (2001) Concentric cracking in tomato – biochemical, physical and anatomical factors. *Vegetable Science* 28, 192–194.

Salgado-Salazar, C., Shishkoff, N., Daughtrey, M., Palmer, C.L. and Crouch, J.A. (2018) Downy mildew: a serious disease threat to rose health worldwide. *Plant Disease* 102(10), 1873–1882.

Sanoubar, R. and Barbanti, L. (2017) Fungal diseases on tomato plant under greenhouse condition. *European Journal of Biological Research* 7(4), 299–308.

Saure, M.C. (2001) Blossom-end rot of tomato (*Lycopersicon esculentum* Mill.) – a calcium or a stress related disorder? *Scientia Horticulturae* 90(3–4), 193–208.

Schutzki, R.E. and Cregg, B. (2007) Abiotic plant disorders, symptoms, signs and solutions. A diagnostic guide to problem solving. *MSU Extension Bulletin E-2996*. Michigan State University, East Lansing, Michigan. Available at: https://www.canr.msu.edu/outreach/uploads/2018-files/

e2996.abiotic%20plant%20disorders%20-%20 symptoms,%20signs%20and%20solutions.pdf (accessed 9 September 2020).

Shear, C.B. (1975) Calcium related disorders and fruits and vegetables. *HortScience* 10(4), 361–365.

Shipp, J.L., Zhang, Y., Hunt, D.W.A. and Ferguson, G. (2003) Influence of humidity and greenhouse microclimate on the efficacy of *Beauveria bassiana* (Balsamo) for control of greenhouse arthropod pests. *Environmental Entomology* 32(5), 1154–1163.

Sikora, E.J. and Chappelka, A.H. (2004) Air pollution damage to plants. *Alabama Cooperative Extension System, Publication ANR-913*. Alabama A&M and Auburn Universities, Alabama. Available at: https://ssl.acesag.auburn.edu/pubs/docs/A/ANR-0913/ANR-0913-archive.pdf (accessed 9 September 2020).

Skog, L.J. (1998) Chilling injury of horticultural crops. *Ontario Ministry of Agriculture, Food and Rural Affairs Fact Sheet*. Ontario Ministry of Agriculture, Food and Rural Affairs, Guelph, Ontario, Canada.

Smith, T. (2012) Fungus gnats and shore flies. Extension Greenhouse Crops and Floriculture Program, University of Massachusetts, Amherst, Massachusetts. Available at: https://ag.umass.edu/greenhouse-floriculture/factsheets/fungus-gnats-shore-flies (accessed 9 September 2020).

Stanghellini, M.E. and Tomlinson, A. (1987) Inhibitory and lytic effects of non-ionic surfactant on various asexual stages in the life cycle of *Pythium* and *Phytophthora* species. *Phytopathology* 77(1), 112–114.

Stanghellini, M.E., Rasmussen, S.L., Kim, D.H. and Rorabaugh, P.A. (1996) Efficacy of non-ionic surfactants in the control of zoospore spread of *Pythium aphanidermatum* in a recirculating hydroponic system. *Plant Disease* 80(4), 422–428.

Sutton, J.C., Sopher, C.R., Owen-Going, T.N., Liu, W., Grodzinski, B., *et al.* (2006) Etiology and epidemiology of *Pythium* root rot in hydroponic crops: current knowledge and perspectives. *Summa Phytopathologica* 32(4). https://doi.org/10.1590/S0100-54052006000400001

Van Lenteren, J.C. (2012) The state of commercial augmentative biological control: plenty of natural enemies, but a frustrating lack of uptake. *BioControl* 57, 1–20.

Volpin, H. and Elad, Y. (1991) Influence of calcium nutrition on susceptibility of rose flowers to gray mould. *Phytopathology* 81(11), 1390–1394.

Waycott, W. (1995) Photoperiodic response of genetically diverse lettuce accessions. *Journal of the American Society for Horticultural Science* 120(3), 460–467.

Weintraub, P.G., Recht, E., Mondaca, L.L., Harari, A.R., Diaz, B.M. and Bennison, J. (2017) Arthropod pest management in organic vegetable greenhouses. *Journal of Integrated Pest Management* 8(1), 1–14.

Yunis, H., Shtienberg, D., Elad, Y. and Mahrer, Y. (1991) Calcium enrichment of fertiliser to reduce gray mould of greenhouse grown eggplant, pepper and cucumber. *Phytoparasitica* 19, 246.

11 Hydroponic Production of Selected Crops

11.1 Introduction

While there is a wide range of potentially profitable crops which can be grown in hydroponics under protected cultivation, greenhouse production is dominated by fruiting crops such as tomatoes, cucumber, capsicum and strawberries, and vegetative species such as lettuce, salad and leafy greens, herbs and specialty crops like micro greens. Cut flower crops such as roses, carnations, chrysanthemums and others are also grown on a commercial scale worldwide. Smaller niche-market crops are becoming increasingly more popular under hydroponic cultivation, both in greenhouses and in plant factories, and include plants such as wasabi (Fig. 11.1), saffron, rhubarb, ginger and turmeric, medicinal species and chicory (Fig. 11.2). The information provided below on a selected range of common hydroponic crops is summarized to give basic procedures for each and an outline of the systems of production.

11.2 Hydroponic Tomato Production

Tomatoes are the crop produced on the largest scale worldwide under greenhouse production, with soilless production methods now dominating the industry. While the majority of greenhouse tomatoes grown are still the traditional round, red beefsteak, a wide diversity of fruit types is now being produced including cocktail, cherry, grape (Fig. 11.3), plum/Italian, cluster types, low acid and heirloom, and in a range of colours from yellow, purple, pink to dark brown; moreover, in countries where this is allowable, organically certified greenhouse tomato crops are also produced. Worldwide, hydroponic tomato crops are produced by growers ranging in size from small backyard operations to large-scale, commercial complexes many hectares in size and utilizing the latest technology while supplying extensive market chains. Some of these operations have greenhouse installations spread over a number of sites, with US and Canadian operations increasingly setting up greenhouses to supply fruit at a lower production cost and to expand supplies at certain times of the year. In Europe, greenhouse tomato production has expanded rapidly in areas such as Almeria on the south coast of Spain, where a mild climate and extensive use of basic plastic greenhouses have seen a large-scale industry develop. While yields are relatively low under these basic plastic structures, the intensive glasshouse industry in the Netherlands, Northern Europe and modern greenhouse production in the USA, Canada and Australasia have seen the rapid adaptation of new technology continually push yields higher.

Modern tomato greenhouses utilize not only hydroponic production methods with precise control over nutrition, but also CO_2 enrichment of the greenhouse atmosphere, supplementary lighting in low-radiation areas, extensive heating, cooling and humidity control, computer-controlled environments and high-technology, energy-efficient greenhouse claddings and designs. Many incorporate water treatments such as RO, ozone and UV sterilization, portable beehives for pollination and complex training systems all designed to make maximum use of greenhouse space, energy inputs and system efficiency. Most greenhouse tomato crops are grown as a long-term vine crop,

DOI: 10.1079/9781789244830.0011

Fig. 11.1. Hydroponic wasabi crop.

Fig. 11.2. Hydroponic chicory production.

Fig. 11.3. Grape tomatoes are one of the diverse ranges of fruit grown hydroponically.

often on a high-wire system with production taking place over an 11-month period; however, a second crop may be intercropped within an existing crop to reduce downtime.

Another cropping option is producing two shorter crops per year with planting occurring in spring and autumn.

Tomato varieties produced hydroponically are largely those which have been bred for greenhouse production and are dominated by F1 hybrids. These have multiple resistances to a wide range of common diseases such as *Fusarium* and *Verticillium* wilts, tomato mosaic virus and many others, while maintaining a highly uniform crop. Fruit characteristics of greenhouse hybrids include firm flesh, thick skin and a long shelf-life which help facilitate handling and transportation with minimal fruit damage. Open-pollinated tomato varieties, such as many of the older heirloom types, can be highly variable in growth, yields and fruiting characteristics, although the seed is less labour intensive to produce and therefore relatively inexpensive.

11.2.1 Hydroponic systems for tomato production

Systems and substrates for hydroponic tomato production are varied; however, the industry is dominated by drip-irrigated slab, pot or bag systems. Substrates such as stone wool, coconut fibre, perlite and peat are the most popular, although locally available and cost-effective alternatives such as sawdust, bark, scoria, pumice, rice hull, LECA and organic wastes or composts have all been used as the basis of some tomato production systems (Fig. 11.4). Substrates may be contained in prewrapped slabs or used to fill bag or bucket systems such as the Dutch bucket (bato bucket). Some systems incorporate sloped gutters or other collection systems under the growing containers to collect the nutrient drainage and channel this to a collection area. Hanging gutters are an increasingly popular system for greenhouse tomato production as they raise the level of the crop up above the floor for convenient access and also allow for a more accurate control of slope for rapid and efficient removal of nutrient drainage. Hydroponic tomato crops may also be grown in aeroponics, NFT and DFT systems, however these are less widely utilized under commercial production. Hydroponic tomato crops may be grown in either open or closed systems, with closed systems which collect and recirculate the nutrient solution having water- and fertilizer-use efficiencies 22% higher compared with open systems (De la Rosa-Rodriguez

Fig. 11.4. Hydroponic tomato crop grown using perlite substrate.

et al., 2020). It has also been reported that recirculation of the nutrient solution can decrease fertilizer costs by 30–40% and water requirements by 50–60% (Portree, 1996).

11.2.2 Tomato propagation

Greenhouse tomato plants are propagated from seed, however at the seedling stage plants may be grafted where this confers a production advantage. Seed is sown into a hydroponic substrate, often small rockwool or other propagation cubes, cell trays or plastic pots and maintained at the optimum temperature of 25°C (Peet and Welles, 2005), with germination occurring within 5–7 days. Once the cotyledons have fully expanded, a complete nutrient solution is applied at an EC of 1.0 mS/cm, increasing to an EC of 2.5 mS/cm in the week before planting out into the hydroponic system. The ideal transplant size is 15–16 cm tall and is as wide as it is tall (Peet and Welles, 2005). Increasing spacing and providing sufficient light assist with the prevention of elongation during the seedling stage and supplementary lighting may be provided during winter in low-light climates. Tomato seedlings may be transplanted out into the production system once they are large enough to handle; however, under commercial production, most are transplanted at the time that the first flower truss has become visible. This limits the time the plants are in the vegetative-only state in the production system. A young plant which is about to flower when transplanted will be setting fruit within 7–10 days after planting under good growing conditions.

Grafting involves the joining of a suitable fruiting variety (the scion) into a rootstock of a different type and is becoming increasingly more common for greenhouse tomato crops. Where bacterial wilt disease occurs, or when growing certain varieties with poor disease resistance, it is beneficial to graft tomatoes on to disease-resistant rootstocks. The process may be carried out by propagation nurseries or directly by the grower, and involves cutting both the rootstock (cut just

below the cotyledons) and scion seedling stems at a 45° then using a grafting clip or tube to secure the cut stem tissues together. After grafting plants require high humidity to help prevent desiccation of the scion tissue (95% relative humidity) for up to a week to facilitate healing and bonding of the graft union.

11.2.3 Tomato environmental conditions

Tomatoes are a warm-season plant with a high light requirement and respond well to CO_2 enrichment. Optimal relative humidity levels are between 60 and 75%, as high humidity promotes diseases such as *Botrytis*. Temperatures between 22 and 26°C are optimal, with a differential in day/night temperature required for balanced growth and strong flowering. Rapidly expanding tomato fruits provide a strong sink strength for photoassimilate and require high light levels to support photosynthesis. A DLI of between 22 and 30 mol/m^2 per day (Torres and Lopez, n.d.; Mattson, 2010) is required for high fruit quality, optimal yields and growth rates, and low-winter-light climates may incorporate the use of supplementary greenhouse lighting to boost the DLI. The most commonly utilized forms of supplementary lighting for hydroponic tomato crops are HPS lamps and increasingly LEDs, which are more energy efficient. LEDs, which do not produce the heat of HPS lamps, can be positioned above, within and to the side of tomato canopies, thus improving light distribution without the risk of burning foliage. Alongside supplementary lighting, CO_2 enrichment of tomato crops is widely used to improve growth and yields particularly in the late autumn, winter and early spring in climates when greenhouse ventilation may be minimized to retain heat. A CO_2 concentration of at least 1000 ppm provides the most benefit and can be supplied as a by-product of heating burners or via other methods.

11.2.4 Tomato crop training systems

Tomato plant density is somewhat dependent on factors such as cultivar, light levels and system of production, but is typically 2.5 to 3.6 plants/m^2 (Peet and Welles, 2005). Most crops are grown as a double-row arrangement and under good light and growing conditions, may allow an extra stem to develop (side shoot) to increase the plant density later in the growing season. Allowing a side shoot to develop under increasing light conditions has also been found to allow more fruit production and greater fruit uniformity on each truss (Ho, 2004). During the cropping cycle, side shoots or 'laterals' are removed weekly once long enough to handle without damaging the main stem. This maintains the indeterminate growth habit and prevents excessive numbers of stems, flowers and fruits developing. A plant which is growing vigorously under ideal conditions will produce 0.8–1 truss and 3 leaves per week with all leaves being removed from below the bottom fruit cluster (Langenhoven, 2018) (Fig. 11.5).

Long-term tomato crops largely use a system of indeterminate growth where the plant continues to grow upwards producing new foliage and fruit trusses until such time as it is 'stopped' via removal of the growing point at the top of the plant. To facilitate this continual upwards growth, a high-wire system is used which can reach up to 3.5 m above the floor. The 'lean and lower' system of indeterminate tomato plant training allows for the growing point to remain at the top of the canopy, while the stem is lowered and supported just above floor level. At the ends of rows, vines are positioned around the row end and back down the next row with care taken to prevent any stem breakages (Fig. 11.6). This brings the ripening fruit lower to allow for easier access by harvesters while maximizing light interception by foliage at the top of the plant. As plants are lowered the leaves are removed from the lower stem to allow stems to be tidily gathered and supported under the plants; this permits maximum air flow around the base of the plants to assist with humidity and disease control. Typically, between 14 and 18 leaves are left on the plant with the canopy maintained at 2–2.5 m in height (Langenhoven, 2018). Plants are supported by twine and plastic clips from the base of the plant to the top, which may, in some cases, reach a maximum canopy height of over 2.5 m. Under certain conditions semi-determinate tomato varieties may also be grown as a hydroponic crop (Fig. 11.7).

Fig. 11.5. Leaves are removed from the plant below the bottom fruit cluster.

Fig. 11.6. Layering of a tomato crop.

11.2.5 Tomato crop steering

The growth and development of greenhouse tomato crops can be manipulated in either a generative or vegetative direction or maintained as balanced growth. Crop 'steering' is carried out via manipulation of the environment and root-zone conditions. A plant with balanced vegetative versus generative growth tends to have dark green foliage, with a stem thickness of approximately 1 cm in diameter at 15 cm below the growing point, and large, strong, closely spaced flower clusters that set fruit well (Peet and Wells, 2005). A plant which is highly vegetative is characterized by strong, rapid leaf development with weaker flower trusses and slow fruit expansion, there may be a slight purplish coloration to the tops of the plant. A generative plant is one with good fruit set but smaller leaves and a thinner head which struggles to produce sufficient assimilate and suffers from carbohydrate starvation, and flowering may occur close to the top of the plant (less than 10 cm below the growing point). Young plants often tend to be overly

vegetative and older plants become more generative. Factors which promote vegetative tomato plants include genetics (some cultivars are more vegetative than others), plant age, greenhouse conditions (plastic claddings give more vegetative plants), moisture (higher moisture retention of the growing medium, lower EC and frequent irrigation favour vegetative development) and warm temperatures combined with high humidity. Factors which favour generative growth are genetics (generative cultivars), plant stress such as high light, high EC or deficit irrigation, and low humidity combined with high temperatures and good light levels. In order to steer crops in either a vegetative or generative direction EC level may be adjusted, with higher ECs promoting generative growth; deficit irrigation may be used to the same effect. Combined with moisture and EC control, the use of temperature DIFs is commonly applied for tomato plant steering. A greater DIF (night temperature much lower than day temperature) will result in a more generative plant, greater DIFs also stimulate fruit set. Lower

Fig. 11.7. Semi-determinate tomato crop.

DIFs (night and day temperature more similar) are used to stimulate vegetative growth.

11.2.6 Tomato pollination and fruit development

Under greenhouse conditions, natural fruit set in tomato crops requires some degree of pollination assistance. Portable bumblebee (*Bombus impatiens*) hives are commonly used for flower pollination where available. These contain fewer than 150 bees which prefer to work close to the nest and generally do not tend to escape through the greenhouse vents. In a greenhouse tomato crop a single bee has the ability to pollinate 450 flowers/h. Greenhouse guidelines are that one worker bee can service 40–75 m^2 which not only saves on labour requirements but also increases yield and quality (Portree, 1996). Bees are also tolerant of conditions such as low temperature, low light and high humidity and the vibrations produced by the bee are at the correct frequency to cause pollen to explode upwards. Cherry or cocktail tomatoes and other small-fruited types require two to three times more hives than beefsteak or round tomatoes due to the increased fruit numbers (Langenhoven, 2018). When bees are in use for tomato crop pollination, care must be taken with pesticides sprays and hives closed up and removed while spraying takes place. Where portable beehives are not commercially available for greenhouse pollination, other methods include the use of hand-held electric pollinators which vibrate the flower truss to release pollen as well as hand tapping or shaking of each plant stem every second day while flowers are open. Poor pollination can be a common problem in tomato crops under certain growing conditions; day temperatures that are either too high or too low (above 34°C or below 13°C) can result in pollen which becomes non-viable and sticky, preventing pollination from occurring.

After pollination and set, fruits expand rapidly and trusses may be pruned where required to obtain a suitable fruit size. The last two fruits on each truss may be removed to achieve a more uniform size, particularly with the production of cluster or 'on the truss' harvested fruit. Tomato fruit may develop a condition termed 'kinking' where the peduncles of the fruit trusses become too weak to support the weight of the fruit and bend or kink downwards (Horridge and Cockshull, 1998). Plastic truss supports including peduncle clamps may be used to prevent the reduction in fruit size which may occur due to kinking.

11.2.7 Tomato crop nutrition

Tomatoes are a much more salinity-tolerant crop than many others grown hydroponically and are less sensitive to water quality. Tomatoes can be grown with a nutrient solution containing as much as 100 ppm chloride without causing any growth problems (Langenhoven, 2018) and will also tolerate moderate levels of sodium in non-recirculating systems. Healthy, mature tomato plants have a large leaf area and can require 2–3 litres of water per plant per day under high light levels, the majority of this water (90%) is used in transpiration. EC levels for hybrid tomato crops are run at levels higher than for other fruiting plants. This is to both restrict excessive vegetative growth and to maintain fruit compositional quality. EC is adjusted for crop stage of growth, season and current environmental conditions and typically run higher under the lower light conditions of winter to assist with quality control and lower under higher temperatures to facilitate water uptake for transpiration.

In hydroponic tomato crops the requirement for potassium is almost equal to that of nitrogen in the early crop stages, from seedling through until anthesis, but after this the requirement for potassium continues to increase with fruit load, while that of nitrogen levels off. While nitrogen is required in large quantities for vegetative growth, potassium is the predominant cation in tomato fruit and has significant effects on fruit quality. The majority of the potassium absorbed during the active fruit expansion stages ends up in fruit tissue, thus higher levels of potassium must be maintained during the flowering and fruiting stages. Crops grown under higher light take up much greater quantities of potassium from a nutrient solution and it is likely that different cultivars also vary in their uptake rates. Commercial tomato growers make regular use of nutrient solution leachate analysis to determine nutrient uptake levels and ratios from their crop as it progresses through the different stages of development. This allows for changes in the nutrient formulation based on actual data on nutrient uptake, which is influenced by growing conditions, plant health, stage of development, genetics and other factors.

11.2.8 Tomato pests and diseases

Hydroponic tomato crops are prone to similar pests and diseases as many other greenhouse fruiting crops, including physiological disorders such as blossom end rot and fungal diseases like grey mould (*Botrytis cinerea*), *Pythium*, *Sclerotinia*, late blight (*Phytophthora infestans*), early blight (*Alternaria solani*), leaf moulds, powdery mildew and various wilt diseases caused by *Verticillium* and

Fusarium. Bacterial diseases of tomatoes include bacterial wilt (*Ralstonia solanacearum*), bacterial spot (*Xanthomonas* spp.) and bacterial canker (*Clavibacter michiganensis*). A number of virus diseases can be problematic in tomato crops and include tomato mosaic virus, cucumber mosaic virus, tomato spotted wilt virus, curly top and a large number of others. Pests commonly occurring in greenhouse tomato crops include greenhouse whitefly (*Trialeurodes vaporariorum*), tobacco whitefly (*Bemisia tabaci*), thrips, two-spotted mite (*Tetranychus urticae*), tomato russet mite (*Aculops lycopersici*), aphids, caterpillar larvae and the tomato psyllid (*Bactericera cockerelli*).

11.2.9 Tomato yields

Hydroponic greenhouse tomato yields vary widely depending on the climate, degree of technology in use, cultivar, growing management and other factors. Yields from high-technology tomato cropping have been steadily increasing, however many basic systems are still relatively low yielding. Langenhoven (2018) reported that potential production from an 11-month cropping period system equates to 70–90 kg/m^2 while a 7- to 8-month cropping system translates to 20–45 kg/m^2.

11.3 Hydroponic Capsicum Production

Capsicums or sweet bell peppers are a hydroponic crop which originates from Central and South America and is grown extensively as a greenhouse crop worldwide. As with tomatoes, capsicum production, which was once dominated by the large, mild, blocky, three- or four-lobed, thick-fleshed fruiting types, has now expanded to include a range of colours, fruit shapes, sizes and pungency levels. Hydroponic capsicum crops may not only include a range of different coloured blocky sweet bell peppers (red, yellow, orange, green, brown, white and even purple types) (Fig. 11.8), but also the smaller-fruited cocktail or 'snack' capsicums (Fig. 11.9),

elongated and cone-shaped types, and hot peppers of varying degrees of heat. Capsicums may be harvested green or left to colour

Fig. 11.8. Coloured sweet bell peppers.

Fig. 11.9. Cocktail or snack capsicums.

before harvesting. Coloured capsicums are sweeter in flavour and generally receive higher returns despite requiring a longer developmental time on the plant.

Most commercial greenhouse capsicum production uses hybrid cultivars which have been bred not only for yields, fruit size and type, but also multiple disease resistances and long shelf-life. Commercial cultivars are largely based on indeterminate types which continually produce new stems, leaves, flowers and fruits throughout the cropping cycle. These indeterminate cultivars require regular pruning to manage and direct growth for maximum fruit loading and productivity. Just as with tomato crops, capsicums require crop management practices that balance and maintain both vegetative and generative growth. Compared with tomatoes and cucumbers, capsicums are a crop which is slow to develop and fruit and most growers produce these on a full-year cycle. Seedlings are raised in a nursery area for the first 6 weeks and are then transplanted into the production system. It takes approximately 20 weeks from sowing seed to the first harvest (Calpas, 2020), with continuous harvesting for 8 months.

11.3.1 Capsicum propagation

Greenhouse capsicum is raised from seed of which there are approximately 120 seeds per gram for the large, blocky, bell pepper cultivars. Seed is sown directly into small propagation cubes, cell trays or pots of sterile growing medium and maintained at a temperature of 25–26°C during the germination stage (Calpas, 2020). Capsicum seedlings emerge within 7–10 days and a dilute nutrient solution at an EC of 0.5 mS/cm is applied, the temperature at this stage is also lowered to 23–24°C. After 2–3 weeks of growth, seedlings sown into small propagation cubes are then transferred into larger blocks and provided with a full-strength nutrient solution at an EC of 2.5 mS/cm under full light with a 24 h average temperature of 22–23°C. Seedlings are spaced wider as growth develops to ensure maximum light interception and prevent the development of overly tall, weak transplants. Young plants are transplanted once they reach a suitable size; this is typically after 6 weeks in the nursery area. Transplants should be 20–25 cm tall and have at least four leaves on the main stem, they may also begin to branch at this stage with roots emerging from the base of the propagation block or pot. No flower development should be visible at the transplant stage as good vegetative vigour needs to be established prior to fruiting. Flower development, particularly of the first crown flower, will inhibit vigorous vegetative shoot development in capsicums and early-set fruit should be removed until a suitable canopy size has formed.

11.3.2 Capsicum systems of production

Greenhouse capsicums are grown at a density of 3–6 plants/m², the plants are then normally trained to give two main stems per plant. In climates with low winter light levels, densities of 5–6 stems/m² are maintained, while high-light climates can support 8–12 stems/m². Maboko et al. (2012) found that a density of 3 plants/m², each pruned to 4 stems/plant, resulted in optimal yields under good light conditions.

Hydroponic production systems for greenhouse capsicum are much the same as those used for other longer-term fruiting crops such as tomato, cucumber and aubergine. Rockwool or coconut fibre slabs, substrate-filled bags, Dutch buckets and similar containers may be used. Capsicums have been grown commercially in DFT and NFT, however these are not widely used systems of production. Commercial capsicum production systems may be open or closed with an increasing use of nutrient recirculation due to economic and environmental concerns over 'drain to waste' open systems. Daytime production temperatures of capsicum are optimal within the 21–24°C range while the minimum temperature threshold for growth is 8–12°C (Ropokis et al., 2019).

11.3.3 Capsicum pollination, fruit set and fruit development

Capsicum flowers are largely self-pollinating, although some cross-pollination does occur, and fruit set can be optimized with the use of electric pollination or portable bumblebee hives. Pollen can remain viable for as long as 8–10 days when temperatures are between 20 and 22°C but can rapidly deteriorate at higher and lower values (Morgan and Lennard, 2000). Capsicum also has capacity to set fruit parthenocarpically (without fertilization), particularly under low temperature conditions (night temperatures of 12–15°C). Failure to set fruit, which can be a common problem in capsicum crops under certain conditions, is partly due to the formation of abnormal or non-viable pollen. Insufficient levels of pollination result in undersized and misshapen fruit and the use of bumblebees can increase fruit weight, width, volume, seed weight and speed of ripening.

Temperature has been found to be the prime factor affecting fruit set with night temperatures being the most critical. The optimum night temperatures are 18–27°C, with little fruit set occurring above 32°C. Conditions after fruit set play a significant role in fruit development. As fruit number per plant increases, the average size per fruit tends to drop with the heavier fruit loading; however, it is possible to restrict fruit set to allow the plant to distribute reserves to a lesser number of fruits which will attain a greater size.

11.3.4 Capsicum training

After transplanting, young plants naturally begin to fork or branch with two or three stems formed; at this point plants are pruned to leave the two strongest stems. Most commercial greenhouse capsicum crops are trained using the 'V' system with two stems per plants selected and maintained with string twisted around each stem and hung from an overhead support (Fig. 11.10). Pepper stems are woody and brittle

and cannot be layered in the way that long-term tomato crops are, thus the stem requires sufficient support to prevent breakages under heavy fruit loading. As each stem develops, flowers are formed in alternate leaf axils. Immediately above each flower bud a lateral shoot forms and these are removed shortly after development to maintain the vigour of the main shoot apex; alternatively, they may be retained under conditions of high vegetative vigour to support higher fruit yields. Another factor to consider with training of capsicum crops is light availability to the area where fruit growth and development are occurring. Leaves supporting developing fruit need to maintain high levels of photosynthesis and should be adequately exposed to sufficient light; however, during periods of high solar radiation fruit require shading by foliage to avoid sunburn.

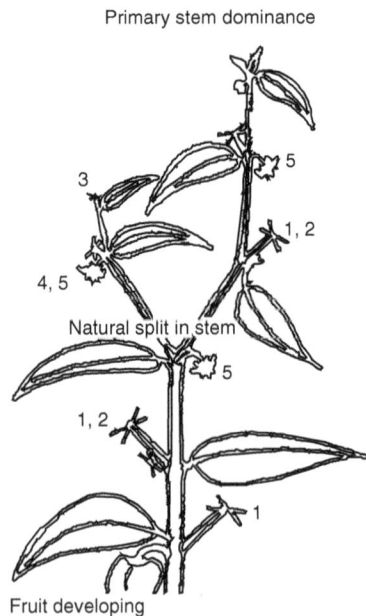

Fig. 11.10. Capsicum training system: 1, lateral stopped to maintain a single stem; 2, flower removed to maintain source strength; 3, stem stopped one leaf past developing fruit; 4, additional flower pollinated to increase fruit load; 5, pollinated flower.

11.3.5 Capsicum crop nutrition

Seedlings are geminated without any nutrient solution irrigation; however, once the cotyledons have expanded, a complete, diluted vegetative nutrient solution is applied which contains equal amounts of nitrogen and potassium. The ratio of nitrogen to potassium during crop growth is determined by factors such as crop stage of development, light levels and other climatic factors. Moderate levels of phosphate are important for flowering and utilization of sugars, while magnesium is required for chlorophyll formation and efficient photosynthesis. Magnesium availability can be limited by excessive applications of potassium, so when increasing potassium levels during fruiting it is often necessary to increase magnesium proportionally (Morgan and Lennard, 2000). Both supply and uptake of calcium are necessary during all stages of crop development and are required for high-quality fruit and maintenance of a vigorous terminal shoot. Capsicum is prone to fruit disorders such as silvering and blossom end rot, which are attributed to a lack of calcium or interruption in the calcium supply or translocation within the plant. Nitrogen levels affect fruit number and fruit weight and thus play an important role in yield determination. With capsicum grown in rockwool at a range of nitrogen levels in solution, Schon et al. (1994) found that plants which received 175 mg N/l produced significantly more fruit than plants receiving lower nitrogen levels. During the fruit development phase, the levels of potassium and phosphate are increased. The nitrogen to potassium ratio may also be adjusted for season with nitrogen levels slightly increased to accommodate higher photosynthetic activity under high light levels, while potassium and nutrient solution EC are increased under lower-light, winter conditions to maintain fruit quality. Potassium levels of 7 mM in the nutrient solution have been found to give maximum yields; however, it is possible to improve certain bioactive compounds of pepper fruits with a higher application of potassium without reducing yield (Botella et al., 2017).

EC levels of the nutrient solution within the root zone can have a significant effect on capsicum yield, growth and fruit quality. High EC levels can reduce yield, fruit size and dry weight and decrease vegetative growth while increasing the occurrence of blossom end rot (Tadesse et al., 1999). Low EC, however, can result in poor-quality fruit and weak, elongated vegetative growth. EC levels are typically run within the range of 2.5–3.5 mS/cm in the root zone for plants once past the seedling stage.

11.3.6 Capsicum pests, diseases and physiological disorders

Capsicum plants are prone to similar pests and diseases as many other fruiting crops such as tomato, cucumber and aubergine. These include greenhouse whitefly (T. vaporariorum), tobacco whitefly (B. tabaci), thrips, two-spotted mite (T. urticae), aphids and caterpillar larvae. Common capsicum diseases are often those associated with high humidity such as B. cinerea and Phytophthora capsici which cause seedling damping off, root rots, stem cankers and lesions. Sclerotinia, Anthracnose, Alternaria, and Fusarium and Verticillium wilts are also potential pathogens of hydroponic capsicum crops. Capsicum crops, like tomato, are prone to being infected by viruses for which aphids are the main vectors. Viruses include tobacco mosaic, tomato spotted wilt virus and cucumber/pepper mosaic virus.

Capsicum is prone to a number of physiological disorders which affect flowering, fruit set, vegetative growth and fruit quality. Flower, flower bud and fruitlet abscission are common problems with larger-fruited capsicum cultivars, particularly those grown in warmer climates. High temperature, low light levels, plant stress, the presence of rapidly growing fruit on the plant and certain pests and diseases all predispose plants to these problems. Fruit quality issues include blossom end rot, a calcium-related disorder, and fruit cracking, silvering and sunscald. Misshapen fruit is another disorder which is common under unfavourable growing conditions. Low temperatures during fruit set

(12–15°C at night) and the absence of seed in fruit result in deformed and often under-sized fruit.

11.3.7 Capsicum harvesting and yields

Capsicum requires between 7 and 9 weeks from fruit set to harvest, however this may be extended during winter. Fruit is harvested when it has obtained at least 80% colour and is cut from the plant with a sharp knife directly through the peduncle. Fruits are graded according to size, with those which are blemished or misshapen classified as reject grade. Yields of capsicum, which is a slow crop to establish, vary considerably between growers based on the level of technology, climate and management. Most crops of coloured capsicums yield between 22 and 32 kg/m^2, with new growers targeting around 23 kg/m^2 (Calpas, 2020) and high-technology year-round crops reaching yields of at least 30 kg/m^2 and often higher.

11.4 Hydroponic Cucumber Production

Hydroponic cucumbers are a rapid-growing, high-yielding crop which require warm growing conditions and high light levels. Traditionally, it has been the 200–250 mm long, thin-skinned, seedless, dark green 'European/Continental or Telegraph' cucumber grown hydroponically in greenhouses worldwide; however, there is a range of different cucumber types in production. These include snack or cocktail 'mini' cucumbers with fruit harvested at 75–100 mm, these are spineless, seedless and several fruits are produced at each node; and the 'Beit-Alpha' types which produce 125–150 mm long, seedless fruits (Fig. 11.11). Both the standard European and mini cucumber cultivars are parthenocarpic and set fruit without pollination. Under greenhouse production it is largely the gynoecious cultivars which are grown, these produce only female flowers so that no pollen is present in the

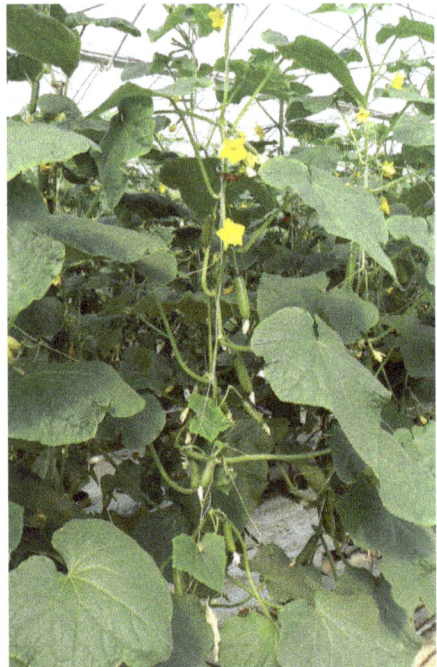

Fig. 11.11. Mini cucumber fruit production.

crop to cause seed to form in the fruit. If pollination does inadvertently occur, fruit will become distorted and bitter flavoured due to the formation of seeds. For this reason, it is essential that pollen carried by bees and other flying pollinators does not enter the greenhouse from outdoors or surrounding field plantings of cucumber crops. Cucumber varieties should be selected which have genetic resistances to powdery mildew, leaf mould and cucumber mosaic virus as well as suitable fruit size and quality characteristics.

11.4.1 Cucumber propagation and production

Seed germinates rapidly, within 2–3 days, at 29°C (Hochmuth, 2012) in free-draining, sterile propagation media and young plants are ready for establishment into the hydroponic system within 2 weeks. Commonly used propagation materials include rockwool cubes, foam blocks, soilless peat or coconut fibre, and perlite substrates in small pots. Once seed has geminated temperatures are lowered to 25–26°C and a diluted, balanced vegetative nutrient solution applied. Once seedlings have developed three or four true leaves, they are transplanted into the production system, and at this stage roots should be seen appearing from the base of the seedling container or cube. Large-fruited cucumbers are grown at densities of 2 plants/m^2 under lower-light winter conditions and 2.5–3 plants/m^2 under high-light, summer conditions. Smaller-fruited types can be grown at densities of 3–4 plants/m^2.

11.4.2 Cucumber environmental conditions

Cucumbers are well suited to warm climates and grow rapidly if provided with sufficient water and nutrients. Drip-irrigated substrate systems are commonly used for production with plants grown in single rows at 30 cm spacing (Fig. 11.12). A wide range of substrates may be used for hydroponic cucumber crops including rockwool, perlite, sawdust, peat, bark, coconut fibre and organic mixes. Gutter systems, slabs, bags, Dutch buckets and bed systems have all been used successfully for cucumber production; growers may produce several successive crops grown in the same substrate. While some cucumber crops are produced in solution culture systems such as NFT or DFT, the large and extensive nature of the root system and extremely high demand for oxygen under warm growing conditions mean that continuous aeration of the nutrient solution is required to maintain a healthy root zone (Hochmuth, 2012).

Cucumbers require warm conditions, with air temperatures in the range of 24–30°C and with a minimum night temperature of 18°C. High light levels are required for good yields and supplementary lighting is beneficial in low-light winter climates. CO_2 enrichment is also highly beneficial for cucumber crops up to a level of 1000 ppm, this increases productivity and is most efficient when used during conditions when

Fig. 11.12. Cucumber production in a drip-irrigated system.

ventilation is reduced. While cucumbers are a warm-season plant, high temperatures in the root zone can have adverse effects on plant growth processes and reduce uptake of water and nutrients. Studies have found that root cooling of hydroponic cucumber crops to 22–25°C as compared with an ambient root-zone temperature of 33°C increased productivity by 71.4 to 74.3% in a recirculating hydroponic system (Al-Rawahy *et al.*, 2019).

11.4.3 Cucumber training and support systems

Cucumbers produce indeterminate growth in a similar way to greenhouse tomatoes, with new foliage, flowers and fruits formed continually. Growth is rapid under ideal temperature and light conditions and plants require regular pruning and training to maintain suitable canopy conditions. The main training method used for greenhouse cucumbers is the 'umbrella' or cordon system. This allows vegetative growth to form as the lower, main stem portion and early flowers or fruitlets are removed during this stage. This allows the plant to produce the vigorous early vegetative growth that is required for maximum fruit production (Hochmuth, 2012). The upper portion of the main stem is then permitted to set fruit in every second axil. When stem growth reaches the top wire, the plants are 'stopped' by removal of the growing point and two sides shoots are permitted to form and continue growth over the wire and back towards the ground. As much fruit as the plant can maintain is permitted to develop on the side shoots (Fig. 11.13).

A second common method of greenhouse cucumber training is the 'V-cordon' where single rows are spaced at 1.3 m with an in-row spacing of 30 cm. Two overhead wires are spaced approximately 70–80 cm apart from each other and plants are trained up strings which are alternately tied to each wire. This means plants are trained at an angle away from the row centre and form a V arrangement down the row. Fruit are pruned in the same way as in the umbrella system with all lateral branches removed as they appear

on the main stem until this reaches the overhead wire. Fruit develops at each node and fruit pruning may be required for large-fruited cultivars to restrict this to one fruit per node; however, vigorous plants may be able to carry two fruits per node. The smaller, snack or Beit-Alpha types are not pruned to a single fruit and will produce several fruits per node (Hochmuth, 2012). Any misshapen fruit should be removed as these will not form a marketable product.

11.4.4 Cucumber crop nutrition

Hydroponic, greenhouse cucumbers are a high-nutrient-demanding crop; however, EC levels are run much lower than for tomatoes because cucumbers are not salinity tolerant. EC levels of 2.2–2.5 mS/cm and pH of 5.6–6.0 are commonly applied and the crop also has a high transpirational water demand under optimal growing conditions. As with tomatoes and other fruiting hydroponic crops, cucumbers require adjustment of the nitrogen to potassium ratio as the plant moves from vegetative to fruiting phases. An N:K ratio of 1:1.5 is suitable for fruiting; however, growers often monitor nitrogen and potassium levels in the nutrient solution leachate draining from the growing system and adjust this ratio based on analysis data and fruit loading levels.

11.4.5 Cucumber harvesting and yields

Cucumbers are a shorter-term crop than tomatoes or capsicums and are usually harvested for a period of 12–14 weeks before crop replacement; however, this varies with cucumber type, cultivar, environment and management. Large European cucumber types are harvested based on size and when the fruit has created a uniform diameter throughout its length. Over-mature fruits start to develop yellowing at the blossom end and a loss in dark green coloration and should not be marketed. Harvesting should be carried out every 2–3 days. During harvest, fruits are cut from the vine and carefully

Stem 'stopped' at top wire
Two laterals continue to hang down from top wire

Fruit carried singly on
every axil of secondary stems

Fruit carried on every
second axil of primary stem

Fruit and laterals removed from lower 50–80 cm

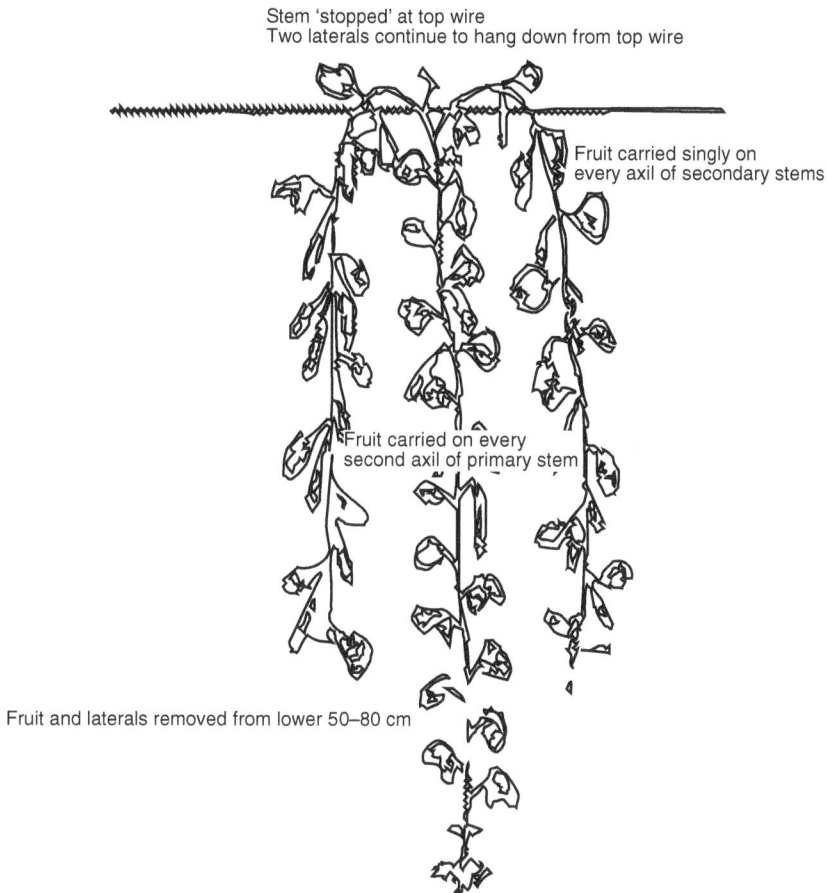

Fig. 11.13. Cucumber training system.

handled to avoid bruising or damage as these are the main factors which reduce shelf-life. Fruits may be graded for size and European cucumbers are shrink wrapped in plastic film after harvest to retain moisture and firmness (Fig. 11.14). Smaller, snack and other cucumber types are often packaged into clamshells, bags, or overwrapped trays with branding to protect the fruit and maintain fruit quality. All cucumber types are at risk of moisture loss postharvest which results in a loss of texture and turgidity; optimum storage temperatures are between 10 and 15°C with 80–90% humidity. Refrigeration should be avoided as this can induce chilling injury which reduces shelf-life. Chilling injury is

characterized by pitting of the fruit surface and a rapid loss of green colour; fungal decay is also common following chilling injury.

Average yields of hydroponic greenhouse cucumbers are in the range 50–65 kg/m² per year (Castilla, 2013), but vary depending on cropping schedule, climate, variety grown and level of technology.

11.4.6 Cucumber pests, diseases and physiological disorders

Cucumber crops are prone to similar pests and diseases as many other fruiting crops. These include greenhouse whitefly (*T. vaporariorum*),

Fig. 11.14. Cucumbers may be shrink wrapped to retain moisture and firmness.

tobacco whitefly (*B. tabaci*), greenhouse thrips (*Heliothrips haemorrhoidalis*), two-spotted mite (*T. urticae),* aphids and caterpillar larvae. Of these, the two-spotted mite is the most damaging and difficult to control and these thrive in the warm environment of a cucumber greenhouse. Mite damage symptoms include mottled, chlorotic leaves resulting from mite feeding damage, followed by the development of fine webbing and this is often initially misdiagnosed as nutrient deficiency or toxicity damage. Common diseases include powdery mildew and viruses, including cucumber mosaic virus which is transmitted largely by the feeding of aphids, with control and prevention of entry of these pests being important steps for virus control. Selecting varieties with resistance to common disease issues largely assists with control.

Cucumber fruits are prone to a number of physiological disorders which downgrade fruit quality. One of the most common is 'crooking' which causes curvature of the fruit, begins at an early stage and remains until maturity. Crooked fruit should be pruned from the plant as soon as they appear as excessive curvature results in unmarketable fruit. Crooking can have a number of causes including impedance of the developing fruit by a leaf or stem, insect feeding on one side of the young fruit or environmental issues such as adverse temperature or nutritional problems. Fruitlet abscission is another common issue and occurs when a high level of fruit loading develops while there is insufficient vegetative growth to maintain the crop load. Sudden changes in temperature or other environmental factors can also result in fruit abscission.

11.5   Lettuce and Other Salad Greens

Hydroponic salad crops are divided into three main types. The first is lettuce, of which there is a wide range of fancy types including butterhead or bib, oak leaf, romaine/cos, red and green coral/frill types ('Lolla Rossa'), 'Salanova®' and other open-leaf types, as well as iceberg or crisphead types. The second type is other salad greens, incorporating endive, watercress, rocket (or arugula), Asian greens such as mitzuna, mustard, spinach, corn salad (or mâché), chard, cress, chicory, kale and micro greens, which are young seedlings of a range of lettuce, herb and other species (Figs 11.15 and Fig. 11.16). Finally, herbs of many varieties are grown hydroponically, the most popular

Fig. 11.15. Hydroponic pak choi produced in a deep flow system.

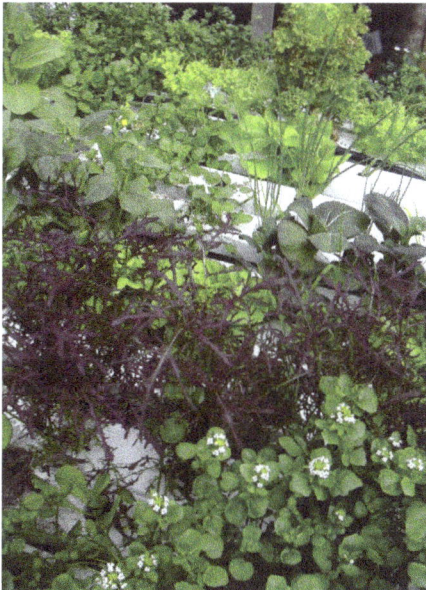

Fig. 11.16. A mixed system of salad greens grown in NFT.

under commercial production being basil and coriander (or cilantro), however this varies by country and local demand.

A wide range of hydroponic systems is suitable for lettuce, herb and salad greens production, with solution culture methods such as NFT and raft/float systems dominating commercial soilless production. Drip-irrigated or tray-based systems may also be used for production of these crops. Substrate and raft culture systems have the advantage that, unlike NFT, moisture is retained around the root systems should an electricity failure or a system breakdown occur. NFT and other solution-grown salad and herb plants have the advantage that the entire plants can be harvested with the root system still attached and marketed in this way. Bundling up the damp roots in plastic and securing with a rubber band prolongs the shelf-life of the plant as compared with cutting at the base. The roots continue to take up moisture and keep the plant turgid after harvest; however, this is not a complete substitute for keeping salad produce under refrigeration postharvest. Potted herbs are also produced hydroponically, often in ebb and flow or modified NFT systems which supply nutrient solution to the base of small substrate-filled pots which are densely sown with herb seed to give a leafy, compact, marketable product.

11.5.1 Lettuce propagation and production

Selection of lettuce cultivars is just as important as with the fruiting plants grown hydroponically and improved lines are continually being bred. European butterhead or bib types such as 'Rex' and 'Elvis' have dominated the butterhead lettuce types for many years and are considered industry standards, while larger-headed cultivars are preferred in some markets. There are many different fancy lettuce cultivars, including heirloom types that often do not have the disease resistance, high degree of coloration or resistance to physiological disorders of the more recently bred, modern cultivars specifically developed for greenhouse production. Lettuce cultivars can be selected based on the local growing climate, varieties suited to cool and low-light winter conditions or warmer-season cropping are available from commercial seed suppliers. During warmer conditions it is advisable to select red varieties which will maintain their colour. Disease resistance is also important when selecting lettuce cultivars: resistances to downy mildew races and to lettuce mosaic virus and the lettuce aphid (*Nasonovia ribisnigri*) are all advantages in some situations. Cultivars which are tolerant to bolting (premature flowering-head formation) and tipburn in spring and summer can be selected for growing in conditions which favour these disorders.

Lettuce seed, like many of the salad greens and herbs, prefers cooler conditions for germination and growth than other common greenhouse crops such as tomatoes and cucumbers. The ideal temperature for lettuce seed germination is 18°C, with the optimal range being 16–18°C (Morgan, 2012). Above 25–26°C germination may be inhibited and the seed becomes dormant, a condition known as 'thermal dormancy'. Growers in tropical climates often chose to germinate lettuce seed inside temperature-controlled facilities at 18°C and transport these into the greenhouse once germination has occurred. Raw lettuce seed may be seeded into the propagation material; however, there is widespread use of pelleted seed which not only facilitates hand sowing, but also is suited to the use of mechanized seeder equipment.

Hydroponic lettuce, herb and many salad greens are propagated in small pots, tubes, containers or blocks of inert substrate such as rockwool or various foam media. Once seedlings have developed the first two leaves, light levels are increased and a dilute nutrient solution applied to provide a suitable level of nutrition at an EC of 0.5–0.6 mS/cm. The EC is then increased to 1.0 mS/cm after the first week of growth until the plants are established into finishing systems. Once roots are seen emerging from the base of the propagation cube or substrate, seedlings are transferred to a production system where light and temperature largely determine the rate of lettuce growth and development. Plant spacing is highly dependent on the type of lettuce being grown with smaller fancy types grown at a higher density than iceberg or crisphead types. A density of 16–25 plants/m^2 is common for most cultivars. Harvest occurs within 3–5 weeks after transplanting, depending on final harvest weight required and growing conditions. Some head lettuce is harvested while relatively young and low in weight at less than 120 g; other markets prefer fully sized, large heads of up to 350 g and these require longer production times and greater plant spacing (Fig. 11.17).

11.5.2 Lettuce environmental conditions

Temperature optimums for greenhouse lettuce production are in the range of 16–22°C, with conditions above 28°C increasing the occurrence of bolting, tipburn, wilting and low plant weight in many fancy lettuce types. Lettuce production in warmer and tropical climates often uses nutrient solution chilling to cool the root zone of the plant to allow acceptable levels of production in conditions which would otherwise restrict growth. Most types of lettuce have a minimum DLI requirement of at least 14–15 mol/m^2 per day (Brechner and Both, 2013); however, densely grown crops and crisphead lettuce benefit from at least 17–18 mol/m^2 per day (Fig. 11.18). Gagne (2019) found that baby leaf vegetables such as lettuce and *Brassica* species were positively affected by increasing DLIs from 6 to 24 mol/m^2 per day, however declining biomass was found with light levels above 24 mol/m^2 per day. Humidity levels for lettuce growth are also important as high humidity can predispose plants to tipburn due to the restriction of transpiration from the foliage that is required to carry calcium out to the leaf tips.

11.5.3 Lettuce crop nutrition

Nutrient solutions for hydroponic lettuce production can vary depending on type and cultivars grown, climate, water supply and production system. Each variety and type may have slightly different nutrient requirements depending on growth rate and genetic

Fig. 11.17. Head size of lettuce is based on market requirements.

Fig. 11.18. Supplementary lighting can be used to increase the DLI for lettuce production.

differences in nutrient ion uptake. Recirculating or closed production systems in particular need regular monitoring and solution analysis to determine how the balance of nutrients may change over time and prevent any deficiencies or accumulation of unwanted elements from occurring. Summer formulations for lettuce typically contain higher nitrogen levels, while winter crops grown under lower light should be provided with a formulation higher in potassium. Lettuce produced in climates which experience high light intensities require higher iron levels than in lower-light regions and growers should avoid using formulae created for low-light/temperature climates and applying these in tropical or subtropical areas (Morgan, 2012). Control over EC, pH and nutrient balance as well as a suitable flow rate are also vital for lettuce production. EC levels depend on the cultivar being grown and the season, with EC levels of 1.8–2.4 mS/cm maintained in winter and 1.4–1.6 mS/cm in summer to assist with rapid uptake of water and tipburn prevention.

In solution culture systems such as NFT and DFT, successful lettuce production requires maintenance of oxygen levels in the root zone (Fig. 11.19). As the temperature of the nutrient solution heats up, less oxygen is held for plant uptake. Reducing the temperature of the solution allows higher oxygen content and improved production in hot/humid climates.

11.5.4   Lettuce pests, diseases and physiological disorders

Lettuce is a quick-turnover crop which is less prone to some of the pests which can plague longer-term greenhouse crops. However, various species of aphids are common at certain times of year and can cause considerable crop damage as well as contamination of the harvested product. Other lettuce pests include the two-spotted mite (*T. urticae*), greenhouse whitefly (*T. vaporariorum*) and various species of thrips, fungus gnat larvae, leaf miners and caterpillar larvae. Less common pests of hydroponic lettuce crops

include nematodes, beetles, and slugs and snails. Common lettuce diseases are similar to many other greenhouse crops and include *B. cinerea* (grey mould) under cool, humid conditions, powdery and downy mildews, *Sclerotinia*, *Anthracnose*, *Alternaria* leaf spot, *Erwinia*, bacterial leaf spot diseases and root rots caused by *Pythium*, *Phytophthora* and *Rhizoctonia*. Lettuce may also become infected with viruses, often spread by insect pest feeding, including lettuce mosaic virus and cucumber mosaic virus.

Physiological disorders of lettuce include premature bolting, tipburn, glassiness, lack of head formation, salinity damage, colour disorders and the development of bitter flavour compounds. One of the most common physiological disorders of lettuce under warm/humid conditions is 'bolting', a premature formation of the flowering stalk. Lettuce which are bolting develop a tall, elongated growth form and if severe, cannot be marketed. Bolting is a physiological disorder

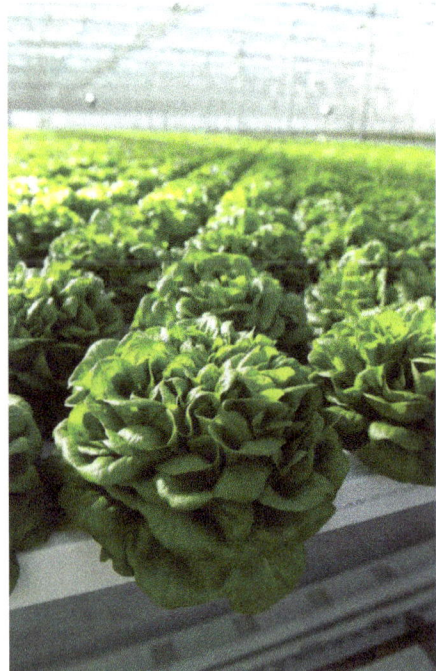

Fig. 11.19. Successful lettuce production requires the maintenance of oxygen levels in the root zone.

with a genetic component and is caused by a number of issues including excessive temperature, particularly root-zone temperature, lack of a differential in day/night temperature, long day length, moisture stress and high EC, root issues and disease, very low light intensity and over-maturity. Bolting can be reduced or eliminated with the use of nutrient solution cooling, which also benefits growth and development. Nutrient solution chilling to 18°C has been shown to dramatically reduce bolting when control treatment temperatures were 28–39°C by day (Jie and Sing Kong, 1998).

11.5.5 Lettuce harvesting and handling

Lettuce should be harvested as soon as marketable weight has been reached, however young plants can be cut for incorporation into baby leaf or salad mixes. Whole lettuce heads should be cut cleanly at the base with a sharp knife as early in the morning as possible to retain turgor and prolong shelf-life. Growers harvesting lettuce or herb products with intact roots may trim these to fit inside packaging if required, roots are then rolled up and placed into the base of a produce bag or clamshell container and tied off to retain moisture. At harvest, the oldest, outer leaves are often removed from heads before packaging into plastic sleeves, clamshells or other bulk boxes. For the production of salad mixes, heads are cut to release individual leaves which are washed, rinsed, dried and processed into pillow packs or larger bulk containers. Lettuce requires rapid chilling for the removal of field heat, and this is essential for maintaining shelf-life and quality postharvest. Lettuce should be chilled to within 1–2°C within 30 min of harvest for maximum product quality.

Lettuce, salad greens and herbs are prone to a number of postharvest issues including fungal and bacterial pathogens such as *Botrytis*, *Erwinia* and others which can cause wet rots and tissue breakdown, particularly under high humidity and where spores were present in the greenhouse. Wilting and loss of turgor can be caused by incorrect temperatures, low humidity and a lack of chilling postharvest. Leaf yellowing due to a breakdown of chlorophyll is often caused by incorrect storage temperatures, insufficient chilling and possible nutrient deficiencies during production.

11.6 Production of Hydroponic Micro Greens

While head lettuce and salad mixes have dominated the market for greens, in more recent years the production of 'micro greens' and 'micro herbs' has become a profitable crop, particularly for smaller growers and those using indoor production facilities. Micro greens are larger than a sprout but smaller than a baby salad leaf and will usually have produced at least two true leaves after expansion of the seedling leaves or cotyledons (Fig. 11.20). Because they are harvested at such an immature stage, seed is sown at a high density to maximize yields from each crop; this also allows the developing seedlings to grow tall and straight with a

Fig. 11.20. Micro greens are larger than a sprout but smaller than a baby salad leaf.

tender, almost blanched stem and bright, well-developed leaves. Micro greens have a short shelf-life which makes them a good prospect for local markets and restaurants as they are best used fresh within 2–3 days of harvest. Micro greens are used as toppings, garnishes and flavourings in salads and feature in many up-market dishes as well as being sold as a high-value product in produce stores and supermarkets.

Micro greens fall into four main categories. The first is the 'shoots and tendrils' such as pea, sunflower and corn shoots which are often used as garnishes, although they all have their own mild flavour. The second is the 'spicy greens' which include rocket, radish, cress and mustards. 'Micro herbs', the third category, include those used not only as garnishes, but also for their characteristic flavour such as parsley, fennel, edible chrysanthemums, coriander, basil, French sorrel, mint, dill, chives and onion, and shisho (*Perilla*). Finally, there are the 'tender greens' which are highly diverse in flavour, leaf size, shape and colour; these include red cabbage, broccoli, spinach, beet (red), tatsoi, mitzuna, amaranth, chard, kale, corn salad, endive, chicory, celery, carrot and lettuce.

Hydroponic micro greens have a distinct advantage over those grown in trays of substrates or soil mixes in that no granular growing medium needs to be used. The very high sowing rate and density of micro greens mean that small particles of substrate can end up in the foliage and since micro greens are not usually washed after harvest, this poses a risk of grit ending up in the final dish. For this reason, hydroponic micro greens are best produced on a thin mat or capillary pad which holds the seed in place and retains some moisture for germination (Fig. 11.21). Paper towel, hessian/burlap sheets, rockwool cubes or sheets, thin kitchen cloth and hydroponic 'micro green pads' can all be used to grow a clean, high-quality crop with little expense.

Seed for micro green production should be obtained which has been specifically produced and packaged for sprout or micro green production; this means the seed will have a low percentage of 'foreign matter', will not have been treated with fungicides or other chemicals and will have been well cleaned. This is particularly important when buying seed for pea, corn or spinach micro greens as the seeds of these species are often coated with fungicide. Seed companies have also now introduced a range of specific micro green cultivars which are an improvement on standard varieties. Many of these feature intensely coloured or modified first leaves, such as some of the radish micro green species and

Fig. 11.21. Hydroponic micro green production.

those grown for pea shoots. Some micro green varieties have seed which is 'mucilaginous' meaning that once wetted the seed forms a thick, gelatine-like layer which holds moisture. Cress and basil are examples of mucilaginous seed and these seed types should not be pre-soaked before sowing. Larger seeds such as wheatgrass, corn and peas may be pre-soaked in warm water for 24 h before sowing, although this step is not essential.

The micro green seed should be weighed out and sown on to a wetted surface as evenly as possible. Use of seed shakers assists with this process. The correct seeding density depends somewhat on the species being grown; however, an approximate guide to seed sowing and yield rates for the commercial production of hydroponic micro greens is given in Table 11.1.

Once the cotyledons (seedling leaves) are visible and are starting to develop chlorophyll the seedling will have exhausted the reserves contained in the seed. At this stage, the young plant is starting to photosynthesize and produce its own assimilate and nutrient ions will be absorbed by the root system. A general-purpose vegetative or seedling nutrient formulation is usually sufficient for micro green production. EC levels are typically run at seedling strength for micro greens (0.5–1.0 mS/cm), although these may be adjusted for season in a similar way to lettuce and herb crops. Nutrient solution needs to be applied regularly and carefully to developing micro greens; this is both to avoid flooding the micro greens and wetting the foliage which encourages fungal diseases, and to ensure fresh nutrient solution is flushed through the root zone, oxygenating and feeding the seedlings. Most hydroponic systems used for micro greens do not use a continual flow of nutrient, but intermittent application followed by a period of drainage with the growing mat/pad or substrate holding sufficient moisture around the roots between irrigations.

11.6.1 Harvesting micro greens

Cutting height is important as high-quality micro greens need a good, clean portion of stem below the leaves, but should not be cut so low as to risk contamination with the growing medium or material the seeds were sown into. This becomes particularly important if a light, loose, granular medium has been used for production as particles can easily be picked up during the harvesting process and contaminate the product. Clean, sharp scissors are suitable for cutting micro greens on a small scale, while larger growers use mechanical harvesters. During warm growing weather, micro greens, just as with herbs and lettuce, are best harvested early in the day when the foliage is cool and most turgid. This will prolong the shelf-life of the packaged product. Some micro greens are shipped out to customers while still growing in trays or cells, prolonging shelf-life and allowing onsite harvesting as required.

11.7 Hydroponic Strawberry Production

Strawberry plants yield and produce well under hydroponic cultivation and the protected environment of a heated greenhouse

Table 11.1. Micro green seed sowing and yield rates. (From Morgan, 2012.)

Seed sowing rates:
- 230 g seed/m^2 for rocket (arugula)
- 460 g seed/m^2 for brassicas[a]
- 3376 g soaked seed/m^2 for wheatgrass

Yields vary for species[b]:
- Rocket can produce 1500 g/m^2 within 3 weeks of sowing
- Brassicas can produce 2000 g/m^2 within 3 weeks at the sowing densities given above
- Many of the smaller, lighter herbs give much lower yields per square metre

[a]Brassicas include red cabbage, broccoli and similar species.
[b]Note that time to harvest is highly dependent on growing conditions such as temperature, light, humidity and nutrition.

allows out-of-season or year-round production in many climates. Many European countries such as Germany, Belgium, Poland, the UK and Spain have been producing large volumes of greenhouse strawberry fruit for decades in climates which prevent outdoor production for much of the year. Strawberry producers in other countries may use greenhouse structures to either extend the cropping season or produce out-of-season fruit when returns are highest, or to avoid some of the production issues that soil-based systems face.

Strawberries are small, but highly productive plants which can be grown as either an annual or longer-term crop and are thus well suited to hydroponic production. Modern cultivars produce their highest yields of good-quality fruit in the first season, making annual cropping systems a more viable option for many growers with crops typically grown for between 5 and 10 months before replacement. Commercially there are two main groups of strawberry plants: the day-neutral and the short-day plants. Day-neutral plants are induced to flower and fruit when temperatures are high enough to maintain growth irrespective of day length. With short-day varieties, floral initiation is triggered by short day lengths (less than 12 h) or by cool conditions. Refrigeration can be given to planting stock to provide the chilling period required for flowering.

11.7.1 Strawberry propagation

Unlike many other hydroponic crops, strawberries are not raised from seed but vegetatively propagated from 'runners' which form on mature plants towards the end of the growing season. Runners from field-grown crops are harvested in autumn or winter and stored bare rooted before being used to establish new crops. These field-grown runners, once washed and given a fungicide application, can be used as planting stock for hydroponic crops; however, there is still a risk of the carryover of crown and root rot diseases using this method. A preferable type of planting material is tip cuttings of

strawberry runners which have been raised as plugs in soilless growing medium, thus have not been in contact with soil. However, supply of this type of soilless planting stock is not always available in many regions and some growers choose to propagate planting stock themselves. Runner plants or runner tip cuttings used for strawberry crop establishment can be stored under refrigeration for an extended period of time, this process also providing the chilling period which is required for floral initiation.

For greenhouse production the largest fruit and highest yields are obtained from runners or plants with a high crown weight and large size. Strawberry propagators in many regions of the world grade their runners or freshly dug plants into size grades, with A and A+ grades being of the largest size. Commercial growers prefer to use larger-grade plug cell transplanting as planting stock as this ensures the crown has fully developed a new root system before transferring to the hydroponic system and has a good store of assimilate for new growth and early fruit development.

11.7.2 Strawberry production systems

Hydroponic strawberries are grown in a diverse range of soilless systems, including solution culture methods such as NFT; however, substrate culture dominates commercial production as this allows support of the plant grown at the correct depth and control over root-zone moisture levels. Strawberry plants are prone to infection by root and crown rot pathogens, particularly in the early establishment phase, and thus control over irrigation is vital for success of this crop. Commonly utilized substrates include rockwool (both slabs and granulated), coco fibre, peat, perlite, pumice, bark, LECA, untreated sawdust, rice hull and organic mixes.

Systems of production for strawberries include standard single-plane cropping, hanging gutters, vertical or 'stacked' pots, columns or staggered systems. Vertical, hanging, column or stacked pot systems of strawberry production aim to utilize greenhouse space efficiently by increasing the plant density as compared with single-plane cropping. While this type of system may seem attractive with greater plant numbers and higher yield potentials, considerable shading of plants on the lower levels of these systems occurs and this method is suited to either high-light cropping areas or where supplementary lighting may be used. Single-level or staggered-level troughs (Fig. 11.22) allow for maximum light interception by all plants and consist of single or multiple levels of troughs filled with substrate or supporting slabs or growing medium such as rockwool or coconut fibre. These are positioned at a convenient height for plant maintenance and harvesting and facilitate air movement up and under the canopy for disease prevention. The nutrient solution is drip irrigated along the rows of troughs and drains from each via a collection channel. Plants are established into the tops or sides of troughs and after developing sufficient foliage, fruit trusses are trained over the edge of each trough so that they are held clear of the foliage and growing substrate.

11.7.3 Strawberry plant density, pruning, pollination and fruit growth

Plant density with hydroponic strawberries varies considerably with the type of production system used and environmental conditions. Low light can significantly lower yields and berry quality, while high light levels can stress plants and cause growth issues. Densities of 12–24 plants/m^2 are common in single-level or staggered trough type systems and this is somewhat cultivar dependent. Strawberry plants require little pruning; however, removal of the lower, older leaves as they age and senesce assists with good rates of air flow around the base of the plant for humidity removal and prevention of fruit and crop rot diseases. Strawberry plants, once established into optimal growing conditions, usually start to flower rapidly. This floral development is often on plants which have insufficient leaf area to support fruit growth and the first few flowers are removed until the canopy is of sufficient size. Once flowers are fully open, greenhouse strawberry crops require pollination assistance for large, high-quality berry production. Strawberry flowers pollinate when pollen produced on the stamen drops on to the receptacle, however this process is assisted by insects, wind and plant movement which cause pollen release. In a greenhouse situation, strawberry pollination assistance is provided by large air blowers which move down the rows directing a blast of air over

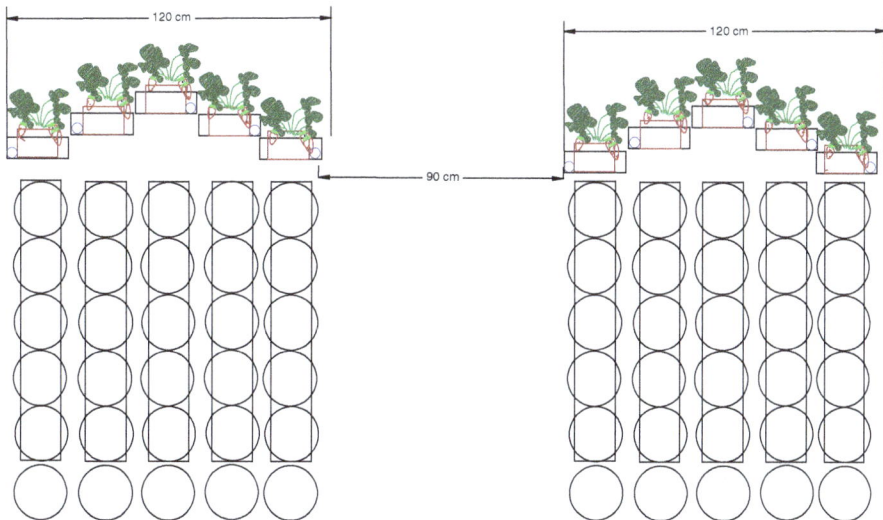

Fig. 11.22. Staggered strawberry planting system.

the flowers. Commercial growers may also introduce portable hives of bees to carry out greenhouse pollination (Fig. 11.23). On a smaller scale, brushing over the tops of the plants will also facilitate pollination. Pollination must be carried out daily as high rates of pollen transfer are required. Insufficient or uneven pollination will result in misshapen and undersized fruit.

11.7.4   Strawberry production environment

Despite being a small, shallow-canopied crop, strawberries have a reasonably high light requirement for good yields and high-quality fruit. Miyazawa *et al.* (2009) found that during the vegetative stage, optimum DLI was 29.3 mol/m^2 per day for strawberry growth at a temperature of 25°C. Temperatures in the range of 16–25°C are suitable for greenhouse strawberry production. Low temperatures slow the emergence of new leaves and extend the fruit development phase, while high temperatures above 30°C both in the root zone and aerial environment tend to inhibit strawberry growth and fruit production severely. Under warmer conditions fruit quality also suffers with a loss in flavour and an increase in softness combined with rapid ripening. Under warm, tropical growing conditions it is possible to manipulate plant growth and fruit development by chilling the root zone to 10°C (Sakamoto *et al.*, 2016). Strawberry crops respond to the use of CO_2 enrichment which gives increases in yields and lessens the time to harvest. Enrichment to levels of 700–900 ppm CO_2 has been found to give an increase in total plant yield of 50–70% (Itani *et al.*, 1999).

11.7.5   Strawberry crop nutrition

As with other hydroponic fruiting plants, strawberries require high levels of potassium

Fig. 11.23. Small, portable hives of bees can be used for pollination assistance.

during the fruiting phase to maintain good fruit quality. During the establishment and early vegetative stage a formulation which provides an N:K ratio of 1:1 may be applied until the time of fruit set on the first inflorescence, with the level of potassium gradually increased as fruit load increases on the crop. N:K ratios for strawberries typically fall within the range 1.0:1.7 to 1.0:2.0 during the fruiting phase (Morgan, 2006). Nitrogen to potassium ratios should be adjusted throughout the crop cycle and monitored through regular nutrient drainage analysis. Plants are prone to iron deficiency due to reduced levels of iron uptake under low temperature conditions.

EC levels for strawberry production vary depending on environmental conditions, stage of growth and system of production. A minimum EC of 1.6 mS/cm is required during the harvest period to maintain good fruit quality from all systems. Under winter low-light conditions, EC levels are run higher at 2.2–3.0 mS/cm to maintain fruit quality with lower levels run in summer and under higher-light conditions to prevent excessive plant stress and assist with water and calcium uptake.

11.7.6 Strawberry pests, diseases and disorders

Greenhouse strawberries are prone to a number of potentially serious pest and disease problems which should be prevented if yields and berry quality are to be maintained. Common pests include greenhouse whitefly (*T. vaporariorum*), tobacco whitefly (*B. tabaci*), greenhouse thrips (*H. haemorrhoidalis*), two-spotted mite (*T. urticae*), aphids and caterpillar larvae. Under warm, dry growing conditions, mites can be a particularly damaging and difficult-to-control pest on strawberries with early control required because crops can be destroyed by this persistent pest. Well-screened greenhouses avoid the problem of birds damaging ripening fruit which is common with field-grown crops, however outdoor hydroponic systems require bird netting for protection.

Diseases of strawberries are often those linked to poor environmental control, particularly overirrigation of the root zone, predisposing the plants to root and crown rot pathogens and high air humidity which is a risk factor for foliar diseases. Many modern cultivars have some resistance to common root rot pathogens such as red stele root rot or red core (*Phytophthora fragariae*) and these should be selected wherever possible. Other root rot pathogens include *Pythium*, *Rhizoctonia*, *Fusarium* and *Verticillium* which may spread through a crop from the initial source of infection. Fruit rots are common where fruit is in contact with damp substrates or when air flow is restricted and humidity levels high. *Botrytis* (grey mould) of fruit can occur under greenhouse conditions, particularly under cooler temperatures, and leather rot (*Phytophthora cactorum*) may also infect ripening fruit.

Most of the physiological disorders of strawberry crops are those which affect fruit size, shape and quality. Malformed berries are commonly caused by poor pollination or damage to the achenes (small seed-like structures) on the surface of the fruit. Incorrect environmental conditions such as temperatures which are too high or too low also contribute to malformed fruit and undersized berries. Overly soft fruit texture is often a problem where crops have been grown under temperatures that are higher than optimal but can also be due to incorrect plant nutrition and low EC levels. Deficiencies in potassium during fruiting is another common cause of soft fruit which does not store well, as is low levels of calcium or restrictions in calcium uptake by the plant.

11.7.7 Strawberry harvest and postharvest handling

Strawberries require daily harvesting as they must be removed from the plant at the correct stage of maturity when at least three-quarters ripe rather than when over-mature and starting to soften, which reduces shelf-life. Allowing fruit to fully ripen on the plant results in a greater flavour and sugar levels than those harvested at the first signs of coloration (Fig. 11.24). Fruit

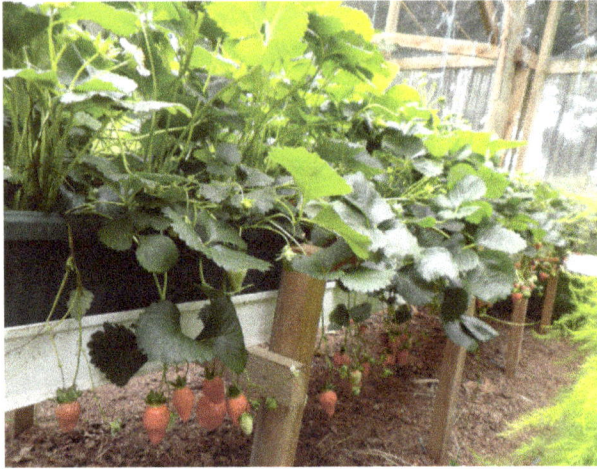

Fig. 11.24. Strawberry fruit at harvest maturity.

harvesting should occur in the morning when temperatures are still cool and the plants turgid. Once harvested berries require cooling to below 5°C as rapidly as possible to prolong shelf-life and fruit quality. Packing for hydroponic strawberries aims to protect the fruit from impacts, crushing and bruising and is often in clear plastic punnets, clamshell packages or small cardboard cartons. Greenhouse growers often grade fruit at the time of harvest since the majority of crop usually falls into the marketable category and this reduces damage caused by double handling. Fruits are harvested from the plant with the calyx intact, often with a length of stalk attached on the largest, high-grade berries.

11.8 Hydroponic Rose Production

Roses, along with carnations and chrysanthemums, are the most cultivated cut flower species under hydroponic production with most production focusing on the hybrid tea and floribunda types, although smaller volumes of spray types are also grown (Fascella, 2009). Red is the colour produced in the highest volume, while other colours such as yellow, pink, orange and white are grown in lesser amounts. Rose production can be carried out year-round with sufficient greenhouse heating through the winter months, with growers often focusing on seasonal markets such as Valentine's Day.

11.8.1 Rose production systems and planting material

Rose planting stock has traditionally been either budded or grafted on to a suitable rootstock; however, more growers, particularly those using hydroponic systems, now use planting stock on its own root system (Reid, 2008). Roses are grown in a wide variety of substrates including rockwool, scoria, coconut fibre, peat, perlite, sawdust, pumice and bark which is contained in large beds, containers, buckets or bags. Studies have found that the addition of coconut fibre to perlite gave higher rose production levels than perlite alone and a mixing ratio of 50–75% coco with perlite proved the most effective (Fascella, 2009; Roosta *et al.*, 2017). Systems may be open, but collect the nutrient leachate which is not reused, or closed where the nutrient solution is recirculated. Drip irrigation is the most commonly used method of nutrient solution application.

11.8.2 Rose plant density, pruning and plant management

Roses are a relatively long-term crop which is replaced on average every 5–6 years as production and quality fall after this time. Most growers renew a certain percentage (often 20–25%) of their crop each year to retain productivity levels. Plant density is dependent on rose type, cultivar and growing system, particularly light levels, and is typically 8–12 plants/m², with a density of 10 plants/m² being usual (Reid, 2008). Planting material is a minimum of 2–3 months old when established into the hydroponic system, which is carried out in winter for bare rooted bushes; those grown in containers can be planted out year-round.

Initial pruning of newly planted rose stock is vital to ensure a suitable structural framework develops while a strong root system establishes rapidly after planting. A suitable plant framework must be established before the first blooms are produced and full production should be achieved within 6–8 months of planting (Reid, 2008). Immediately after planting, pruning is carried out to promote the development of vigorous bottom shoot breaks, which are the structural shoots arising from the base of the plant. Once bottom shoot breaks have occurred, these are pruned back to 30–55 cm above the base of the plant. Further pruning is carried out throughout the productive life of the plant as required to promote bottom shoot breaks. A process of shoot 'bending' may also be used to promote shoot breaks. Once flowering has begun, disbudding is often required as most of the large-flowered varieties produce at least some lateral auxiliary buds. Pinching or cutting out lateral buds allows a single, main apical bud to produce a large, high-quality bloom.

11.8.3 Rose growing environment

Roses are a long-term crop which requires suitable temperatures for both yield and bloom quality. The broad production temperature range is 17–25°C with a relative humidity of 70–75%. Minimum temperature is considered to be 15°C while the maximum is 32°C. Temperatures below 15°C result in an increased occurrence of flower deformation including the production of 'bull head' flowers. High temperatures produce smaller flowers which may have fewer petals and with diluted coloration. Roses require high light levels for optimal production and supplementary lighting to maintain a 16 h day length is used in some winter climates to boost yields. Flower production is generally higher in summer than winter due to longer day length and higher DLIs. Many hydroponic rose producers use winter heating to maintain year-round production and force the blooms, however others may 'rest' the plants during winter if heating is not economical. Rose crop yields are dependent on plant age, cultivar, density and growing environment, with commercial operations obtaining between 225 and 350 stems/m² per year (Harrison, 2009; Reddy, 2015).

11.8.4 Rose crop nutrition

Roses require balanced nutrition for continued shoot growth and production of high-quality blooms. EC levels are dependent on growing conditions, with lower EC run in summer to assist with sufficient water uptake and increased levels in winter for plant strengthening. EC levels of 2.0–2.6 mS/cm are commonly used for hydroponic roses with pH in the 5.8–6.0 range. Higher levels of potassium are applied once the plant has entered the flower production stage to assist with bloom quality. Potassium to calcium ratios are also important for the production of cut flower roses. High levels of potassium compete for the uptake of calcium which is essential for bloom quality and yields. A suitable K:Ca ratio was determined to be 6:4, which reduced the antagonistic effects of potassium and calcium, and also improved postharvest quality and longevity of the cut flowers (Ouchbolagh et al., 2015). Apart from the essential macro- and microelements contained in the nutrient solution, some rose growers incorporate the use of silicon supplements, often as potassium metasilicate, to assist with plant growth and disease resistance. These supplements have a high pH however, which must be countered with the use of additional acid.

11.8.5 Rose pests, diseases and disorders

Roses are prone to similar pests as many other greenhouse hydroponic crops; however, in warm, dry climates and under summer cropping, mites and thrips can be particularly challenging to control. Mites cause bronzing of the foliage followed by the appearance of fine webbing on the undersides of leaves and in the leaf axils. Mites are difficult to control due to widespread chemical resistance to many of the insecticide compounds available. Aphids, thrips, whitefly and caterpillars are other common pests of greenhouse roses; these are often seasonal and greenhouse screening assists with the prevention of infestations. Roses are prone to a number of diseases, most of which are more common under humid conditions and facilitated by insufficient ventilation and air flow around the plants. Using drip irrigation and avoiding overhead wetting of the leaves or the formation of condensation assist with disease prevention. *Botrytis* flower blight which infects the bloom tissue can occur under high-humidity, cool-temperature conditions but is not often a problem in summer. Powdery mildew, which produces white spores, usually on the upper sides of the leaves, stems and flower buds, and can cause distorted growth of new shoots, is a common issue with greenhouse roses in many climates. Downy mildew may also occur and causes leaf drop and dark brown markings to appear on the upper leaf surfaces. Control of mildew requires prevention of moisture forming on the plants, sufficient air flow and humidity control as well as the use of protectant and curative fungicide products.

11.8.6 Rose harvesting

Harvesting of rose blooms requires removal from the plant at the correct position on the stem in order to promote the development of more flowering shoots. A new flowering shoot will start to grow from a bud in the axil of a leaf underneath a harvested flower stem. Flowering stems should be cut just above the second five-leaflet leaf or if stems are long and strong, cut to the third five-leaflet leaf (Reid, 2008), which will leave three more successive cuts from that particular stem. The most vigorous shoots will then develop from the leaf axils of leaves with five leaflets.

Sharp secateurs are used for harvesting roses, which should be carried out in the morning when temperatures are coolest and once any condensation has dried on the flowers. Harvesting at the correct stage of maturity is vital as those harvested at an immature stage will not open fully and may develop a bent neck. Roses should be harvested when one or two sepals have relaxed and the outer petals have started to loosen. Once harvested, roses are immediately placed into water and cooled to 1–4°C as rapidly as possible to maintain a long vase life. Blooms are graded based on stem length, lower leaves are removed and floral preservative treatments may be given before shipment.

References

Al-Rawahy, M.S., Al-Rawahy, S.A., Al-Mulla, Y.S. and Nadal, S.K. (2019) Effect of cooling root zone temperature on growth, yield and nutrient uptake in cucumber grown in hydroponic system during summer season in cooled greenhouse. *Journal of Agricultural Science* 11(1), 47–52.

Botella, M.A., Arevalo, L., Mestre, T.C., Rubio, F., Garcia-Sanchez, F., *et al.* (2017) Potassium fertilization enhances pepper fruit quality. *Journal of Plant Nutrition* 40(2), 145–155.

Brechner, M. and Both, A.J. (2013) *Hydroponic Lettuce Production Handbook*. Controlled Environment Agriculture Program, Cornell University, Ithaca, New York. Available at: https://cpb-us-e1.wpmucdn.com/blogs.cornell.edu/dist/8/8824/files/2019/06/Cornell-CEA-Lettuce-Handbook-.pdf (accessed 10 September 2020).

Calpas, J. (2020) Production of sweet bell peppers. Government of Alberta. Available at: https://www.alberta.ca/production-of-sweet-bell-peppers.aspx (accessed 10 September 2020).

Castilla, N. (2013) *Greenhouse Technology and Management*, 2nd edn. CAB International, Wallingford, UK.

De la Rosa-Rodriguez, R., Lara-Herrera, A., Tri-jo-Tellez, L.I., Padilla-Bernal, L.E., So-lis-Sanchez, L.O. and Ortiz-Rodriguez, J.M. (2020) Water and fertilizer use efficiency in two hydroponic systems for tomato production. *Horticultura Brasileira* 38(1), 47–52.

Fascella, G. (2009) Long-term culture of cut rose plants in perlite based substrates. *Floriculture and Ornamental Biotechnology* 3(1), 111–116.

Gagne, C.G. (2019) *The Effects of Daily Light Integral on the Growth and Development of Hydroponically Grown Baby Leaf Vegetables*. Report presented to Cornell University, Ithaca, New York. Available at: https://ecommons.cornell.edu/bit-stream/handle/1813/69362/Gagne_Charles.pd-f?sequence=1 (accessed 10 September 2020).

Harrison, D. (2009) A focus on roses. *Greenhouse Canada*, 5 May 2009. Available at: https://www.greenhousecanada.com/a-focus-on-roses-1666/ (accessed 10 September 2020).

Ho, L.C. (2004) The contribution of plant physiology in glasshouse soilless culture. *Acta Horticulturae* 648, 19–25.

Hochmuth, R.C. (2012) Greenhouse cucumber production. *Florida Greenhouse Vegetable Production Handbook, Volume 3*. University of Florida Extension, Gainesville, Florida. Available at: https://edis.ifas.ufl.edu/pdffiles/CV/CV26800.pdf (accessed 10 September 2020).

Horridge, J.S. and Cockshull, K.E. (1998) Effect on fruit yield of bending ('kinking') the peduncles of tomato inflorescences. *Scientia Horticulturae* 72(2), 111–122.

Itani, Y., Hara, T., Phum, W.N., Fujime, Y. and Yoshida, Y. (1999). Effects of CO_2 enrichment and planting density on the yield, fruit quality and planting density and absorption of water and mineral nutrients in strawberry grown in peat bag culture. *Environmental Control in Biology* 37(3), 171–177.

Jie, H. and Sing Kong, L. (1998) Growth and photosynthetic responses of three aeroponically grown lettuce cultivars (*Lactuca sativa* L.) to different root zone temperatures and growth irradiances under tropical aerial conditions. *Journal of Horticultural Science and Biotechnology* 72(3), 173–180.

Langenhoven, P. (2018) Hydroponic tomato production in soilless culture. *Indiana Horticulture Congress, February 13, 2018*. Purdue University Extension Service, West Lafayette, Indiana. Available at: https://ag.purdue.edu/hla/fruitveg/Presentations/Hydroponic%20Tomato%20Production%20in%20Soilless%20Culture_February%2013,%20

2018_Petrus%20Langenhoven.pdf (accessed 10 September 2020).

Maboko, M.M., Du Plooy, C.P. and Chiloane, S. (2012) Effect of plant population, stem and flower pruning on hydroponically grown sweet pepper in the shadenet structure. *African Journal of Agricultural Research* 7(11), 1742–1728.

Mattson, N. (2010) *Greenhouse Lighting*. Cornell Greenhouse Horticulture, Cornell University, Ithaca, New York. Available at: http://www.greenhouse.cornell.edu/structures/factsheets/Greenhouse%20Lighting.pdf (accessed 10 September 2020).

Miyazawa, Y., Hikosaka, S., Goto, E. and Aoki, T. (2009) Effects of light conditions and air temperature on the growth of everbearing strawberry during the vegetative stage. *Acta Horticulturae* 842, 817–820.

Morgan, L. (2006) *Hydroponic Strawberry Production*. Suntec NZ Ltd, Tokomaru, New Zealand.

Morgan, L. (2012) *Hydroponic Salad Crop Production*. Suntec NZ Ltd, Tokomaru, New Zealand.

Morgan, L. and Lennard, S. (2000) *Hydroponic Capsicum Production*. Casper Publications Pty Ltd, Narrabeen, New South Wales, Australia.

Ouchbolagh, A.J., Ajirlo, S.A., Tabatabaei, S.J. and Hassani, R.H. (2015) Effects of K/Ca ratios of nutrient solution on some physiological traits and cut flower vase life of rose cultivars (Capitol and Magic Red). *Journal of Plant Physiology and Breeding* 5(2), 19–28.

Peet, M.M. and Welles, G. (2005) Greenhouse tomato production. In: Heuvelink, E. (ed.) *Tomatoes*. CAB International, Wallingford, UK, pp. 257–304.

Portree, J. (1996) *Greenhouse Vegetable Production Guide*. British Columbia Ministry of Agriculture, Fisheries and Food, Abbotsford, British Columbia, Canada.

Reddy, J. (2015) Rose cultivation in greenhouse guide. *AgriFarming*, 24 June 2015. Available at: https://www.agrifarming.in/rose-cultivation (accessed 10 September 2020).

Reid, A. (2008) Greenhouse roses for cutflower production. *Bulletin 4738*. Department of Agriculture and Food, Perth, Western Australia, Australia. Available at: https://researchlibrary.agric.wa.gov.au/bulletins/164/ (accessed 10 September 2020).

Roosta, H.R., Bagheri, V. and Kian, H. (2017) Effect of different planting substrates on vegetative and physiologic characteristics and nutrient content of rose (*Rosa hybrida* var. Grandgala) in hydroponic system. *Journal of Science and Technology of Greenhouse Culture* 7(28), 27–39.

Ropokis, A., Ntatsi, G., Kittas, C., Katsoulas, N. and Savvas, D. (2019) Effects of temperature and grafting on yield, nutrient uptake and water use efficiency of a hydroponic sweet pepper crop. *Agronomy* 9(2), 110.

Sakamoto, M., Uenishi, M., Kiyamoto, K. and Suzuki, T. (2016) Effect of root zone temperature on the growth and fruit quality of hydroponically grown strawberry plants. *Journal of Agricultural Sciences* 8(5), 122–131.

Schon, M.K., Peggy Compton, M., Bell, E. and Burns, I. (1994) Nitrogen concentration affects pepper yield and leachate nitrate-nitrogen from rockwool culture. *HortScience* 29(10), 1139–1142.

Tadesse, T., Nichols, M.A. and Fisher, K.J. (1999) Nutrient conductivity effects on sweet pepper plants growth using a nutrient film technique. *New Zealand Journal of Crop and Horticultural Science* 27(3), 229–237.

Torres, A.P. and Lopez, R.G. (n.d.) Commercial greenhouse production: measuring daily light integral in greenhouses. *Publication HO-238-W*. Purdue University Extension Service, West Lafayette, Indiana. Available at: https://www.extension.purdue.edu/extmedia/ho/ho-238-w.pdf (accessed 10 September 2020).

12 Plant Factories – Closed Plant Production Systems

12.1 History and Background

A plant factory is an indoor, enclosed, crop cultivation system where the growing environment is precisely controlled to maximize production. This control over all aspects of plant growth includes light, temperature, humidity, air movement, CO_2 and nutrition, which is largely achieved through soilless (hydroponic) cultivation techniques. The term 'plant factory' has, in the past, included high-intensity production systems such as greenhouses reliant on natural light only or natural light supplemented with horticultural lighting, as well as those utilizing only artificial lighting. More recently, 'plant factory' has come to represent a 'closed plant production system' (CPPS) or been termed a 'plant factory with artificial lighting' (PFAL) where no natural sunlight is required. Plant factories may also be referred to as 'indoor vertical farms' or 'indoor vertical cropping' in the USA and other countries, based on the multilevel nature of the systems which aim to maximize growing space. With a high level of environmental control, plant factories can produce vegetables two to four times faster than by typical outdoor cultivation (Luna-Maldonado *et al.*, 2016) and, with the use of vertical systems, optimize yields per unit of floor area. In addition to higher levels of production, plant factories are becoming increasingly sustainable as less water, fertilizers, pesticides and labour are consumed during cultivation (Hu *et al.*, 2014).

The use of totally enclosed growing systems is not new; 'growth cabinets', 'phytotrons' and CELSS (controlled ecological life-support systems) as part of the space development programmes of NASA and others have been in use by researchers for a number of decades. However, in recent years this technology has been advanced, adopted and rapidly expanded by both commercial and smaller-scale growers to produce a wide range of crops in locations previously unsuited to horticulture. The history leading to the development of modern plant factories began in the late 1940s and 1950s with the first greenhouses utilizing control of lighting, temperature, humidity and CO_2. One of the earliest to develop these was the Earhart Plant Research Laboratory at the California Institute of Technology, such a research facility was termed a 'phytotron' (Hirama, 2015). These types of greenhouses were adopted for use by a number of research facilities through the 1950s and 1960s, and by the 1970s, Hitachi Ltd of Japan was the first in the world to begin test runs with plant factory technologies (Hirama, 2015). At the same time space programmes began developing CELSS and by 1990 NASA had begun research into zero-gravity plant production systems. During the 1980s large-scale plant factories using natural sunlight had become widespread and incorporated into greenhouse horticulture. These installations relied on soilless cultivation techniques and supplementary lighting when required, alongside automated control over other plant growth factors such as temperature, air flow, ventilation, humidity and CO_2 enrichment. In more recent times, the development of highly energy-efficient lighting systems utilizing LED technology has been a major contributor to the investment in large-scale, exclusively indoor, CPPSs in many countries. LED technology allowed precise control not only over light intensity and duration, but also light spectrum while at the same time being significantly more energy efficient than older forms of artificial crop lighting.

By 2017, approximately 211 plant factories were operating across Japan cultivating

© L. Morgan 2021. *Hydroponics and Protected Cultivation* (L. Morgan)
DOI: 10.1079/9781789244830.0012

lettuce, herbs, tomatoes, strawberries and other crops (METI, 2017). Japan has also established a non-profit organization, the Japan Plant Factory Association (JPFA), founded in 2010. The main objective of the JPFA is 'to contribute to concurrently solve the global food, environmental, energy and resource issues of the 21st century through R&D and dissemination of sustainable plant factory systems, enabling resource saving, environmental conservation, high-quality/profitable/stable plant production' (JPFA, 2020). Many Asian countries, including Taiwan and Korea, have PFAL industries which are rapidly developing with research supported by universities, institutes and private companies. The annual domestic market of the PFAL sector in Korea was worth nearly US$600 million in 2016 (Kozai and Nui, 2016c). In the USA vertical farming has been a developing technology for a number of years, with large-scale, totally enclosed plant factories utilizing only artificial lighting being established close to large cities such as Chicago (Kozai and Nui, 2016c). Advancements in transgenic technology have seen large-scale plant factories established in North America with the specific purpose of producing pharmaceutical proteins in plants grown under highly controlled conditions (Goto, 2016). Many plant factory installations have been set up in Europe, including the UK, Belgium and the Netherlands, for both research and commercial production.

12.2 Advantages of Plant Factories

The development of plant factories has been in response to a number of environmental, population and food supply issues. In Japan, plant factory development and investment have been in response to a general concern about the declining rate of domestic vegetable supply and how the problem of heavy dependence on imports might be remedied (Dickie, 2014). Plant factories have potential to help address the ever-increasing problems of poor harvests due to weather events, global warming, lack of agricultural employees and food shortages due to overpopulation

(Hirama, 2015). A lack of suitable horticultural land and global issues with water supply and quality for crop production also limit large-scale expansion of field-based food production. With an ever-increasing population worldwide, significantly more food must be produced from the same or less area of land and technological advances are seen as the major method to achieve this. It is estimated that 90% of the global increase in food production required to keep pace with population growth needs to come from higher yields and increased cropping intensity and only 10% from the expansion of productive land (FAO, 2009). Plant factories which use vertically stacked systems to grow produce at extremely high densities are seen as one way of increasing yields per unit area of land, coupled with using sites which are not suitable for traditional cropping methods anyway (Fig. 12.1). Outdoor factors such as arid climates with excessive heat and humidity, low-temperature winter regions, pollution and poor natural light levels, which may limit or prevent the production of certain fresh produce such as salad vegetables outdoors, are other main reasons for the adoption of plant factory technology. Inside the environmentally controlled climate of a plant factory, unfavourable outdoor conditions play no role in yields or productivity reductions. Well-insulated plant factories can also be more thermally efficient than greenhouses in some climates. Certain situations, such as cooling to counter rapid temperature increase in the crop due to sunlight penetration into the greenhouse structure or the requirement for high levels of heating in cold winter climates, use considerable energy resources. An example of the adaptation of plant factory technologies in harsh climates is the indoor crop production facilities in Antarctica. Not only do these installations allow fresh fruit and vegetables to be produced for those personnel resident on the ice, but also their use of soilless culture instead of importing soils or growing mixes limits the risk of accidental introduction of non-native species (Bamsey et al., 2015). To date, a total of 46 plant production facilities have been in operation at one time or another

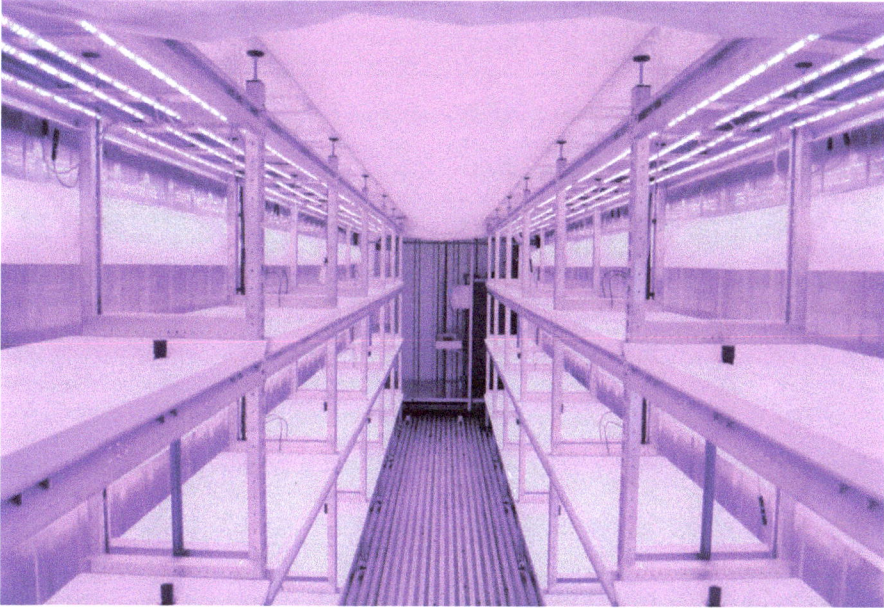

Fig. 12.1. The FarmFlex container farm developed by Urban Crop Solutions. (Image courtesy of Urban Crop Solutions, https://urbancropsolutions.com, accessed 11 September 2020.)

in Antarctica with nine onsite installations currently operating in nine different Antarctic stations (Bamsey *et al.*, 2015).

Not being reliant on outdoor climatic conditions means seasonality is eliminated from plant factory cropping cycles. Crops are produced with uniform productivity and reliable yields every week of the year. Being highly predicable in output ensures markets do not experience sudden gluts or shortages of fresh produce and can maintain a stable price year-round for consumers. Plant factories, if installed and operated correctly, are also highly efficient in water usage compared with outdoor field cropping. Most plant factories operate a system of soilless culture which continually recirculates the nutrient solution past the root systems, thus providing an efficient method of water usage and fertilization through precise control of plant nutrition. Apart from efficient water use, plant factories can be operated to prevent any risk of fertilizer runoff, degradation of the soil and pollution of natural waterways. Due to the nature of the enclosed and highly protected environment, use of pesticides, fungicides and other crop protective compounds can be kept to a minimal level, if required at all.

Food safety is another increasing concern among both horticultural producers and the general public, with salad vegetables and fresh fruit in particular being responsible for a number of produce recalls and food-related disease outbreaks over recent years. While there can be a number of different causes for food safety issues in fresh produce, many cases are linked back to issues with biological contamination in the field, with water supplies or by human handling. Plant factories can eliminate many of these risks by producing crops in a fully enclosed, clean, hygienic environment with treated water supplies, minimal human handling and exclusion of other sources of potential contamination such as birds, rodents, domestic or wild animals, incorrectly processed organic fertilizers and other pollutants. Shelf-life of plant factory crops may also be considerably longer than those of crops grown outdoors through prevention of contamination with postharvest spoilage pathogens. The bacterial load of plant factory produce has been reported as being less than 300 CFU/g, which is 1/100 to 1/1000 that of field-grown produce (Kozai and Niu, 2016a).

A fully enclosed environment where air, water quality and all inputs are carefully controlled is also suitable for the production of highly contamination-sensitive crops. These include a range of medicinal and pharmacological plants which are being increasingly grown using plant factory installations and technology.

Apart from food security issues, plant factories have the advantage of being set up anywhere there is a reliable source of sufficient energy and water and are not reliant on finding suitable cropping land. This means disused buildings such as warehouses or offices, underground facilities, basements, rooftops, empty inner-city lots and urban disused land can be turned into efficient plant factories providing fresh produce in the immediate vicinity of high-population centres, consumers and markets. Storage, energy usage, time and shipping costs to move fresh produce from the production field into populated areas are significantly reduced with use of urban-based plant factories and is seen as one of the main advantages of such installations. Freshness and quality of produce, particularly of highly perishable items such as salad greens, herbs, leafy vegetables and soft fruits, can be maintained when production takes place close to urban centres with minimal transport distances. The production of fresh produce within urban areas is also seen as an important step for reducing food waste generated when perishable fruits and vegetables must to be transported large distances. Considerable losses in quality and damage may be incurred when highly perishable fruits and vegetables are moved from field to market, requiring storage, handling and several steps in the postharvest chain before reaching the final consumer. Plant factories may also be static or mobile installations, allowing flexibility of location. Plant factories have been developed using shipping containers and other modular units that can be deployed as necessary to supply fresh produce in disaster-affected areas or where food shortages may occur. These types of mobile plant factories may also

be employed onsite by restaurants, hotels, resorts and other businesses, as well as individual households, that require a regular and guaranteed supply of safe, fresh production onsite.

12.3 Criticisms of Plant Factories

The increasing development and expansion of plant factories worldwide have met with some criticism of both the installations and the type of produce grown. Much of this is focused round the perceived high energy requirements, level of capital investment, cost of production and lack of profitability with these facilities, despite many becoming increasingly energy efficient. Simulation studies in the continental USA climate have indicated that naturally lit greenhouses use less energy and have a smaller carbon footprint than plant factories producing equivalent yields (Harbick and Albright, 2016). Other criticisms are that plant factory crops are in some way 'unnatural' or 'artificial', lacking in nutrition and flavour as compared with those produced in the field, utilizing sunlight or more traditional greenhouse methods. A study carried out in Japan into consumer awareness and opinions of plant factory vegetables indicated a high interest in food safety with particular concern for radioactive contamination, a priority for freshness over price, and a perception that factory-produced vegetables were 'sanitary but expensive' (Kurihara et al., 2014). Consumers in Japan considered a 20–40% premium for factory-produced over outdoor-produced vegetables to generally be acceptable, however many of those surveyed still preferred cultivation with sunlight (Kurihara et al., 2014). Another issue is the fact that plant factory crops produced using standard hydroponic methods and nutrient solutions cannot be classified or certified as organic and questions regarding the sustainability of such production methods have arisen.

Other concerns are that urban plant factories are using valuable land which would better serve other industries more suited to industrial or residential sites and that the range of crops that can be grown is restricted and limited to leafy vegetables, herbs and a few other minor crops. With the increasing use of robotics and automation in plant factories, the loss of horticultural industry jobs is another factor – highly efficient plant factories may aim for high volumes of production without crops ever being handled by workers. However, skilled personnel will still be required to design, build, run and maintain plant factories. Many of these criticisms are somewhat unfounded and advancing technology and increasing skill of operation managers are likely, in the future, to negate the issues of produce quality and energy consumption. However, re-education of the public and improving the image of plant factories may take additional promotion before consumers become fully aware of the potential benefits of production of fresh, high-quality and safe produce in their own communities.

12.4   Costs and Returns

One of the main criticisms of plant factories is the high capital cost compared with greenhouse or field-based operations. An example of a cost–benefit analysis of plant factories in Japan is outlined by Kozai and Niu (2016b). This indicated that in 2014, the initial cost of a PFAL installation in Japan with a production system consisting of 15 vertical tiers (50 cm between tiers) was US$4000/m^2, approximately 15 times the cost of an environmentally controlled greenhouse utilizing natural sunlight. Yields from such a PFAL installation growing leaf lettuce would result in an income of US$2100–2400/m^2 per year on the Japanese wholesale market (3000 leaf lettuce heads/m^2, each weighing 80–100 g, per year). Compared with estimated yields from outdoor field production of 32 heads/m^2 per year and 200 heads/m^2 per year from greenhouse systems, PDFL productivity is 100 times that of the open field and 15 times that of the greenhouse (Kozai and Niu, 2016b). Apart from the initial investment in a plant factory installation, production cost percentages stated in

this example from Japan were: 25–30% for electricity which included lighting, 25–30% for labour, 25–35% for depreciation and 20% for all other input costs (fertilizer, seeds, water, office, lighting consumables, etc.) (Kozai and Niu, 2016b). Future projections are that developing technology will see an overall decrease in the capital investment required for plant factory installations, associated equipment and running costs, alongside improvements in cultivars, yields and productivity.

12.5   Domestic and Other Small-Scale Plant Factories

Along with large, intensive, commercial enclosed plant factories which aim to produce high volumes of crops for the wholesale market, smaller installations are gaining in popularity for a range of domestic uses. Urban residents who do not have access to an outdoor area for gardening may adopt plant factory technology for small-scale production. Apartment dwellers may install scaled-down, fully automated hydroponic systems, along with lighting indoors, to grow a wide range of plants, both food and ornamental crops for aesthetic value. A number of domestic plant production systems can now be purchased as kit sets or fully operational units designed for use in homes and other domestic situations. Innovations in home plant factory system designs have aimed to use a wide range of hydroponic techniques to deliver the nutrient solution to plant roots. These include aeroponics and 'ultrasonic nebulizers' which produce vibrations that turn the nutrient solution into a fog-like cloud allowing the roots to absorb more oxygen and nutrients for growth (Li et al., 2014). These systems are designed to not only improve the growth efficiency of vegetables as compared with more traditional hydroponic methods and systems, but also allow mini plant factories to be incorporated more seamlessly into interior design features.

Other environments where small-scale plant factories have been installed include supermarkets, cafés, restaurants, hospitals, offices, schools, shopping centres, recreational centres, aged care homes, tourism ventures, and many others. These may be used to grow fresh produce onsite for sale to the final consumer or used in onsite food preparation and are often on display for customers to observe the cropping process. In offices and similar environments, plant-factory type installations may simply be used to grow ornamental plants to beautify and provide relaxation for workers. In some instances, certain types of plants can be grown to improve air quality inside buildings and assist in the removal of toxins.

Apart from small, static, domestic and semi-commercial plant factory installations, mobile enclosed growing units have also been developed. Many of these types of self-contained systems have been designed based on 20–40 ft shipping containers which can be deployed to areas of need as required. This includes regions which may have suffered natural disasters and require a temporary source of fresh produce or sites where no suitable buildings are available to site a permanent facility. Shipping-container-based mobile plant factories typically contain vertical systems of three to five tiers, a basic hydroponic system, lighting for each tier, heating and cooling, along with air-circulation fans and automated control of the environment. The main crops grown in mobile plant factory units include lettuce, salad and other compact vegetables, micro greens, sprouts, herbs and other small plants with a high crop turnover rate.

12.6 Crops Produced Including Pharmaceuticals

Horticultural crops most suited to production in indoor plant factories that only utilize artificial light sources are those of high value, a rapid growth rate and those which are restricted in height and compact. Since plant factories are reliant on high planting density rates and the use of vertical multi-tiered racks, short plants fit within this system while still permitting sufficient air flow through the crop. Plants which are 30 cm or less in height at maturity allow for production in multilevel growing racks where the distance between culture beds is 40–50 cm (Kozai and Niu, 2016a). Requirement for a relatively low level of light to complete the production cycle is another important variable for plant factory production as energy costs for lighting are relatively high compared with greenhouse facilities utilizing only natural light. Leafy vegetables such as lettuce, mustard, spinach and herbs have a relatively low light requirement (approximately 200 µmol/m^2 per s) as compared with crops such as tomatoes which require approximately 800 µmol/m^2 per s (Yamori et al., 2014). Further suitability is based on other crop characteristics such as a high proportion of the plant being saleable as in the case of lettuce, salad greens and herbs and where high-quality, clean product is a major advantage. Plant factories may also be used to produce transplants – small seedlings or other plant material which is raised to later be grown on in a different environment. Nursery plugs, grafted and non-grafted tomato, cucumber and aubergine, propagation of a wide range of small, indoor, high-value ornamental or food crop plants, tissue culture explants, cuttings and other transplants are well suited to plant factory multilevel systems where the precise control of the environment, hygienic conditions and high density make this economical.

The use of enclosed indoor plant factories to produce high-value medicinal and pharmacological crops represents an opportunity for increasing health benefits from plants and is an aspect of this industry which seems to have a great deal of potential (Luna-Maldonado et al., 2016). Such crops require strict hygiene and quality control as well as manipulation of environmental conditions to maximize levels of active compounds within the plant. An example of this is the use of highly advanced plant factories for 'molecular farming' as an efficient method for producing vaccines and therapeutic proteins in plants. An automated plant-based vaccine production factory is currently being used to consistently grow and make proteins using a fully robotic system and vacuum infiltration of vectors into tobacco plants (Fraunhofer Center for Molecular Biotechnology, 2013). Production of transgenic plants in fully enclosed facilities has the further advantage of reducing the risk of gene flow compared with outdoor, field-grown crops (Yamori *et al.*, 2014).

12.7 Vertical or Multilevel Systems, Including Moveable Systems

Plant factories using only artificial lighting are often sited in urban areas where land is expensive and limited, thus must be used as efficiently as possible. Furthermore, plant factories make use of fully environmentally controlled installations which are costly to set up and run. These factors make it necessary to use high plant densities in order to achieve an economic return from investment. The majority of totally enclosed plant factories using only artificial lighting are able to grow crops in vertical, multilevel systems which contain a number of planted tiers. Each tier, for many of the types of crops grown such as small vegetables, is 40–60 cm distance from the base containing the irrigation system to the top where light is provided. In installations such as those based in high stud warehouses and factory buildings, this can represent a large number of tiers which maximizes production per unit of floor area. Many of the more advanced plant factories also incorporate the use of moveable conveyors and production systems, reducing the requirement for pathways and access space within the crop (Fig. 12.2).

Crop production inside modern plant factories is often highly automated and aims to minimize the use of expensive labour to carry out plant cultivation processes. Propagation areas may use automated seeding equipment to precisely sow seed into sheets or trays of sterile growing medium such as foam, oasis, rockwool, peat, coconut fibre or similar hydroponic substrate. These hold the seed in place during the germination phase, providing moisture and oxygen, and facilitate the handling of the young seedling during transplanting into the hydroponics growing system. During the germination

phase, seeded trays are kept at optimal temperatures for rapid establishment and once cotyledons have expanded light is provided for seedling development. When seedlings have reached a predetermined size, typically two or three true leaves for lettuce and similar salad crops, the individual seedlings are separated and transplanted into the production system, grown in multilevel tiers each with an array of light fixtures. Many systems rely on conveyors to move seedlings out to production areas and fill the tiered production tray systems as this is a labour-intensive, repetitive task which is rapidly being taken over by robotic technology. During the crop growth cycle plants are respaced in some plant factory systems to maximize growing area and are subsequently harvested after a set number of days. Automated or manual harvesting may be carried out depending on the level of technology used in each plant

Fig. 12.2. Use of conveyor systems maximizes space in an automated plant factory. (Image courtesy of Urban Crop Solutions, https://urbancropsolutions.com, accessed 11 September 2020.)

factory. During this stage plants are removed from the hydroponic production system, root systems are removed and discarded, and any older, pale, damaged or unmarketable leaves are trimmed from each plant. For some hydroponics crops such as high-value lettuce and herbs, the hydroponic root system may be left attached to the plant, wrapped moist in plastic and sold as a 'living product'. Leaving the bare, wrapped and moist root system on hydroponic salad or herb plants increases the shelf-life of the produce by providing moisture for the plant. This assists in preventing postharvest wilting and desiccation. Plants may then be immediately packaged for retail sale or further processed into salad mixes and other products for wholesale distribution. Due to the clean and hygienic nature of the enclosed plant factory environment, crops such as lettuce may not need to be washed postharvest. Salad crops, leafy vegetables and herbs are typically cooled to 0–2°C after harvest to maintain shelf-life before shipping out to markets.

12.8   Crop Nutrition in Plant Factories

The majority of plant factory systems utilize soilless, hydroponic methods of cultivation; however, a small number may use organic substrates, potting mixes and other media. Plant factory hydroponic cultivation systems have the advantage of being clean, hygienic (the water and nutrient solution may be disinfected if required), lightweight, water efficient and provide optimal control over nutrition via use of balanced nutrient solutions. Composition of hydroponic nutrient solutions varies depending on crop, system, stage of growth, water quality and other factors, however all contain the essential elements for growth: nitrogen, phosphorus, potassium, calcium, magnesium, sulfur, iron, manganese, boron, zinc, copper and molybdenum. Highly soluble, high-grade fertilizers are dissolved into water to create concentrated stock solutions which are further diluted to working strength for use in the hydroponic systems. These concentrates

are used to increase the EC (a measure of the concentration of ions) of the nutrient solution recirculating within the plant factory system as required on a regular basis. pH and EC are precisely controlled in plant factory hydroponic systems to maximize growth and nutrient uptake at each stage of plant growth. EC levels for many lettuce, salad greens, leafy vegetable and herb crops used in plant factory systems range from 0.8 to 2.2 mS/cm. The EC level of the nutrient solution can have significant effects on growth and quality of different crops in plant factories. While small, fancy hydroponic lettuce types are typically run at lower EC levels between 0.8 and 1.5 mS/cm, crisphead or iceberg lettuce requires higher values for optimal growth and head quality. Leafy Asian vegetables such as red mustard and pak choi grown in plant factories have been found to develop higher levels of ascorbic acid (vitamin C) at EC levels between 2.0 and 2.5 dS/m (Lee *et al.*, 2012). pH of the recirculating nutrient solution may be lowered with the use of acids such as nitric or phosphoric acid and increased with potassium hydroxide.

A number of different types and modifications of hydroponic systems are used in plant factories; these include NFT, where a thin film of the nutrient solution flows along the base of the growing channels which contain the plant root systems. Float, raft or deep flow technology, water culture systems, or modifications of these are also used in a number of plant factories. In these systems the plants are supported above a pond or tank of nutrient solution which contains the root systems. The nutrient solution in these types of systems is typically aerated and circulated to maintain DO levels around the root systems. Ebb and flow (flood and drain) type hydroponic systems are a commonly used method of nutrient delivery in plant factories. In these systems, trays containing the plants supported in containers are flooded with nutrient on a regular basis, then drained of nutrient solution. Plants such as lettuce, salad greens, leafy vegetables, strawberries and herbs may be supported in small cups, pots or tubes allowing the root systems to access the nutrient solution, while holding the plant upright and in place. Other systems utilize thin sheets of inert growing medium such as rockwool into which seed is sown and which retains the nutrient solution between irrigations. A number of systems use aeroponic methods where the root systems are suspended in chambers and nutrient solution misted or fogged either continually or on an intermittent basis. This allows maximum aeration of the root system and a number of modified systems have been developed based on aeroponic principles. The majority of the hydroponics systems currently in use in plant factories for commonly grown crops such as lettuce, herbs and leafy greens recirculate the nutrient solution in order to optimize water and fertilizer usage. Recirculation of the nutrient solution allows precise control of the pH and EC directly around the root systems. Most plant factories use fully automated nutrient control systems where sensors monitor factors such as solution pH, EC, temperature and DO. Required adjustments to the pH or strength of the nutrient solution are then carried out by automatic doser units controlled by computer or electronic systems. A further level of nutritional control is carried out by monitoring the composition of the nutrient solution via laboratory analysis for each element. This allows adjustments to the nutrient formulation to be calculated based on current plant uptake ratios. More recent innovations in nutrient solution compositional monitoring and control in high-technology plant factories are based on the development of new ion-selective sensors. These sensors continually detect levels of calcium, potassium, magnesium, nitrate and ammonium with use of ion-selective electrodes positioned in the recirculating nutrient solution (Kozai and Niu, 2016c).

High-quality water supplies for plant factories are essential for hydroponic production. Many facilities use RO to produce demineralized and sterile water to make up the plant cultivation nutrient solutions. Some facilities recycle water collected by dehumidification of the plant growing environment, which can become overly humid due to plant transpiration. In order to maintain the high quality and maximum food safety aspect of plant factory cultivation, use of clean water is essential and prolongs the usable life of the nutrient solution. Plant factory facilities may use local agricultural laboratories for analysis of the recirculating nutrient solution or have onsite equipment to measure individual elements on a regular basis and allow immediate solution adjustment if required. By precisely monitoring, adjusting and controlling the nutrient solution, water and fertilizer efficiency can be maximized with minimal waste of these inputs. Along with water supplies, recirculating hydroponic nutrient solutions may undergo treatment or disinfection such as with UV radiation, heat, by ozonation, filtration or other processes to prevent microbial contamination which can pose a food safety and plant pathogenic risk.

Recent research into crop nutrition in plant factories has been aimed at further improving fertilizer-use efficiencies. Since some nutrients such as nitrogen, phosphorus and potassium may be absorbed in excess (luxury absorption) than is needed for maximum productivity, new concepts of fertilizer application have been studied which aim to prevent this issue. A quantitative nutrient management method with low nutrient concentrations has been developed which is a feasible and resource-saving method for hydroponic vegetable production in plant factories (Maneejantra et al., 2016). Another aspect of crop nutritional control to manipulate plant quality is the development of cultivation systems which produce leafy vegetables with low potassium content. These low-potassium vegetables may be consumed by dialysis patients who must restrict potassium intake. By growing plant factory vegetables with no potassium application during the latter half of their growth period, the foliar potassium content was reduced 30–40% as compared with normal plant growth (Ogawa et al., 2012). Production of vegetables with low nitrate content is another possibility with hydroponic nutrition, avoiding the issue of some soil-grown crops which may contain high nitrate levels (Ohashi-Kaneko, 2016).

12.9 Plant Factory Environments

One important feature of totally enclosed plant factories with artificial lighting is the precise control over the growing environment which can be achieved in order to maximize growth rates and crop turnover. PFAL often refers to plant production systems housed in warehouse-like structures which are thermally insulated and almost airtight (Kozai, 2013). Thermal insulation of both roof and walls increases the energy efficiency of heating and cooling and is particularly important when the external environment experiences extremes of heat or cold. Inside these structures, the plant factory runs as a 'closed production system' with minimal flow between the external and internal environments. The closed environment nature of many plant factory installations includes a requirement for strict hygiene and prevention of contamination from outdoors. This may include the use of double-door entries into growing areas, foot baths, air showers or hot water showers, protective clothing for workers, sterilization of floors and other hard surfaces, and disinfection of water supplies. Filtration of any incoming air, particularly where air pollution is an issue in urban areas, may also be carried out.

Heating and cooling equipment are an essential component of plant factory environmental control, the degree of which is dependent on the outside local climate and seasonal differences in temperatures. Cooling is achieved with use of air-conditioner units which also act to dehumidify the air and assist with air circulation. Water condensed from air-conditioner units is collected to be reused within the system. Air movement across the plants growing in high-density, multilevel plant factory systems is also precisely controlled. It is essential to remove the stale and humid 'boundary layer' of air which forms close to the leaf surface and to replenish CO_2 levels around the stomatal openings for rapid rates of photosynthesis. Without sufficient air flow which promotes transpiration from the leaf and calcium incorporation into cells in the leaf tips, plant factory crops such as lettuce

are prone to develop the physiological disorder tipburn. Stable, horizontal, 24 h airflow rates above 0.28 m/s have been found to effectively reduce tipburn occurrence in plant factory lettuce crops by increasing the calcium content of the foliage and enhancing transpiration in the centres of the plants (Lee et al., 2013). Combined with precise temperature control, air flow through high-density plant factory cropping systems is essential for normal growth and prevention of physiological disorders.

Along with recirculating soilless culture systems, other aspects of the production environment such as temperature, humidity and air flow, CO_2 and lighting are typically controlled via computer systems. Control over these environmental elements is performed by using fixed set points, which may be matched to individual growth stages (Yamori et al., 2014). A range of environmental control systems has been developed and adapted to plant factory situations; while many of these are designed to operate onsite, some are web based and can be adjusted by the grower remotely (Dongxian et al., 2007). Environmental control systems for plant factories typically employ a range of sensors which gather microclimate information in various positions around the crop – these include variables such as air, leaf and root zone or nutrient solution temperature, humidity, light and CO_2 levels (Moon et al., 2012). Other systems may incorporate sensors or devices which monitor plant growth via weight or image processing. These sensors feed continual information back to the central, computerized system controller which adjusts heating and cooling, light levels, CO_2 dosing, irrigation programmes and composition of the hydroponic nutrient solution (EC, pH and DO).

One significant advantage of fully enclosed growing systems is the ability to precisely control the level of CO_2 in the air surrounding the crop. The benefits of CO_2 enrichment to well above ambient on crop growth, yields and earliness of harvest are well known; however, fully enclosed plant factories do not experience the loss of CO_2 through ventilation required in greenhouses, so are highly CO_2 efficient. CO_2 enrichment to levels of 800–1500 ppm inside plant factories allows plants to become highly photosynthetically efficient and increases overall crop productivity.

Precise environmental control inside fully insulated and enclosed plant factories allows operators to provide the exact conditions required for maximum growth rates. While this may differ from installation to installation, a typical environmental programme for hydroponically cultured lettuce would be a 16 h photoperiod at 24/20°C (day/night), 75% relative humidity, CO_2 level of 900 μmol/mol and photon flux density of 210 μmol/m^2 per s under red/blue LED lighting (Lin et al., 2013).

12.10   Lighting

Energy-efficient and effective artificial lighting is the technology which is largely responsible for the current rapid development of plant factories. Plant factories in the early 1990s were reliant on the use of HPS or MH lamps, the standard method of providing supplementary horticultural lighting in greenhouses. While HPS and MH lighting can provide the intensity required for completely enclosed production systems, these lamp types also produce a large amount of heat and must be positioned well above the crop being grown to prevent leaf burn. HPS and MH lighting systems also have a high energy requirement and thus the cost of running such lamps in a plant factory situation where no sunlight is available becomes a major running expense. Towards the end of the 1990s the use of fluorescent horticultural lamps became the more preferred method of lighting plant factory crops due to their higher PAR (wavelength 400–700 nm) output per Watt (Kozai et al., 2013). Fluorescent lamps also had a considerably lower heat output than older HPS or MH forms of artificial lighting and allowed the use of multilevel stacked and other vertical rack systems to be developed, thus increasing plant density per unit of area considerably. The most significant development in plant factory lighting came with the technology

improvements in LEDs which has resulted in highly advanced lighting systems. These LED systems can be used not only to provide sufficient illumination for photosynthesis, but also to control other important aspects of plant growth and development (Fig. 12.3). LED technology was a major breakthrough in the commercialization of plant factories as these highly energy-efficient lamps can produce the necessary light intensity for production at a far lower cost than HPS, MH or fluorescent lamps.

Apart from being highly energy efficient, LED lighting systems have a number of other advantages for plant factory installations. LEDs have a long operating life and do not require the regular replacement of bulbs other forms of horticultural lighting do. HPS, MH and fluorescent bulbs develop a gradual fall-off in light output over time which negatively affects plant growth unless these are regularly replaced based on their operating lifespan. LEDs can be positioned directly above the crop production plane with minimal distance between the tops of the plants and the LED source, as they produce minimal heat. Despite these advantages, one of the most promising aspects of LEDs is the manipulation of light wavelengths to benefit plant growth and this remains an area of intensive current research. In the past, the use of artificial lighting has focused on providing those wavelengths of light which are used in photosynthesis; however, plant responses to light are considerably more complex and LEDs allow flexibility with regard to wavelength and spectrum control. The ratio of red to blue light emission from LEDs has been shown to have different effects on plant growth. It has been found that the optimal light colour or wavelength for plant growth differs among species (Kim et al., 2006) and that it is necessary to determine the optimal light source for various species to maximize plant yields (Zhang et al., 2015). Various studies into the use of LEDs for plant factory lighting have revealed just how complex species or even cultivar interactions with wavelengths actually are. The use of white LEDs has been reported as being the most appropriate for romaine (cos) lettuce growth as compared

Fig. 12.3. LED lighting in the FarmPro container farm. (Image courtesy of Urban Crop Solutions, https://urbancropsolutions.com, accessed 11 September 2020.)

with either red or blue LEDs (Zhang *et al.*, 2015). Lettuce grown under exclusively red LEDs developed more leaves than that grown under blue LEDs (Yanagi *et al.*, 1996), however this lighting was not suitable for growth of spinach or radish (Yorio *et al.*, 2001). Many recent studies have indicated that combinations of white, red and blue LEDs have the most effect on photosynthesis and productivity and that the ratio of red to blue LEDs in lighting arrays can be customized to specific plant factory crops. Apart from photosynthesis, the spectral quality of LEDs can have dramatic effects on crop anatomy and morphology as well as nutrient uptake and pathogen development (Massa *et al.*, 2008). As more research is carried out into the effects of wavelength, this is likely to be a continuing technological development in plant factory design.

Other developments in plant factory lighting systems have focused on improving efficiency, minimizing wastage and accelerating growth under LED fixtures. High-density plant production systems with overhead lighting fixtures can have issues with leaf senescence on the bottom or outer leaves owing to shading by the upper leaves and by neighbouring plants, resulting in foliage beneath the plant canopy suffering from low light conditions (Terashima *et al.*, 2005). Lower leaves receiving insufficient light degrade, become yellowed and must be removed at harvest, resulting in increased wastage and reduced marketable weight. Studies have shown that providing light to both the adaxial and abaxial sides of a leaf can increase photosynthesis and different light colours have varying effects on leaf ageing and breakdown (Terashima, 1986; Causin *et al.*, 2006; Soares *et al.*, 2008). The use of supplementary LED lighting from underneath the plant in high-density lettuce crops has been shown to improve marketable leaf fresh weight, nutritional quality of plants, prevent senescence of the lower leaves and reduce both crop wastage and labour requirements (Zhang et al., 2015).

While LED lighting systems have become the focus of considerable research and provide advantages in the form of energy efficiency, low heat output, long usable lifespans and flexibility in spectral output, the main disadvantage of this technology is currently the high capital cost of such installations. Compared with fluorescent lighting systems, LED arrays are considerably more expensive to install and are a large contributor to the high establishment cost of plant factories. As technology develops, the cost of LED lighting systems is likely to decrease and the long-term saving in energy costs of running LED systems compared with other forms of lighting somewhat mitigates the initial capital cost of installation.

12.11 Environmental Control and Plant Quality in Plant Factories

While many studies have been carried out into the effect of environmental factors within plant factories on growth rates, yields and overall productivity, improvements in compositional quality are another potential advantage of this method of production. By consistently growing crops with higher nutritional status or other health benefits such as improved antioxidant properties, plant factory vegetables may receive premium prices, greater consumer acceptance and maximize the economic efficiency of such installations. Furthermore, the production of high-quality crops based on phytochemicals is a possible strategy for accelerating the practical use of plant factories (Lee et al., 2013). One of the most promising advancements in plant factory research has been into the relatively new adaptation of LED lighting technology. Not only do LEDs have a long lifespan, a high energy-conversion efficiency and low electric consumption, they can also be operated at a defined wavelength (Lee et al., 2014). Certain combinations of red and blue LEDs are known to be an effective source for plant growth and development; however, wavelength also influences a wide range of other compositional quality factors. The spectral quality of LED lights is the quantity of different wavelengths emitted by the light source and perceived by photoreceptors within the plant (Lin et al., 2013). Combinations of red, blue and white (broad-spectrum light) LEDs have been investigated to determine how different ratios of these may affect various aspects of plant quality. LED wavelengths differentially affect the metabolic systems of many vegetables, resulting in changes in soluble sugars and a number of other bioactive compounds. While red and blue LED combinations may result in optimal levels of yields and productivity, incorporation of a broader spectrum with the use of white LEDs, alongside red and blue, has been shown to have positive effects. These included improvements in development, nutrition, appearance and the edible quality of lettuce plants (Lin et al., 2013). Other studies have taken this principle a step further and investigated the effects of a wider light spectrum on plant quality. This included examining the role of UV wavelength ranges consisting of UVA (400–315 nm), UVB (315–280 nm) and UVC (280–100 nm) on the levels of valuable, health-promoting phytochemicals in plant factory crops. Lettuce plants exposed to UVA or UVB were found to have an increased concentration of polyphenols, including anthocyanin and flavonoids (Tsormpatsidis et al., 2010). UV light treatment can also be used to improve crop quality in terms of phenolic compounds and antioxidant properties in red leaf lettuce grown in closed-type plant production systems (Lin et al., 2013). More research into this field is continuing as these effects are likely to be species dependent.

12.12 Automation and Robotization

Plant factories are characterized by high-density and rapid-turnaround crops such as lettuce and other small vegetables. Traditionally under protected cultivation, each plant undergoes a number of handling stages, from seeding to transplanting into the production system, manual inspection, spacing out of plants as they develop, through to harvesting, trimming and packaging. The labour requirements for high-intensity plant factories where all tasks are carried out manually are high and a significant cost to the installation. Furthermore, these types of tasks are highly repetitive, may involve working in uncomfortable conditions or reaching up to hazardous elevated areas (Ohara *et al.*, 2015). Automation of the growing system in plant factories has been a long-term goal of researchers with increasing developments being seen in the field of robotics. Large, commercial plant factory installations where all tasks are carried out via robotic technology are close to becoming a reality. A Japanese company in Kameoka, Kyoto Prefecture is aiming to have industrial robots carry out the majority of cultivation tasks inside a vast indoor factory. Robots in this facility will be used to boost production from 21,000 to 50,000 lettuces per day and will perform all tasks apart from seeding which is still to be carried out by workers (McCurry, 2016). Robots and automation technology being developed for plant factory installations are seen as a way of filling labour shortages created by an ageing Japanese population and to improve efficiency, as it is estimated the use of robots will reduce labour costs by half (McCurry, 2016). Use of robotics and full automation is also considered a way of maintaining bacteria-free, clean cultivation by preventing contamination from being carried in from outside (Ohara *et al.*, 2015). Automated systems which control the robotic labour force may also control the environment within the plant factory including light, temperature, humidity and CO_2 levels. Plant factory robotics consist of a simple conveyor equipped with custom-made robotic arms able to handle and transplant sensitive young seedlings without causing damage. In Europe, the 'Urban Crops' automated plant factory has developed similar technology where only robots have access to the production trays in a completely closed, controllable growth environment.

12.13 New Innovations

Many new innovations in the design and development of plant factory installations focus on automation, robotics and improved methods of plant monitoring and assessment. One such example is a new automated and continuous plant weight measurement system developed for use in plant factories and research facilities employing CPPSs. Continuous monitoring of plant growth using high-technology, non-destructive methods that record plant weight during the growing cycle alongside an imaging system provides valuable data for hydroponic plant monitoring and researchers (Chen *et al.*, 2016). By being able to quantify plant weight accumulation in response to changes in the plant factory growing environment, researchers are better able to assess how developments in lighting, nutrition, CO_2, temperature and humidity may affect a given crop. Other automatic control systems measure growth stages of crops by utilizing a digital image-processing technique and aim to provide the optimal environmental conditions according to each stage of crop growth (Hwang *et al.*, 2014). Image-processing technology may also be used to measure and predict growth and determine the correct timing of harvest (Kim *et al.*, 2013).

Automation and robotics are perhaps one of the most intriguing aspects of plant factory development. While robots are currently being used in a number of existing plant factories for certain tasks, fully automated installations will become increasingly common as a way of reducing production costs and increasing efficiency. Robotic technology reduces or eliminates the requirement for staff to carry out repetitive, often highly physical tasks within an enclosed cropping environment. Alongside robotics and automation, one of the main factors determining the ongoing success of plant factories will be with increases in energy efficiency and lower capital investment, particularly with lighting equipment, and reductions in running costs. Increases in productivity and profitability to ensure large-scale plant factories are economically viable along with increased consumer acceptance of plant factory produce are important factors under investigation in a number of countries.

Increasing the efficiency of plant factory production is an ongoing area of research which includes not only improvements in energy efficiency, but also productivity. Plant breeding to produce cultivars more highly suited to enclosed plant factory production systems and environments is one aspect of plant productivity currently being developed. Another is the manipulation of plant growth to yield more of the marketable portion of the crop while minimizing the waste component, which is typically the root system and any leaves that require trimming or removal at harvest. Controlling root weight while not restricting the growth of the aerial portion of the plant is a key cultivation technique in plant factory production cost management (Kozai and Niu, 2016b).

Improving the flavour and nutritional quality of crops produced in commercial plant factories has been identified as important in order to improve overall consumer perception of produce grown in this way. Manipulation of the growing environment including LED lighting spectra, improved cultivars, precise nutritional control and other factors are being investigated to determine their effects on flavour, aroma, volatiles, vitamin and mineral content, bioactive and pharmaceutical compounds in the crops being grown. In the future, plant factories may be promoted as much for the proven high and consistent compositional quality of the vegetables produced as for their freshness and availability compared with outdoor, field-grown crops.

References

Bamsey, M.T., Zabel, P., Seidler, C., Gyimesi, D. and Schubert, D. (2015) Review of Antarctic greenhouse and plant production facilities: a historical account of food plants on the ice. *Presented at 45th International Conference on Environmental Systems*, 12–16 July 2015, Bellevue, Washington. Available at: https://core. ac.uk/download/pdf/31017763.pdf (accessed 11 September 2020).

Causin, H.F., Jauregui, R.N. and Barneix, A.J. (2006) The effect of light spectral quality on leaf senescence and oxidative stress in wheat. *Plant Science* 171(1), 24–33.

Chen, W.-T., Yeh, Y.-H.F., Liu, T.-Y. and Lin, T.-T. (2016) An automated and continuous plant weight measurement system for plant factory. *Frontiers in Plant Science* 7, 392.

Dickie, G. (2014) Q&A: Inside the world's largest indoor farm. *National Geographic*, 19 July 2014.

Dongxian, H., Po, Y. and Benhai, Z. (2007) Web-based environmental control system for closed plant factories. *Transactions of the Chinese Society of Agricultural Engineering No. 12.* Available at: https://dl.sciencesocieties.org/publications/tcsae/abstracts/2007/12/2007.12.030 (accessed March 2017).

FAO (Food and Agriculture Organization of the United Nations) (2009) Global agriculture towards 2050. *How to Feed the World 2050 High-Level Expert Forum*, Rome, 12–13 October 2009. Available at: http://www.fao.org/fileadmin/templates/wsfs/docs/Issues_papers/HLEF2050_Global_Agriculture.pdf (accessed 11 September 2020).

Fraunhofer Center for Molecular Biotechnology (2013) Automated plant factory for the production of vaccines. *Research News*, June 2013. Available at: https://phys.org/news/2013-06-automated-factory-prodution-vaccines.html (accessed 26 Septmeber 2020).

Goto, E. (2016) Production of pharmaceuticals in a specially designed plant factory. In: Kozai, T., Niu, G. and Takagaki, M. (eds) *Plant Factory: An Indoor Vertical Farming System for Efficient Quality Food Production*. Elsevier, London, pp. 193–200.

Harbick, K. and Albright, L.D. (2016) Comparison of energy consumption: greenhouses and plant factories. *Acta Horticulturae* 1134, 285–292.

Hirama, J. (2015) The history and advanced technology of plant factories. *Environmental Control in Biology* 53(2), 47–48.

Hu, M.-C., Chen, Y.-H. and Huang, L.-C. (2014) A sustainable vegetable supply chain using plant factories in Taiwanese markets: a Nash–Cournot model. *International Journal of Production Economics* 152, 49–56.

Hwang, J., Heong, H. and Yoe, H. (2014) Study on the plant factory automatic control system according to each crop growth step. *Advanced Science and Technology Letters* 49, 174–179.

JPFA (Japan Plant Factory Association) (2020) Homepage. Available at: https://npoplantfactory.org/en/ (accessed 26 September 2020).

Kim, H.H., Wheeler, R.M., Sager, J.C., Goins, G.D. and Norikane, J.H. (2006) Evaluation of lettuce growth using supplemental green light with red and blue light-emitting diodes in a controlled environment – a review of research at Kennedy Space Centre. *Acta Horticulturae* 711, 111–119.

Kim, M.H., Choi, E.G., Baek, G.Y., Kim, C.H., Jink, B.O., *et al.* (2013) Lettuce growth prediction in plant factory using image processing technology. *International Federation of Automatic Control Proceedings* 46(4), 156–159.

Kozai, T. (2013) Plant factory in Japan: current situation and prospectives. *Chronicals of Horticulture* 53(2), 8–11.

Kozai, T. and Niu, G. (2016a) Introduction. In: Kozai, T., Niu, G. and Takagaki, M. (eds) *Plant Factory: An Indoor Vertical Farming System for Efficient Quality Food Production*. Elsevier, London, pp. 3–6.

Kozai, T. and Niu, G. (2016b) Role of the plant factory with artificial lighting in urban areas. In: Kozai, T., Niu, G. and Takagaki, M. (eds) *Plant Factory: An Indoor Vertical Farming System for Efficient Quality Food Production*. Elsevier, London, pp. 7–32.

Kozai, T. and Niu, F. (2016c) PFAL business and R&D in the world: current status and perspectives. In: Kozai, T., Niu, G. and Takagaki, M. (eds) *Plant Factory: An Indoor Vertical Farming System for Efficient Quality Food Production*. Elsevier, London, pp. 35–66.

Kurihara, S., Ishida, T., Suzuki, M. and Maruyama, A. (2014) Consumer evaluation of plant factory produced vegetables. *Focusing on Modern Food Industry (FMFI)* 3, 1–8.

Lee, J.G., Choi, C.S., Jang, Y.A., Jang, S.W., Lee, S.G. and Um, Y. (2013) Effects of air temperature and air flow rate control on the tip burn occurrence of leaf lettuce in a closed-type plant factory system. *Horticulture, Environment and Biotechnology* 54(4), 303–310.

Lee, M.-J., Son, J.E. and Oh, M.-M. (2014) Growth and phenolic compounds of *Lactuca sativa* L. grown in a closed-type plant production system with UV-A, -B or -C lamp. *Journal of Science, Food and Agriculture* 94(2), 197–204.

Lee, S.G., Choi, C.S., Lee, J.G., Jang, J.A., Nam, C.W., *et al.* (2012) Effects of different EC in nutrient solution on growth and quality of red mustard and pak-choi in plant factory. *Korea Agricultural Science Digital Library*. Available at: http://agris.fao.org/agris-search/search.do?recordID=KR2015005510 (accessed 11 September 2020).

Li, Z., Paskevicius, A., Nagase, A., Ono, K. and Watanabe, M. (2014) Study of new technology supporting design of home use plant factory. *Bulletin of Japanese Society for the Science of Design* 61(3), 45–50.

Lin, K.-H., Huang, M.-Y., Huang, W.-D., Hsu, M.-H., Yang, Z.-W. and Yang, C.-M. (2013) The effects of red, blue and white light emitting diodes on the growth, development and edible quality of hydroponically grown lettuce (*Lactuca sativa* L. var. capitata). *Scientia Horticulturae* 150, 86–91.

Luna-Maldonado, A.I., Vidales-Contreras, J.A. and Rodriguez-Fuentes, H. (2016) Editorial: Advances and trends in development of plant factories. *Frontiers in Plant Science* 7, 1848.

McCurry, J. (2016) Japanese firm to open world's first robot-run farm. *The Guardian*, 2 February 2016. Available at: https://www.theguardian.com/environment/2016/feb/01/japanese-firm-to-open-worlds-first-robot-run-farm (accessed 11 September 2020).

Maneejantra, N., Tsukagoshi, S., Lu, N., Supaibulwatana, K., Takagaki, M. and Yamori, W. (2016) A quantitative analysis of nutrient requirements for hydroponic spinach (*Spinacia oleracea* L.) production under artificial light in a plant factory. *Journal of Fertilizers and Pesticides* 7(2). https://doi.org/10.4172/2471-2728.1000170

Massa, G.D., Kim, H.-H., Wheeler, R.M. and Mitchell, C.A. (2008) Plant productivity in response to LED lighting. *HortScience* 43(7), 1951–1956.

METI (Ministry of Economy, Trade and Industry, Japan) (2017) What is a plant factory? Available at: http://www.meti.go.jp/english/policy/sme_chiiki/plantfactory/about.html (accessed 11 September 2020).

Moon, A., Lee. S. and Kim, K. (2012) Design of a component-based plant factory management platform. *Presented at Service Computation 2012, The Fourth International Conferences on Advanced Service Computing*, Nice, France, 22–27 July 2012.

Ogawa, A., Eguchi, T. and Toyotuku, K. (2012) Cultivation methods for leafy vegetables and tomatoes with low potassium content for dialysis patients. *Environmental Control in Biology* 50(4), 407–414.

Ohara, H., Hirai, T., Kouno, K. and Nishiura, Y. (2015) Automatic plant cultivation system (automated plant factory). *Environmental Control in Biology* 53(2), 93–99.

Ohashi-Kaneko, K. (2016) Functional components in leafy vegetables. In: Kozai, T., Niu, G. and Takagaki, M. (eds) *Plant Factory: An Indoor Vertical Farming System for Efficient Quality Food Production*. Elsevier, London, pp. 177–185.

Soares, A.S., Driscoll, S.P., Olmos, E., Harbinson, J., Arrabaca, M.C. and Foyer, C.H. (2008) Adaxial/abaxial specification in the regulation of photosynthesis and stomatal opening with respect to light orientation and growth with CO_2

enrichment in the C4 species *Paspalum dilatatum*. *New Phytologist* 177(1), 186–198.

Terashima, I. (1986) Dorsiventrality in photosynthetic light response curves of a leaf. *Journal of Experimental Botany* 37(3), 399–405.

Terashima, I., Araya, T., Miyazawa, S.I., Sone, K. and Yano, S. (2005) Construction and maintenance of the optimal photosynthetic systems of the leaf, herbaceous plant and tree: an eco-development treatise. *Annals of Botany* 95(3), 507–519.

Tsormpatsidis, E., Henbest, R.G.C., Battey, N.H. and Hadley, P. (2010) The influence of ultraviolet radiation on growth, photosynthesis and phenolic levels of green and red lettuce: potential for exploiting effects of ultraviolet radiation in a production system. *Annals of Applied Biology* 156(3), 357–366.

Yamori, W., Zhang, G., Takagaki, M. and Maruo, T. (2014) Feasibility study of rice growth in plant factories. *Journal of Rice Research* 2, 119. https://doi.org/10.4172/jrr.1000119

Yanagi, T., Okamoto, K. and Takita, S. (1996) Effects of blue, red and blue/red lights of two different PPF levels on growth and morphogenesis of lettuce plants. *Acta Horticulturae* 440, 117–122.

Yorio, N.C., Goins, G.D., Kagie, H.R., Wheeler, R.M. and Sager, J.C. (2001) Improving spinach, radish and lettuce growth under red light-emitting diodes (LEDs) with blue light supplementation. *HortScience* 36(2), 380–383.

Zhang, G., Shen, S., Takagaki, M., Kozai, T. and Yamori, W. (2015) Supplemental upward lighting from underneath to obtain higher marketable lettuce (*Lactuca sativa*) leaf fresh weight by retarding senescence of outer leaves. *Frontiers in Plant Science* 6, 1110.

13 Greenhouse Produce Quality and Assessment

13.1 Background – Produce Quality and Testing

Producers of greenhouse crops have long been focused on yields as economic returns are largely dependent on marketable volumes of produce. However, in more recent years, consumers have become increasingly aware of other factors such as compositional quality, flavour and the health attributes of fresh produce. 'Quality' of greenhouse and hydroponic produce implies suitability for a particular purpose or the degree to which certain set standards are met. Aspects of produce quality may encompass sensory properties (appearance, texture, taste and aroma), nutritive values, chemical constituents, mechanical properties, functional properties and defects (Abbott, 1999). While greenhouse and hydroponic methods can produce higher yields and produce quality, soilless culture systems do not automatically result in the production of high-quality vegetables (Gruda, 2008). Fruit and vegetable production, being part of a larger biological system, is prone to a high degree of variability as factors such as growing environment, crop management and genetics all play a role in produce quality determination.

Quality standards and testing methods have been developed for most commercial crops to help ensure consumers receive produce of a suitable standard. These quality standards can range from basic grading for removal of damaged produce and for size, shape, weight and overall appearance, to analytical testing for compositional factors such as acidity, volatiles, dry matter, starch and sugars, toxins, vitamins and minerals, and others. With increasing competition globally within the greenhouse and horticultural industry and consumers' growing interest in high-quality, nutritious produce, compositional testing has

become more widespread, both at a grower and postharvest level. While appearance is still vitally important in marketing and for consumer appeal at the point of sale, focus on other aspects of fruit and vegetable quality is becoming more prominent. These include not only organoleptic qualities of flavour and aroma but also nutritional factors such as vitamins and minerals, fibre content and levels of phytoactive compounds such as antioxidants. Compositional quality also encompasses food safety and shelf-life concerns such as the presence of nitrates in leafy greens, microbial contamination and naturally occurring toxins with implications for human health and foodborne illness.

Crop quality testing is becoming more frequent and technologically advanced as rapid, accurate and portable testing equipment and methods are developed. These types of tests serve as an invaluable tool for not only greenhouse producers looking to maximize returns from a crop, but also for scientists and researchers whose objective is to improve compositional quality through a range of genetic improvements and production practices. Guiding crop improvement with quality assessment has long been a process carried out by producers, starting with the first crops to become domesticated several thousand years ago. These types of early testing may have been simple observations of obvious characteristics such as yields, size and appearance with selection of seed for the next generation based on these important factors. However, through the process of crop domestication and improvement, other types of testing began to emerge depending on which traits were considered the most valuable to agricultural societies. These may have included postharvest storage characteristics, such as produce which could be stored for longer and suffer less postharvest decay

DOI: 10.1079/9781789244830.0013

and losses before the days of modern refrigeration, freezing or controlled atmosphere (CA) storage. Other basic quality compositional testing before the modern age would have included sensory evaluation by humans. Long before any analytical equipment was developed to detect sugar levels (soluble solids), pH, acidity, flavour, texture and aromatic volatiles, humans were tasting and evaluating new crops and genetic strains for flavour and eating quality. These tests obviously led to the selection of open-pollinated fruits in particular with higher sugar levels, less seed and more pleasing organoleptic qualities. While taste panels and consumer acceptability assessment are still a component of produce evaluation, analytical testing is seen as a more precise and predictable indicator of specific quality characteristics. More recent developments in produce quality assessment have focused on developing sensors for automated, real-time, non-destructive instant quality measurement systems which can evaluate each piece of produce as it passes over a postharvest line, giving instant feedback on the quality of a crop.

Compositional quality testing of an increasing number of attributes is being more widely used than ever before by plant breeders looking to improve not only crop yields, but also many other internal quality factors of growing importance. Growers are also making increasing use of portable, accurate, rapid quality assessment instruments and equipment to determine factors such as harvest maturity, overall crop quality and the effect of production variables and crop management practices.

13.2 Components of Crop Quality

Quality is often highly specific to the crop being evaluated and industry standards are well publicized for many of the commonly grown fruit, vegetable and flower crops under commercial cultivation. For most fruits and some vegetables, internal quality factors such as sugar levels (Brix or total soluble solids (TSS)), flavour, aroma, nutritional value, degree of maturity, absence of biotic and non-biotic contaminants, titratable acidity which contributes to the sugar/acid balance, firmness and texture have long been part of regular internal quality assessment programmes. External quality aspects including size, shape, uniformity, colour, ripeness, freshness and overall presentation are the main components assessed by the consumer when selecting fruit and vegetables. Quality variables of importance vary not only between different crops but also within a species. Specific quality attributes are often well quantified and described with published industry standards, as produce of different quality grades often receives different prices on wholesale and retail markets.

13.3 Quality Improvement

Since a wide range of genetic and production factors plays a role in final crop quality determination, the role these variables have on influencing different aspects of quality has been intensely studied for most species. Seed companies carry out extensive testing of a wide range of cultivars and hybrids which result from their breeding programmes to determine how genetics has positively influenced quality, as well as yield and other factors. Increased emphasis these days has been on improvements in the nutritional quality of horticultural produce, including vitamin, mineral, fibre, antioxidant and other bioactive compound concentrations, as well as improving flavour in terms of sugars and acids. Important compositional quality factors such as flavour, shelf-life and firmness are linked to the dry matter content or soluble solids of many fruits and vegetables and since these tests are relatively simple and inexpensive to carry out, they are often used in the crop improvement process. Many crop management and production factors can influence TSS and percentage dry matter, thus testing these parameters allows an assessment of how changes in production might influence produce quality at harvest. Brix measurement to determine soluble solids is often used to provide feedback information on crop management practices which affect compositional quality. Under green-house production, variety selection and crop maturity are controllable factors affecting Brix readings. Irrigation level and fertiliza-tion, however, may be managed to improve Brix in some crops. Genetic differences in soluble solids levels between cultivars of the same crop are common, therefore variety se-lection is one of the most important and dir-ect methods of shaping percentage Brix and crop quality (Kleinhenz and Bumgarner, 2013). Growers often must carry out the process of variety soluble solids evaluation themselves as the local growing environ-ment interacts with plant genetics to deter-mine final Brix measurements.

13.4 Cultural Practices to Improve Greenhouse Produce Quality

13.4.1 Nutrient solution electrical conductivity levels, salinity and deficit irrigation

In some fruiting crops such as tomatoes, re-duced water availability during fruit devel-opment can increase fruit compositional quality in terms of soluble solids, but this is often at the expense of fresh weight and yields and requires precise control and moni-toring. Deficient irrigation practices, high EC and salinity via application of sodium chlor-ide are all methods of improving the compos-itional quality of hydroponic tomato fruit; however, these same practices are highly det-rimental to many other greenhouse crops and will negatively affect produce quality. Deficit irrigation is more commonly practised in field crops, although it can be applied in hydro-ponics with careful management. Withhold-ing irrigation and allowing a certain percentage of 'dry down' in the root zone is a practice which can be carried out in tomato crops to improve compositional quality of fruit via re-duced water uptake (Banjaw et al., 2017).

Application of salinity, via the addition of sodium chloride to the nutrient solution, is another method of compositional quality improvement in some hydroponic crops, most notably tomatoes. Many studies have

reported the positive effects of salinity on hydroponic tomato crops in terms of fruit compositional quality variables such as con-sumer-evaluated sweetness and overall fla-vour, TSS (Brix), total acids and titratable acidity (Peterson et al., 1998; Qaryouti et al., 2007; Azarmi et al., 2010; Zhang et al., 2016). Zhang et al. (2016) reported that yield of tomato is significantly reduced at salinity levels of 5 dS/m and above, where a 7.2% yield reduction per unit increase in salinity can be expected, and that the only parameter which is positively affected by salinity is fruit quality. While hydroponic to-matoes have been extensively studied with regard to salinity and compositional quality, these benefits have also been found in other crops. Similar positive results have been obtained in sweet pepper, cucumber, auber-gine, watermelon and courgette (zucchini squash) with increases in dry matter, sugars, starch and vitamin C contents (Sonneveld and van der Burg, 1991; Savvas and Lenz, 1994; Colla et al., 2006; Rouphael et al., 2006). Salinity application to strawberry crops has been reported to increase soluble solids con-tent and flavour as well as increasing the levels of antioxidants, ascorbic acid and anthocyanin concentration (Keutgen and Pawelzik, 2008; Cardenosa et al., 2014). Positive benefits have also been found in capsicum crops with moderate levels of sal-inity where fructose, glucose and myo-inosi-tol fruit concentrations increased relative to a control treatment (Rubio et al., 2009).

The stress of salinity via addition of so-dium chloride has been attributed to a com-bination of factors including possible toxicity of Na^+ and Cl^- ions, reduction in the uptake of other essential ions and a change in the osmotic potential in the root zone causing water deficit conditions (Zhang et al., 2016). Yield reductions under high salinity treat-ments are common; however, applications of silicon have been shown to assist with salin-ity tolerance in some crops by restricting the translocation of Na^+ and Cl^- into young foli-age (Savvas et al., 2007). Under increased salinity applied to improve fruit quality in pepper plants, the application of a higher concentration of calcium in the nutrient solution increased marketable yield and reduced the occurrence of blossom end rot (Rubio *et al.*, 2009). Khayyat *et al.* (2007) also found that application of additional calcium as $CaSO_4$ increased both fruit yield and quality (acidity, TSS and vitamin C) under salinity treatments as compared with controls where only salinity was applied.

Higher EC of the nutrient solution in soilless crops such as tomatoes and capsicum has also been proven to increase Brix levels as well as influencing aromatic volatiles and overall fruit flavour (AlHarbi *et al.*, 2014; Rosadi *et al.*, 2014). Application of increased EC differs from salinity treatments as all macroelements are increased, rather than the application of sodium chloride alone and the potential for toxicity this may create. In solution culture NFT systems, it has been found that increasing the nutrient solution EC from only 2 to 4 mS/cm resulted in an increase in tomato fruit quality in terms of fruit flavour and furthermore that even small increases in soluble solids (Brix) or titratable acidity can be detected by the majority of consumers (Morgan, 1996). Further to improvements in dry matter percentage, high EC treatments (up to 10 dS/m) applied to hydroponic tomato crops have been found to increase levels of vitamin C, lycopene and β-carotene (the precursor of vitamin A) in fresh fruit by up to 35% (Krauss *et al.*, 2006). The application timing of increased EC levels has also been studied and indicates that delaying this treatment until 4 weeks after anthesis did not change tomato fruit quality improvement of a 30–40% increase in dry matter under an EC of 4.5 mS/cm as compared with an EC of 2.3 mS/cm (Wu and Kubota, 2008). Other crops have been found to respond to higher EC levels in a similar way. Caruso *et al.* (2011) found that strawberry fruit quality was improved when the nutrient solution concentration was increased to 2.2 mS/cm in a winter-grown crop. In a study carried out on hydroponic pepper crops, Amalfitano *et al.* (2017) reported that fruit antioxidant concentrations, soluble solids, dry matter residue and sugar content increased when a nutrient solution at EC of 4.4 mS/cm was applied as compared with lower EC levels.

While increases in the compositional quality of tomato fruit are a positive outcome of applying high EC levels, the drawbacks of this technique have been reported as a reduction in fruit size, higher occurrence of blossom end rot and a loss in yield (Nichols *et al.*, 1994; Morgan, 2004). Losses in yield potential under high EC treatment are likely to be due to an increased concentration of phloem sap entering the fruit, thus less water accumulates in fruit tissue and fruit size is subsequently reduced (Morgan, 2008). Methods of reducing yield losses under salinity or high EC treatment used to improve fruit compositional quality have been evaluated. These include strengthening of the root system and reducing plant stress via high rates of oxygenation in the root zone (Morgan, 2008), application of silicon (Stamatakis *et al.*, 2003) and inoculation with beneficial microorganisms such as mycorrhizal fungi (Sellitto *et al.*, 2019). Use of arbuscular mycorrhizal fungi (AMF) has been found to improve plant performance under salinity stress conditions in crops such as tomato, cucumber and lettuce; the suggested mechanism is that these fungi improve salt tolerance by assisting water and nutrient uptake, increasing osmotic regulation and photosynthesis rates, and improving water-use efficiency (Porcel *et al.*, 2012). Thus, inoculation of greenhouse crops with AMF has been shown to help overcome some of the negative effects on crop growth of increased salinity used to improve the compositional quality of fruit such as tomatoes and reduce yield losses. Sellitto *et al.* (2019) found that the highest values of most fruit quality indicators in tomato fruit, mineral elements, antioxidant compounds and activity were recorded under an EC of 6.0 mS/cm with the addition of AMF formulations as inoculants.

13.4.2 Calcium and potassium and compositional quality

Calcium and potassium are the two elements most associated with the compositional quality of hydroponic fruit and vegetables.

A deficiency in either calcium or potassium can have significant effects on appearance and compositional quality. Calcium deficiency is common in many fruit and vegetable crops and while usually plentiful in the nutrient solution or growing medium, is often the result of inadequate translocation of calcium within the plant (Olle and Bender, 2009). Calcium uptake and movement within plants are driven by transpiration from the foliage, thus calcium deficiency and disorders often occur when humidity levels restrict transpiration and thus calcium distribution to the ends of the transpiration steam. Calcium deficiency is particularly severe where high salinity in the root zone is present. Symptoms of calcium deficiency or induced calcium issues within the plant include tipburn (marginal necrosis) in lettuce, chervil, onion, fennel, Chinese cabbage and other cabbages (Olle and Bender, 2009), blossom end rot of fruits such as tomatoes and many others, plant stunting, leaf chlorosis, death of terminal buds and root tips; root growth may be inhibited as well. Calcium is not highly mobile within the plant and older leaves are typically unaffected by deficiency, which is most pronounced on the growing tissues and growing points. Due to its role in strengthening cell walls, calcium deficiencies may lead to increased susceptibility to certain plant diseases (Kennelly et al., 2012). Nutritional interactions which may reduce the uptake of calcium by the root system and induce deficiency symptoms include high levels of the ammonium form of nitrogen or luxury uptake of potassium which competes with calcium uptake.

Potassium deficiency can be particularly damaging in many greenhouse crops as it affects not only yields, but also fruit quality. Symptoms of potassium deficiency include interveinal chlorosis, leaf necrosis, leaf margin dieback, leaf curling and browning, low fruit solids and reduced shelf-life. Potassium is a mobile element and symptoms show first on older foliage. Potassium is generally found in sufficient quantities in most hydroponic formulations, however deficiencies in this element may be induced by high levels of competing cations such as calcium (Ca^{2+})

and ammonium (NH^+_2) in the root zone or where certain environmental conditions limit uptake.

Particularly for fruiting crops such as tomatoes, cucumbers, aubergine and melon, high levels of potassium are strongly linked to compositional quality. Fanasca et al. (2006) found that a high proportion of potassium in the nutrient solution increased levels of antioxidants (especially lycopene) in tomato fruit. In hydroponic strawberries grown in a coconut fibre/perlite medium, the highest values of fruit pH, EC, TSS/titratable acidity ratio, vitamin C content, ellagic acid and colour were found in fruit grown with a nutrient solution containing K:Ca ratio of 1.4 (Haghshenas et al., 2018).

13.5 Environmental Conditions and Produce Quality

Environmental conditions play a significant role in determining the quality of greenhouse crops, not only affecting the physiological processes and produce appearance, but also the internal quality such as sugars, acids, flavour, vitamins and secondary plant compounds (Gruda, 2009). Under greenhouse and hydroponic production growers have considerably more control over the environment than with crops grown in the field, however cost can become a factor when significant levels of heating/cooling or lighting are required. The main environmental factors which are known to influence produce quality include light and temperature, with cultural practices such as moisture levels, EC, salinity, plant stress, genetics and nutrition interacting with climatic conditions to determine the final quality of a crop.

13.5.1 Light levels and produce quality

Light and rates of photosynthesis, which determine assimilate levels available for growth and importation into fruit, are essential factors in produce quality. Low light is common in many climates during certain seasons and high-density cropping puts further pressure on supplying sufficient light for the production of produce with optimal compositional quality. Symptoms of insufficient light are straightforward to identify on most crops and include elongated growth, etiolated seedlings, pale, weak stems, pale foliage coloration, reduced growth rates, lower yields, poor-quality fruit often with low solids levels and weakened plants which are more prone to disease infection. Low light may be common under shorter day lengths in winter, under protected cultivation where twin-skin plastics are used for heat retention and in nursery situations. Apart from a reduction in growth rate and lower yields, for high-light vine crops such as tomatoes, melons and capsicum a low DLI also restricts fruit quality in terms of sugars, fruit dry weight and flavour. If DLI falls particularly low, flower and fruitlet drop may occur as the plants fail to produce sufficient assimilate for reproductive growth. Plants under low light may also have plant 'balance' altered, with tall, weak, thin or excessive vegetative growth and minimal fruit set and growth.

Greenhouse tomato, aubergine and capsicum are high-light-requiring crops (Fig. 13.1) – far greater DLIs are required for these plants than for many other crops growing under protected cultivation such as nursery seedlings, many potted indoor plants, lettuce and herbs. High-light crops are those which typically require a DLI of 20–30 mol/m^2 per day (Morgan, 2004). Since tomato and capsicum tend to be grown at a relatively high density (thus plant shading occurs), maintaining the correct light levels becomes vital for production. The optimum DLI for hydroponic lettuce (butterhead), a crop with lower light requirement, is recommended to be about 14–16 mol/m^2 per day; however, a mature tomato crop needs an estimated DLI minimum of at least 22 mol/m^2 per day for good production (Morgan, 2012). Some studies report tomato crops may need DLI as high as 30 mol/m^2 per day (Dorais, 2003).

Incoming light levels are determined by a combination of factors including naturally occurring radiation levels, day length, loss of light through greenhouse claddings, use of screens and shading, and factors intrinsic to the production system. The use of vertical hydroponic systems to maximize greenhouse area is one factor which can play a significant role in light optimization by crops. Vertical systems are characterized by high plant density and are most commonly used for smaller plants such lettuce, herbs and

Fig. 13.1. High-light-requiring crops include aubergine, tomato and capsicum.

strawberries; however, they can result in insufficient light in the lower levels of the system. Strawberry fruit compositional quality is particularly prone to low light issues with reductions in TSS (Brix) and flavour reported. In vertical strawberry systems a number of studies have found that fruit compositional quality is negatively affected in the lower regions of the system (Karimi et al., 2013). Lower levels of some vertical strawberry systems had only 30% of the light compared with upper levels and this reduced fruit Brix significantly by 2° (Ramirez-Arias et al., 2018). Ramirez-Gomez et al. (2012) found a similar result, with strawberry fruit Brix reduced by more than 2° from the top to the base of vertical production systems. Consideration of light penetration and levels, plant density, use of supplementary lighting and matching vertical systems to high-light climates are all factors which can limit compositional quality losses in fruiting crops.

While insufficient or excess light can damage growth, reduce yields and create physiological symptoms, fruiting crops in particular may be prone to cosmetic damage such as scorch, sunscald and other forms of cosmetic surface damage. Excess light may cause a range of symptoms depending on species and adaptation to radiation levels. Plants which have been grown under shaded or protected conditions, such as transplants raised under cover in a nursery situation, may suffer damage if not gradually hardened off to full sunlight levels. Such damage often shows as scorched or bleached foliage with a growth check until acclimatization has occurred. Persistent excessive light damage is more common in shade plants and often results in stunted, hard and dark-coloured growth. Some plants, when exposed to excessively high and damaging light levels, will develop strategies to avoid or limit cell damage. This may include leaf rolling, leaf drooping or orienting leaves upwards to restrict the amount of foliage surface area exposed to the damaging radiation levels. High light intensity can also lead to physiological disorders of fruiting crops which significantly reduce quality; these include sunscald injury and uneven ripening of tomatoes and other fruit which are brought on by direct effects

of light on fruit and an interaction with high temperature (Adegoroye and Jolliffe, 1987). Apart from light intensity, radiation spectrum can also play a role in compositional quality of hydroponically grown fruit and vegetables. In hydroponic lettuce, blue and UV irradiation increased the anthocyanin content of the foliage (Johkan et al., 2011), thus giving a more intense coloration in red varieties.

13.5.2 Temperature and produce quality

Many plants are negatively affected by high temperature conditions and may stop growing, develop wilt during the warmest part of the day and suffer root dieback if extreme temperature conditions persist for an extended length of time. Plants may also exhibit foliar symptoms such as leaf scorch, tipburn, premature leaf drop, leaf bleaching and abnormal coloration. Fruit may develop disorders such as blossom end rot of tomatoes and capsicum or fruit skin disorders such as blistering, crazing, cracking, solar yellowing or other discoloration. Fruiting crops such as tomatoes, capsicums and strawberries are particularly prone to damage to surface and internal tissues by combinations of high temperatures and intense solar radiation (Hall, 2001). Some species react with a lack of fruit set caused by poor pollen viability under high temperatures. Heat injury in some species may also damage floral development so that plants do not produce flowers (Hall, 2001).

Exposure of chilling-sensitive plants to low temperatures causes disturbances in a number of physiological processes including water regulation, nutrition, photosynthesis, respiration and metabolism, with the degree of symptoms dependent on both temperature and duration of exposure to chilling conditions. Chilling injury symptoms include wilting or water-soaked areas, surface lesions, water loss and desiccation or shrivelling, internal discoloration, tissue breakdown, failure of fruit to ripen or uneven or slow ripening, accelerated senescence and ethylene production, shortened storage or shelf-life, compositional changes, loss of growth or sprouting capacity, and increased decay due to leakage of plant metabolites (Skog, 1998).

Air temperature not only plays a significant role in the rate of plant growth and yields, but also in compositional quality. Low temperatures have been shown to accelerate anthocyanin and chlorophyll production in red leaf lettuce giving a more intensely coloured product (Gazula et al., 2005). The ascorbic acid (vitamin C) content of strawberry fruit has been found to increase when plants are produced at low temperatures (Wang and Camp, 2000), while fruit coloration decreases at high temperatures (Ikeda et al., 2011).

13.5.3 Nutrient solution chilling

Under high temperature conditions such as those often experienced in tropical greenhouses, root-zone cooling via application of a chilled nutrient solution has proven to be beneficial for a number of different crops. This has often been attributed to the fact that the most significant temperature influencing the growth of some plants is that in the root zone (Niam and Suhardiyanto, 2019). Use of nutrient solution chilling in NFT, DFT and aeroponics systems for cool-season crops such as lettuce to maintain commercially acceptable growth, quality and yields in tropical climates has been widely accepted (He et al., 2001). Chilling the nutrient solution down to as low as 16–18°C allows cool-season vegetables to crop well at ambient air temperatures which are often well above optimal for these crops (28–36°C). Without nutrient chilling, the root zone usually warms to the level of the air and this gives numerous growth problems including slow growth, lack of heart formation, bolting, tipburn and low marketable yields. However, more recent research has indicated other crops also benefit from root-zone cooling and that this can also play a role in fruit quality. He et al. (2019) reported that root-zone cooling (from 21–29°C to 10°C) of hydroponic cocktail tomato crops

grown on rockwool in summer improved fruit quality in terms of sugars, vitamin C and lycopene contents. Under high ambient temperatures (33.7°C), root-zone cooling to 24.7°C in young tomato plants was found to increase the relative growth rate of roots and shoots as well as increasing calcium and magnesium uptake (Kawasaki et al., 2013). Cooling of the nutrient solution to 22°C as compared with an ambient solution temperature of 33°C improved the levels of DO in both the fresh nutrient tank and drain nutrient solution in a perlite-grown hydroponic cucumber crop under high-temperature growing conditions in Oman (Al-Rawahy et al., 2019). Furthermore, nutrient solution chilling improved all growth, production and quality attributes in hydroponic cucumber crops in both summer and winter growing periods under high air temperatures (Al-Rawahy et al., 2019). Hydroponically grown strawberries in a deep flow system with roots exposed to low temperatures (10°C) produced higher biomass than those exposed to ambient conditions (20°C) (Sakamoto et al., 2016). Hydroponically grown red leaf lettuce produced with a low root-zone temperature of 10°C as compared with 20°C for 7 days resulted in leaves which contained higher concentrations of anthocyanin, phenols, sugar and nitrate, suggesting that such low-temperature treatments activate antioxidant secondary metabolic pathways (Sakamoto and Suzuki, 2015) which would further improve compositional quality.

13.6 Genetics and Produce Quality

Genetics plays an essential role in the quality of greenhouse-grown produce with plant breeders now focusing not only on yields, disease resistance and shelf-life, but also the compositional quality of many fruits and vegetables. This incorporates flavour quality via improved sugar and acids, as well as higher levels of antioxidant compounds such as lycopene concentration in tomato fruit. The final quality achieved from a particular cultivar is, however, dependent on the interaction between plant genetics, the growing environment and crop management, thus the grower cannot rely on genetics alone to produce a high-quality crop. Evaluating a number of different suitable cultivars is advisable for growers as an ongoing process to determine which genetics are best matched to the local growing conditions and system.

13.7 Quality Testing and Grading Methods

13.7.1 Colour analysis

The development of natural pigments such as carotenoids, anthocyanins, flavonoids, betalains and chlorophylls in fruits and vegetables determines the final colour which often changes through the process of ripening. Colour is often the first impression made on consumers when selecting fresh fruits and vegetables and is thus vitally important for quality and overall acceptance. Colour sorting is carried out in produce which is harvested within a range of maturity such as tomatoes and other fruits where skin colour changes as fruit ripens and is therefore regarded as a quality index. Colour charts are typically used for manual colour assessment of fruit such as tomatoes and peppers. Colour indices may also be assessed by using optical methods to measure skin colour that is based on the level of chlorophyll degradation or the development of colour pigments. Postharvest, colour imaging processing has become an important feature of many fruit graders and different models exist based on the fruit being graded. Fruit bruising and other surface defects may also be assessed via electronic graders using colour imaging. Defective or bruised regions of the fruit surface have a different spectral reflectivity from healthy tissue, and this can be used to detect and remove fruit from the grading line. Grading using machine vision is achieved in six steps: image acquisition, ground colour classification, defect segmentation, calyx and stem recognition, defects characterization and finally fruit classification into quality classes (Leemans et al.,

2002). Colour grading or optimal sorting using image vision technology may be carried out in combination with size, shape and defect classification as the fruits pass over the grading line. Computer optical imaging and grading is being utilized on an increasing number of crops. The major advantages of this system is the rapid throughput of produce, a high degree of quality control and assessment, reduced labour costs, minimal errors in grading and a reduction in the reliance on human involvement in tedious, repetitive and subjective tasks (Narendra and Hareesh, 2010).

13.7.2 Total soluble solids (Brix) testing

Refractive index (Brix) measurement is one of the most common quality assessment tests carried out on a wide range of fruit and vegetables (Fig. 13.2). Percentage Brix readings are an indication of TSS content, which is representative of the sweetness of the produce. Brix may be used as both a quality control test and also to determine the level of ripeness or maturity, particularly of fruit. As starch is transformed into sugars during the ripening process, Brix levels can be used as an assessment of ripeness stage which allows growers to predict or determine when fruit are of sufficient quality for harvest. Since many other production factors also determine final product TSS levels, Brix readings are also used by growers and researchers in crop improvement processes. Management practices such as cultivar selection, fertilization, irrigation control, environmental conditions and plant health all play a role in final TSS levels and Brix measurement is typically used to determine variables which will improve compositional quality.

Percentage Brix (TSS) is measured with a refractometer in 0.1% graduations. A small sample of fruit juice or extract is placed within the refractometer and the determination of TSS is based on the capacity of sugars in the juice to deviate light. TSS assessment using refractometers is a long-standing practice and may be carried out using hand-held, manual refractometers and an external light

Fig. 13.2. Brix testing is a common compositional quality assessment.

source such as sunlight, or with digital battery or mains-operated models in a laboratory situation. Hand-held refractometers are operated by placing a few drops of the juice sample to be tested on to the prism plate and closing the lid. The meter is then turned towards a strong light source and the eyepiece focused until a clear image can be seen. The scale on the image can then be read where the line between the light and dark regions forms, this gives a reading in percentage soluble solids. Digital models simply require a sample of juice or other plant extract to be placed into the well and the soluble solids reading will be displayed. Basic refractometers may not automatically compensate for temperature, thus for readings taken outside the standard 20°C range, reference to the International Temperature Correction Table (usually supplied with basic Brix meters) is required. Electronic refractometers often have temperature compensation built in or can be calibrated to provide accurate readings at a given temperature.

While Brix is relatively straightforward to measure with a refractometer, the samples to be tested must be representative of the crop and prepared in the correct way to ensure accuracy. Squeezing juice from a single cut fruit on to a refractometer plate will not give a suitable assessment of the entire crop or batch of fruit as variability always exists within a crop. A sample of at least ten fruits which are free from blemishes, pest and disease damage and are representative of the size and stage of maturity of the crop should be collected for Brix testing. Juice can then be extracted from all the fruits and blended for testing; this process varies depending on the species under evaluation. Tomato and other fruit may be put through a domestic juicer and the sample allowed to settle for a few minutes or be filtered before Brix determination. In the field, where juice may be squeezed directly from the fruit, this should be taken from a number of different samples and the mean result recorded. However, where the fruit is directly squeezed into the refractometer for measurement, this may not be completely representative of the Brix level of the entire fruit as different tissues produce varying levels of TSS. Hand-squeezed juice often results from certain regions of the whole fruit more than others, i.e. the locules in tomatoes. If Brix measurements are being used as part of an evaluation into the effect of different production techniques or cultivars on Brix levels, then a basic comparison of samples from the same testing date can be used. However, if Brix measurements are to be used to determine crop ripeness for harvesting or overall compositional quality, then TSS baseline guides should be consulted to determine where the samples lie within the ideal range for each crop. USDA crop grading standards for cantaloupe and watermelon state that a minimum soluble solids level of 11% is rated as 'very good internal quality', while 9% is 'good internal quality' for cantaloupe. Examples of the typical soluble solids (°Brix) values for a number of other crops have been published (see Table 13.1).

13.7.3 Dry weight percentage

Dry matter content is a component of compositional quality often used during assessment of many fruits and vegetables. There is a good correlation between the dry matter content and organoleptic (flavour) characteristics in a wide range of produce and generally a higher content of solids means improved quality and taste (Camelo, 2004). Dry matter determination is a straightforward test which involves removal of all moisture from a sample. The accepted method for determination of percentage dry matter is dehydrating the sample in a (vacuum) oven at 70°C until consecutive weighings made at 2 h intervals show that a constant weight has been reached (OECD, 2009). For this process a sample of at least 15 fruits, free of defects and of the same stage of ripeness is taken at random and prepared by cutting and weighing of the sample. The sample is placed in the oven with an air flow of 70°C

Table 13.1. Typical soluble solids (°Brix) for selected crops. (From Kleinhenz and Bumgarner, 2013.)

Crop	°Brix
Cucumber	2.2–5.4
Pepper	6–9
Processing tomatoes	4.7–6.0
Greenhouse tomatoes	3.8–5.0
Cherry tomatoes	5–8
Fresh market tomatoes	3.5–5.3
Courgette	2.4–6.0

for a number of hours until no further weight loss occurs. Once dry, the sample is reweighed immediately after removal from the oven and percentage dry weight calculated based on the fresh and dry weight recordings for each sample. A similar process can be carried out using a microwave oven to increase the speed of desiccation. Samples are placed in a microwave of 800 W for 4 min initially, then for further 1 min intervals, weighing between each time period until no further weight loss occurs (OECD, 2009).

13.7.4 Acidity and pH

As with TSS, acidity quantification in fruit and vegetable samples can be used as both an indicator of ripeness and as a quality assessment variable. The ratio of sugar to acid in produce contributes significantly to consumer acceptance and overall flavour perception and is an important aspect of quality control. During the early stages of the ripening process the sugar-to-acid ratio tends to be low, thus immature fruit taste unacceptably sour. As ripening progresses fruit acids begin to be degraded and sugar levels increase, resulting in a higher sugar-to-acid ratio which gives many fruits their characteristic ripe flavour. Over-ripeness results in very low levels of fruit acids and a lack of overall flavour acceptability. Acidity or sourness can be determined by sensory evaluation or analytical tests of either pH or more accurately determination of titratable (total) acidity. pH meters can be used to determine the pH level of correctly prepared fruit and vegetable samples.

Determination of total acidity involves titration of the sample of fruit or vegetable juice against sodium hydroxide solution and measuring the volume of sodium hydroxide required to reach an end point pH of 8.1. For determination of titratable acidity, at least ten randomly selected fruits of the same stage of ripeness, free from defects and damage are prepared by homogenization or juicing to create a liquid sample which is then filtered to remove solid particles. The clear, filtered extract is then used for titration against 0.1 M NaOH solution and the volume of NaOH required to reach pH 8.1 is recorded for each replicate. To determine the final percentage of acid in the sample being evaluated, the correct multiplication factor must be used. Some produce contains more than one type of acid and it is the primary acid that is tested during titration. For citric acid, a multiplication factor of 0.0064 (1 ml of 0.1 M NaOH is equivalent to 0.0064 g of citric acid) is used and for tartaric acid (grapes), a factor of 0.0075 is used. Percentage acid is then calculated by multiplying the amount of titrate × the acid factor × 100, then dividing by the volume of juice sample used (in millilitres), usually this is 10 ml (OECD, 2009). This value of percentage acid can then be used, along with Brix (TSS) of the same sample, to determine the sugar-to-acid ratio of the produce being evaluated.

13.7.5 Flavour quality – aroma and taste

Flavour is a vital aspect of the compositional quality of food crops and is made up of a combination of the sensations of taste, aroma, texture and pressure. Determining flavour quality is a complex process and there is no one single compound in any fruit or vegetable that is solely responsible for its taste or aroma. Volatile aromatic compounds are perceived through smell, while taste receptors in the mouth contribute to the overall flavour experienced by the consumer. Taste is divided into five basic categories: salty, sour, sweet, bitter and umami. Umami is a difficult-to-describe taste which is associated with salts of amino acids and nucleotides (Yamaguchi and Ninomiya, 2000). Flavour of some types of produce can be linked to groups of related compounds; for example, onions contain characteristic sulfur compounds, bananas contain isoamylacetate and celery distinctive phthalides, while a number of volatiles convey specific aromas in muskmelons and tomatoes (Barrett *et al.*, 2010). When evaluating the flavour of fruits and vegetables it is not just the desirable characteristics that are under assessment. The development of 'off flavours', which may be produced by a number of biochemical reactions including oxidation, is often evaluated as they constitute an important aspect of compositional flavour quality.

13.7.6 Sensory evaluation of compositional quality

Sensory evaluation of fruit and vegetables is used to determine variables such as flavour, aroma and, to a lesser extent, texture. While instrumental measurements can be used to precisely determine compositional quality characteristics such as sugar or acid levels, these do not always give a completely accurate indication of the overall taste or degree of acceptability as experienced by consumers. Sensory evaluation is widely used in the horticultural industry, often alongside laboratory analysis and compositional testing, for a number of different purposes. It is often used by plant breeders to develop new cultivars with improved compositional quality in terms of flavour perception by consumers or to determine how various changes in production factors and crop management practices impact final flavour. Postharvest storage conditions and packaging methods may be evaluated as these contribute to the final eating quality of fresh produce. Sensory evaluation may also be used by marketers to determine what characteristics consumers prefer in fresh intact fruits and vegetables or in those which have been minimally processed such as fresh-cut product lines. With the greenhouse fresh-cut industry growing at a rapid rate and quality maintenance being a considerable priority with these higher-value products, sensory evaluation has become an important aspect of this industry. While with intact fruits and vegetables sensory quality is largely attributed to production factors and stage of ripeness, with fresh-cut produce the effect of tissue damage, oxidation and treatments to prolong shelf-life can all change flavour before the visual appearance deteriorates. Thus, sensory evaluation is an important tool for monitoring how flavour and texture change postharvest in a wide range of fruit and vegetable products.

Quality assessment of ripe fruit using sensory evaluation panels may also be used in the marketing and distribution chains to determine the effect on stage of maturity at harvest. It is common for some fruits to be harvested before they have reached final eating maturity, and this can affect the compositional quality of the produce when tasted by the final consumer. Maturity at harvest plays a central role in flavour development, particularly for climacteric fruits where ripening is regulated by ethylene (Mattheis and Fellman, 1999). Fruit harvested at a stage which is too immature often do not achieve characteristic flavour due to the development of only low levels of essential compounds. Early harvest to prolong shelf-life and maintain texture during handling and transportation can come at the expense of flavour and this can be assessed with the use of postharvest sensory evaluation. For example, tomatoes picked at a relatively immature stage and allowed to ripen are less sweet, more sour, less 'tomato-like' and have more 'off flavour' compared with vine-ripened fruit (Kader et al., 1977).

Sensory evaluation testing of horticultural produce and processed products is divided into two main categories. Analytical sensory measurements use a small number of panellists, often trained to recognize and quantify certain characteristics, to detect differences (difference tests) between samples or to describe the product (descriptive analysis) (Barrett et al., 2010). Affective sensory measurement is used to determine overall consumer preference or acceptance of produce being evaluated. Affective sensory evaluation requires the use of a much larger number of panellists who are untrained and these often include focus groups, surveys and point-of-sale consumer testing. Affective consumer tests are used to determine what consumers prefer, whereas analytical sensory testing can be used to identify small differences in quality between similar samples (Barrett et al., 2010). Results from consumer panel evaluation can be highly variable and require large numbers of participants to obtain useful data; however, they are relatively easy to run and do not require extensive training of panel

members. Trained taste panels provide a higher degree of accuracy regarding differences between samples than consumer testing, however if panellists are not trained correctly results can be highly variable.

Of the analytical test methods of sensory evaluation, the most commonly used testing procedures are the difference test and the descriptive analysis test. Difference tests, including the triangle test, provide panellists with three samples of the fruit or vegetable being evaluated and ask panellists to identity which sample is different from the other two. A more basic example of difference testing is where panellists are provided with two samples and asked if there is any detectable difference between the two. Since panellists of these types of difference tests are not trained, at least 30, but preferably 50 or more participants are required. The samples in difference tests must all be of the same size, colour, degree of ripeness and overall appearance and presented in a way in which the product would normally be consumed. The testing environment, as with all controlled sensory evaluation procedures, must be controlled to minimize distractions and prevent panellists communicating with each other during the evaluation process.

Descriptive sensory analysis uses trained panellists to describe differences between presented samples of produce. Panellists are initially selected based on their ability to perceive slight differences in flavours and aromas and then put through a process of training and calibration to ensure the panel results will be accurate and precise (Barrett et al., 2010). During the testing procedure trained panellists are asked to rate certain flavour attributes by placing a mark on a line from low to high for values such as sweetness, sourness, bitterness, salty, flavour intensity, umami and aroma, or alternatively to give a range from 1 to 10 for each attribute. Other minor attributes assessed are dependent on the fruit or vegetable under evaluation but may include descriptors such as 'astringent', 'rotten' or 'off flavours', 'fermented', 'green' or 'grassy', 'musty/earthy' or 'floral' (Bett, 2001). Apart from flavour and aroma, texture may also be assessed by trained taste panellists who have sufficient experience to differentiate between different aspects of tissue firmness. The texture attributes of produce can be divided into four areas: surface properties, first-bite properties, chew down and after-swallowing properties (Meilgaard *et al.*, 1999). Some of the most commonly tested texture attributes assessed during sensory evaluation are crispness, hardness, juiciness or moisture release (Bett, 2001). Crispness is an assessment used in both fruits and vegetables as it as an indicator of freshness or the amount of moisture loss which may have occurred postharvest and is related to turgor pressure within the cells. Crispness is relatively easy to assess and is determined by the amount of force and noise with which the sample breaks when compressed with molar teeth (Meilgaard *et al.*, 1999). Juiciness can indicate that ageing or dehydration has occurred or simply be an evaluation of the juice content of different fruit tissue samples and is assessed by the amount of moisture perceived in the mouth. Minor attributes which may also be assessed during texture perception are chewiness, which is the number of chews required to prepare for swallowing, and mouth coating, which is the amount of residue left in the mouth and on teeth surfaces after swallowing (Meilgaard *et al.*, 1999).

Samples used in sensory evaluation procedures must be prepared correctly for accurate results to be obtained. This includes careful selection of fruit and vegetable samples which are uniform, of the same degree of ripeness, particularly colour (unless sample colour is being masked with use of lighting) and prepared in the same way. Produce should be washed, dried and cut into uniform and even bite-sized pieces so that appearance is similar between samples. Even slight differences in preparation, size or colour between samples can influence panellists' flavour perception. Samples should be evaluated at room temperature to allow maximum release of volatile aromatic compounds. Panellists should be instructed to smell each sample first to assess aroma before tasting and evaluating for flavour and texture profiles. Analysed results from sensory evaluation may be used as a stand-alone

quality assessment; however, these may also be correlated with results from analytical quality tests on the same samples, such as TSS (often linked to sweetness), dry weight (linked to overall flavour quality, firmness and shelf-life) and titratable acidity (may be linked to sourness in some fruits and vegetables). Sensory evaluation is considered an important indicator of compositional quality in many fruits and vegetables used by researchers, growers and those involved in produce processing, marketing and sales.

13.7.7 Volatiles testing – aroma

Fruit and vegetables at maturity develop a specific blend of volatile compounds that characterize aroma and also contribute to the overall flavour quality of the produce. Most ripe fruits produce a large number of volatile compounds in minute quantities; however, these can be perceived by human olfaction and sensory evaluation is often used to assess differences in aromatic factors. The human olfactory system can detect aromas at levels of only a few parts per billion and is therefore an important part of quality assessment for aroma differences. Fruit volatile composition involves a wide array of chemicals from various classes including alcohols, aldehydes, esters, ketones and terpenes which may be present in intact fruit (primary volatiles) or only produced once tissues have been ruptured (Valero and Serrano, 2010). Primary volatiles contribute largely to the quality of intact fruit and vegetables, while secondary volatiles are more important when contributing to the eating quality of produce where they are a vital aspect of overall flavour perception.

While a large number of compounds has been identified in fruit and vegetables, only a limited number have been categorized as having a role in flavour determination although these contribute greatly to the compositional quality of fresh and processed produce. In tomato fruit, over 400 volatile compounds have been detected; however those which contribute significantly to the overall distinctive tomato aroma are *cis*-3-hexenal, hexanal, *trans*-2-hexenal, hexanol, *cis*-3-hexenol, 2-isobutylthiazole,

6-mehyl-5-hepten-2-one, geranylacetone, 2-phenylethanol, β-ionone, 1-penten-3-one, 3-methylbutanol, 3-methylbutanal, acetone and 2-pentenal (Tandon *et al.*, 2000).

Volatile compounds in fresh fruit and vegetables can be quantified using either whole intact fruits or various extracts which are collected using headspace techniques and analysed directly or concentrated using various trapping technologies (Valero and Serrano, 2010). Distillation and solvent extraction methods can also be used to collect these for analysis. Once extraction has been carried out the volatile compounds can be identified and measured using gas chromatography–mass spectrometry (GC/MS). More recent developments in the detection and quantification of the volatiles which produce aroma have included the 'electronic nose'. Currently there are several commercially available electronic noses which are able to perform volatile analysis based on different technologies (Vallone *et al.*, 2012). One such device consists of ultrafast gas chromatography coupled with a surface acoustic wave sensor which carries out the aroma analysis with three major steps: headspace sampling, separation of volatile compounds and detection (Vallone *et al.*, 2012).

are either water- or fat-soluble, thus the determination of levels of these is carried out using chemical methods following their extraction in either water or lipid media. Most vitamin levels can be determined following extraction using high-performance liquid chromatography (HPLC), while mineral analysis is carried out using atomic absorption spectroscopy (AAS) (Barrett *et al.*, 2010). Fibre content, another compositional quality factor of growing interest among consumers, is determined by loss on incineration of dried residue remaining after digestion of a sample with dilute sulfuric acid and sodium hydroxide (Meloan and Pomeranz, 1980).

Nutritional compositional quality of fruits and vegetables can vary considerably and is determined by a number of factors including genetics, crop management, growing environment, plant nutrition and occurrence of pest and diseases. Stage of maturity at harvest and postharvest handling conditions also contribute to nutritional quality as tissue changes during senescence result in vitamin contents decreasing. Fruits and vegetables which are damaged or cut for packaging and processing also experience losses in vitamin levels which lower the compositional quality of the final consumer product.

13.8 Nutritional Quality

Horticultural produce is a major source of fibre and carbohydrates, and also a wide range of essential vitamins and minerals required in the human diet. The nutritional quality of various fruits and vegetables is dependent on the concentration of vitamins such as C, B-complex (thiamine, riboflavin, B_6, niacin, foliate), A and E, minerals and beneficial phytochemicals such as polyphenolics, carotenoids (e.g. lycopene and β-carotene) and glucosinolates (Barrett *et al.*, 2010). Vitamin C is one of the most widely tested for in fruit and vegetable samples, as it is not only one of the most recognized and valued by consumers, but also prone to being degraded due to heat, oxygen, light, tissue ageing and damage. Nutrients in fruit and vegetables

13.9 Biologically Active Compounds

Greenhouse-grown fruits and vegetables not only contain vitamins and minerals but also a wide range of other compounds beneficial to human health. These compounds or their metabolites that have been described as 'functional' may help to prevent diseases like cancer, have a protective effect on cardiovascular problems, are neutralizers of free radicals, reduce cholesterol and hypertension, and prevent thrombosis, besides other beneficial effects (Camelo, 2004). Phytochemicals commonly found in fruit and vegetables include the terpenes (carotenoids such as lycopene), phenols (blue, red and purple colours found in aubergines, berries and grapes), lignans (broccoli) and thiols (sulfur compounds found in garlic, onion, leek and other alliums, cabbages and other cruciferous vegetables) (Camelo, 2004). Concentration of lycopene, an important carotenoid in the human diet, is considered a quality parameter of fruits and vegetables. Lycopene is present in tomato, watermelon, pink grapefruit, red guava and red-fleshed papaya; however, tomatoes and tomato products constitute the main source of lycopene intake at around 85% of total lycopene consumed (Garcia and Barrett, 2006). Lycopene has taken on a role as a quality constituent in fruits and vegetables due to its potential health benefits, protecting against oxidative damage implicated in the pathogenesis of several human chronic diseases (Garcia and Barrett, 2006). Assessment of lycopene content in greenhouse tomatoes is used to determine which crop production practices increase the levels of this important phytochemical. Lycopene determination is also extensively used by plant breeders in the process of developing high-pigment hybrids to not only improve the red coloration of tomato fruit, but also develop tomato products which possess higher concentrations of lycopene for health benefits.

To analyse for lycopene content in tomatoes and other fruit, samples are prepared by pulping and juicing, then given a hot break treatment to 95°C which inactivates enzymes. Samples are then analysed for lycopene content by spectrophotometric methods after lipid-soluble components have been extracted with a solution of ethanol and hexane. HPLC may also be used to analyse samples for lycopene and other phytochemicals once extracted. Lycopene can also be determined by measuring the intensity of red coloration (Anthon and Barrett, 2005) with use of a colorimeter. Results for lycopene determination are expressed in milligrams of lycopene per kilogram of fresh tomato juice and typically ranges from 55 to 181 mg/kg for processing tomatoes (Garcia and Barrett, 2006). Since lycopene content changes with fruit maturity, samples obtained for analysis must be of the same stage of ripeness and not held in storage before assessment.

13.10 Texture and Firmness Quality Assessment

Texture and firmness are compositional quality parameters related to the sense of touch when the produce is either handled or consumed. The texture of fruits and vegetables is due to turgor pressure and the composition of individual plant cell walls and the middle lamella that holds individual cells together (Barrett *et al.*, 2010). Assessment of firmness or texture may be carried out to determine degree of ripeness of fruit, to determine the correct stage to harvest, to assess losses in textural quality during storage or handling or to determine eating quality. Unlike flavour, texture and firmness are relatively simple to accurately assess with instrumental methods, either basic hand-held meters or more advanced laboratory testing. The most commonly used instruments for evaluation of the firmness or textural qualities of produce apply a large deforming force such as puncture or compression and are therefore destructive tests. The puncture test carried out with a hand-held penetrometer has been used by growers and in laboratory testing for many years and is still a useful and relevant test. Penetrometer testing involves pushing a probe into the surface of a fruit to a depth that causes the skin to puncture, with the resultant reading recording the force required. Penetrometers can be obtained to cover different ranges of pressure suitable for measuring either soft or harder types of fruit depending on the variety and stage of ripeness of the produce being tested (OECD, 2009). Penetrometers should be ideally bench mounted or solidly fixed to ensure the pressure applied to the produce is steady and at a constant angle (i.e. vertically downwards). While hand-held penetrometers are often used in the field to assess harvest ripeness, care must be taken to ensure a smooth and uniform application pressure is used when taking readings (OECD, 2009). As with all methods of quality assessment, firmness testing should be carried out on a representative sample of at least ten fruits of similar size and degree of maturity randomly selected from the crop. Sampled fruits should be free of defects, pest and disease damage and be fully representative of the crop being tested.

Compression testing is similar to puncture testing, with a flat plate, larger than the diameter of the fruit to be tested, being used to obtain a compression value. Non-destructive tests for produce texture and firmness have been developed which use vibration techniques. Using vibration responses in the frequency range of 20–10,000 Hz it was deemed possible to separate fruit by maturity and textural properties (Barrett *et al.*, 2010).

13.11 Microbial Quality and Food Safety

The presence of microbial and other contaminants has become a major component of the concern over fresh produce quality among consumers due to a number of serious food-related disease outbreaks involving fruit and vegetables. Contamination from heavy metals, pesticides and other agrochemicals, parasites, naturally occurring toxic compounds or unwanted cross-contamination with genetically modified organisms are all aspects of produce quality which may be tested for if required. Microbial contaminants implicated with food safety include bacteria such as *Salmonella* spp., *Escherichia coli*, *Listeria monocytogenes* and *Shigella* spp., toxins produced by *Clostridium botulinum* and others, as well as the hepatitis A virus and parasites such as *Entamoeba histolytica* and *Giardia lamblia* (Camelo, 2004). The presence of microorganisms implicated in food safety may be identified using testing methods such as culture media, immunoassay and polymerase chain reaction (PCR). Culture methods involve the use of either solid or liquid culture media to grow the microbial target organism, if it is present on the produce being tested. Not only do cultural methods show if a particular organism is present, they can also provide information on the number of organisms present through the use of plate counts. Immunoassay uses antibodies to detect specific proteins that are unique to the target microorganism. A common example used commercially is the enzyme-linked immunosorbent assay (ELISA), which indicates the presence of particular microorganisms in samples being tested. PCR methods are a rapid and sensitive test which recognizes pieces of DNA or RNA that are unique to the target microorganism. PCR tests are usually presence or absence tests, but some can also be quantitative (United Fresh Produce Association Food Safety & Technology Council, 2010).

Microbial testing is one aspect of food safety quality assessment which may be used as part of an overall programme to improve crop production and handling practices. Ensuring the safety of fresh produce as part of a quality assurance programme has become of international importance and systems such as GAP and HACCP have been developed to guarantee that food is not exposed to any type of contamination that could put health at risk (Camelo, 2004).

13.12 Mycotoxins and Contaminants

The most significant naturally occurring contaminants which pose a food safety risk and crop quality issue are those typically synthesized by fungi, termed 'mycotoxins'. Mycotoxins are secondary metabolites produced by microfungi that are capable of causing disease and death in humans and other animals (Bennett and Klich, 2003). Mycotoxin testing for detection in food and animal feeds is routinely carried out where large-scale commercial production of susceptible crops occurs. The methods used for detection range from commercial, portable test kits using well assays and strip tests, to laboratory analyses which are specific for a range of commonly occurring mycotoxins. ELISA is an antibody-based assay that is often used to detect mycotoxins, with ELISA kits available for aflatoxins, deoxynivalenol, fumonisins, ochratoxins and zearalenone (Schmale and Munkvold, 2009). Food safety testing laboratories use mycotoxin detection and quantification methods such as HPLC and GC/MS. Both HPLC and GC/MS require trained staff and the correct laboratory equipment; however, these are highly sensitive compared with ELISA test kits and often have detection limits which exceed 0.2 ppm. HPLC separates a mixture of compounds on a stationary column using a carrier solvent such as methanol or acetonitrile, the mycotoxins are detected and quantified in the sample as they pass through a specific detector. GC/MS separates a mixture of compounds on a stationary column using a carrier gas such as helium, and the mycotoxins are detected and quantified using a mass spectrometer (Schmale and Munkvold, 2009).

13.13 Heavy Metals and Chemical Contamination

Examples of environmental contaminants which pose a food safety risk include heavy metals, polychlorinated biphenyls (PCBs), dioxins (polychlorinated dibenzodioxins and dibenzofurans), persistent chlorinated pesticides (DDT, aldrin, dieldrin, heptachlor, mirex, chlordane), brominated flame retardants (mainly polybrominated diphenyl ethers), polyfluorinated compounds, polycyclic aromatic hydrocarbons (PAHs), perchlorate, pharmaceutical and personal care products, and haloacetic acids and other water disinfection by-products (Mastovska, 2013). Most chemical contaminants implicated in crop food safety are typically present at very low concentrations (parts per million or less) and require correct identification and quantification by trained staff in food safety laboratories. MS is a commonly used method of detection and identification of chemical contaminants including pesticides and other agrochemicals.

Most developed countries have well-established guidelines for maximum allowable heavy metal concentrations in food crops. The European Commission Directive 1881/2006 specified maximum levels for cadmium, tin, mercury and lead in a range of different food items, while in the USA, the FDA enforces levels for a number of substances including cadmium, lead, mercury and others. Laboratory analytical methods are used to detect and quantify the levels of heavy metals in food samples to ensure regulatory compliance and maximum food safety; these tests include the use of inductively coupled plasma–mass spectrometry (ICP-MS) and AAS, among others.

13.14 Naturally Occurring Toxins

Levels of naturally occurring toxins exceeding those recommended in food safety standards represent a significant loss in quality of a crop and may result in the produce being discarded or downgraded for alternative uses. Examples of naturally occurring toxins include excessive nitrates and nitrites in some leafy vegetables and oxalates in rhubarb and spinach (Kader, 1992). Rhubarb contains compounds such as oxalic acid which is found in particularly high levels in the leaves, this makes the foliage too toxic for consumption. Rhubarb stems contain much lower levels of oxalates and can be safely eaten, generally in a cooked state. Oxalic acid toxicity can cause a number of symptoms such as cardiac, respiratory, gastric and neurological effects. Cucurbit plants such as courgette and cucumber may develop naturally occurring cucurbitacins, particularly when produced under poor growing conditions or when over-mature. Cucurbitacins are bitter-tasting compounds which may cause gastric disorders in sensitive people or if consumed in large quantities.

13.15 Nitrate in Leafy Greens

Nitrate content is an important quality characteristic of vegetables (Santamaria, 2006) that has implications for human health and regulations are set in many countries which producers must meet (Schnitzler and Gruda, 2002). Nitrate may harm the health of consumers as it can be converted to nitrite, causing methaemoglobinaemia, or carcinogenic nitrosamines (Blom-Zandstra, 1989; Afzali and Elahi, 2014). Since an estimated 80% of nitrate enters the body through fruit and vegetable consumption, assessment of nitrate in fresh produce is used to determine when levels may be excessive based on acceptable daily intake recommendations set by various health organizations. The US Environmental Protection Agency reference dose for nitrate is 1.6 mg nitrate-N/kg body weight per day (equivalent to about 7.0 mg nitrate/kg body weight per day) (Mensinga et al., 2003).

Hydroponic production allows the use of certain techniques to limit final foliar nitrate levels and thus produce a high-quality product. Nitrate content in vegetables can be controlled to a certain extent by production practices, thus nitrate testing of leafy vegetables in particular is an important quality control aspect for growers and researchers looking at ways of keeping nitrate levels below set thresholds. Nitrate content in fresh produce is derived from naturally occurring sources such as decaying organic matter and manures, as well as from the use of synthetic nitrogen fertilizers. Other factors which contribute to nitrate uptake and accumulation in vegetable tissues are genetic factors (such as species and cultivar) and environmental factors (humidity, moisture levels, temperature, irradiance and photoperiod) (Santamaria, 2006). Nitrogen fertilization factors (such as application rate and chemical forms) and light intensity are the main variables that influence nitrogen content in vegetables, particularly in the period immediately before harvest. Under conditions of light deficiency, considerable amounts of nitrate cannot be metabolized in plant tissue, thus resulting in high nitrate content in the harvested products (Schnitzler and Gruda, 2002). Vegetable species known to be accumulators of high levels of nitrate are some of the most commonly grown hydroponically, belonging to the families of *Brassicaceae* (rocket, radish, mustard), *Chenopodiaceae* (beetroot, Swiss chard, spinach) and *Amaranthaceae*; but also *Asteraceae* (lettuce) and *Apiaceae* (celery, parsley) (Santamaria, 2006). Rocket (arugula) has been found to be the highest nitrate-accumulating vegetable as it absorbs nitrate rapidly and nitrate concentrations in the leaves can be much higher than in the growth medium (Santamaria, 2006).

Nitrate compositional analysis has been used to determine production methods which can limit nitrate accumulation in the foliage of crops such as lettuce, spinach and rocket. In hydroponic systems this includes precise control over nitrate application rates during production and removal or limiting of nitrate levels around the root zone for a short period before harvest. Studies with hydroponic methods have found that conversion to a nitrogen-free nutrient solution several days before harvest resulted in a significant reduction of nitrate content to levels below the critical for lettuce. Similar findings for endive and celery report that eliminating 90% of the nitrogen in nutrient solutions one week prior to harvest reduced nitrate content by half in endive and by 56% in the leaves of celery (Martignon et al., 1994) Nitrate analysis is also used by plant breeders to examine potential nitrate accumulation between different varieties and develop cultivars with fewer propensities for this problem.

References

Abbott, J.A. (1999) Quality measurement of fruits and vegetables. *Postharvest Biology and Technology* 15(3), 207–225.

Adegoroye, A.S. and Jolliffe, P.A. (1987) Some inhibitory effects of radiation stress on tomato fruit ripening. *Journal of the Science of Food and Agriculture* 39(4), 297–302.

Afzali, S.F. and Elahi, R. (2014) Measuring nitrate and nitrite concentrations in vegetables, fruits in Shiraz. *Journal of Applied Science and Environmental Management* 18(3), 451–457.

AlHarbi, A.R., Saleh, A.M., Al-Omran, A.M. and Wahb-Allah, W. (2014) Response of bell pepper (*Capsicum annuum* L.) to salt stress and deficit irrigation strategy under greenhouse conditions. *Acta Horticulturae* 1034, 443–450.

Al-Rawahy, M.S., Al-Rawahy, S.A., Al-Mulla, Y.A. and Nadaf, S.K. (2019) Influence of nutrient solution temperature on its oxygen level and growth, yield and quality of hydroponic cucumber. *Journal of Agricultural Science* 11(3), 75–60.

Amalfitano, C., del Vacchio, L., Somma, S., Cuciniellom, A. and Caruso, G. (2017) Effects of cultural cycle and nutrient solution conductivity on plant growth, yield and fruit quality of 'Friariello' pepper grown in hydroponics. *Horticultural Science* 44(2), 91–98.

Anthon, G.E. and Barrett, D.M. (2005) *Lycopene Method Standardization. Report to the California League of Food Processors*, Sacramento, California.

Azarmi, R., Taleshmikail, R.D. and Gikloo, A. (2010) Effects of salinity on morphological and physiological changes and yield of tomato in hydroponics systems. *Journal of Food, Agriculture and Environment* 8(2), 573–576.

Banjaw, D.T., Megersa, H.G. and Lemma, D.T. (2017) Effect on water quality and deficit irrigation on tomato yield and quality: a review. *Advances in Crop Science and Technology* 5(4). https://doi.org/10.4172/2329-8863.1000295

Barrett, D.M., Beaulieu, J.C. and Shewfelt, R. (2010) Colour, flavour, texture and nutritional quality of fresh-cut fruits and vegetables: desirable levels, instrumental and sensory measurement, and the effects of processing. *Critical Reviews in Food Science and Nutrition* 50(5), 369–389.

Bennett, J.W. and Klich, M. (2003) Mycotoxins. *Clinical Microbiological Review* 16(3), 497–516.

Bett, K.L. (2001) Evaluating sensory quality of fresh-cut fruits and vegetables. In: Lamikanra, O. (ed.) *Fresh-Cut Fruits and Vegetables: Science, Technology, and Market*. CRC Press, Boca Raton, Florida. Available at: https://sceqa.files.wordpress.com/2012/05/frutas-y-vegetales-ciencia-y-tecnologia.pdf (accessed 26 September 2020).

Blom-Zandstra, M. (1989) Nitrate accumulation in vegetables and its relationship to quality. *Annals of Applied Biology* 115(3), 553–561.

Camelo, A.F.L. (2004) The quality in fruits and vegetables. In: Manual for the preparation and sale of fruits and vegetables from field to market. *FAO Agricultural Services Bulletin 151*. Food and Agriculture Organization of the United Nations, Rome. Available at: http://www.fao.org/3/y4893e/y4893e08.htm#bm08 (accessed 11 September 2020).

Cardenosa, V., Medrano, E., Lorenzo, P., Sanchez-Guerreo, M., Guevas, F., *et al.* (2014) Effects of salinity and nitrogen supply on the quality and health related compounds of strawberry fruits (*Fragaria* × *ananassa* cv. Primoris). *Journal of the Science of Food and Agriculture* 95(14), 2924–2930.

Caruso, G., Villari, G., Melchionna, G. and Conti, S. (2011) Effects of cultural cycles and nutrient solutions on plant growth, yield and fruit quality of alpine strawberry (*Fragaria vesca* L.) grown in hydroponics. *Scientia Horticulturae* 129(3), 479–485.

Colla, G., Rouphael, Y., Cardarelli, M. and Rea, E. (2006) Effect of salinity on yield, fruit quality, leaf gas exchange and mineral composition of grafted watermelon plants. *HortScience* 41(3), 622–627.

Dorais, M. (2003) The use of supplemental lighting for vegetable crop production: light intensity, crop response, nutrition, crop management, cultural

aspects. *Canadian Greenhouse Conference October 9, 2003*. Available at: https://www.agrireseau.net/legumesdeserre/documents/cgc-dorais2003fin2.pdf (accessed 11 September 2020).

Fanasca, S., Colla, G., Maiani, G., Venneria, E., Rouphael, Y., *et al.* (2006) Changes in antioxidant content of tomato fruits in response to cultivar and nutrient solution composition. *Journal of Agricultural and Food Chemistry* 54(12), 4319–4325.

Garcia, E. and Barrett, D.M. (2006) Assessing lycopene content in California processing tomatoes. *Journal of Food Processing and Preservation* 30(1), 56–70.

Gazula, A., Kleinhenz, M.D., Strreter, J.G. and Millar, A.R. (2005) Temperature and cultivar effects on anthocyanin and chlorophyll B concentrations in three related Lollo Rosso lettuce cultivars. *HortScience* 40(6), 1731–1733.

Gruda, N. (2009) Do soilless culture systems have an influence on product quality of vegetables? *Journal of Applied Botany and Food Quality* 82(2), 141–147.

Haghshenas, M., Arshad, M. and Nazarideljou, M.J. (2018) Different K:Ca ratios affected fruit colour and quality of strawberry 'Selva' in soilless system. *Journal of Plant Nutrition* 41(2), 243–252.

Hall, A.E. (2001) *Crop Responses to Environment*. CRC Press, Boca Raton, Florida.

He, F., Thiele, B., Watt, M., Kraska, T., Ulbrich, A. and Kuhn, A.J. (2019) Effects of root cooling on plant growth and fruit quality of cocktail tomato during two consecutive seasons. *Journal of Food Quality* 2019, 3598172.

He, J., Lee, S.K. and Dodd, I.C. (2001) Limitations to photosynthesis of lettuce growth under tropical conditions: alleviation by root zone cooling. *Journal of Experimental Botany* 52(359), 1323–1330.

Ikeda, T., Yamazaki, K., Kumakura, H. and Hamamoto, H. (2011) The effects of high temperature and water stress on fruit growth and anthocyanin content in pot grown strawberry (*Fragaria* × *ananassa* Duch. cv. Sachinoka) plants. *Environmental Control in Biology* 49(4), 209–215.

Johkan, M., Shoji, K., Goto, F., Hahida, S. and Yoshihara, T. (2011) Effect on green light wavelength and intensity of photomorphogensis and photosynthesis in *Lactuca sativa*. *Environmental and Experimental Botany* 75, 128–133.

Kader, A.A. (1992) Postharvest technology of horticultural crops, 2nd edn. *Publication 3311*. University of California, Division of Agriculture and Natural Resources, Davis, California.

Kader, A.A., Stevens, M.A., Albright-Holton, M., Morris, L.L. and Algazi, M. (1977) Effect of fruit ripeness when picked on flavor and composition in fresh market tomatoes. *Journal of the American Society for Horticultural Science* 102, 724–731.

Karimi, F., Arunkumar, B., Asif, M., Murthy, B.N.S. and Venkatesha, K.T. (2012) Effect of different soilless culture systems on growth, yield and quality of strawberry, cv. *Festival. International Journal of Agricultural Science* 9(1), 366–372.

Kawasaki, Y., Matsuo, S., Suzuki, K., Kanayama, Y. and Kanahama, K. (2013) Root-zone cooling at high air temperature enhances physiological activities and internal structures of roots in young tomato plant. *Journal of the Japan Society of Horticultural Science* 82(4), 322–327.

Kennelly, M., O'Mara, J., Rivard, D., Millar, G.L. and Smith, D. (2012) Introduction to abiotic disorders in plants. *The Plant Health Instructor.* https://doi.org/10.1094/PHI-I-2012-10-29-01

Keutgen, A.J. and Pawelzik, E. (2008) Quality and nutritional value of strawberry fruit under long term salt stress. *Food Chemistry* 107(4), 1413–1420.

Khayyat, M., Tafazoli, E., Eshghi, S., Rahemi, M. and Rajaee, S. (2007) Salinity, supplementary calcium and potassium effects on fruit yield and quality of strawberry (*Fragaria ananassa* Duch.). *American-Eurasian Journal of Agriculture and Environmental Science* 2(5), 539–544.

Kleinhenz, M. and Bumgarner, N.R. (2013) Using °Brix as an indicator of vegetable quality: linking measured values to crop management. *Ohio State University Extension Fact Sheet HYG-1651.* Ohio State University, Columbus, Ohio. Available at: http://ohioline.osu.edu/factsheet/HYG-1651 (accessed 14 September 2020).

Krauss, S., Schnitzler, W.H., Grassmann, J. and Woitke, M. (2006) The influence of different electrical conductivity values in a simplified re-circulating soilless system on inner and outer fruit quality characteristics of tomato. *Journal of Agricultural and Food Chemistry* 54(2), 441–448.

Leemans, V., Magein, H. and Destain, M.F. (2002) On-line fruit grading according to their external quality using machine vision. *Biosystems Engineering* 83(4), 397–404.

Martignon, G., Casarotti, D., Venezia, A., Schiavi, M. and Malorgio, F. (1994) Nitrate accumulation in celery as affected by growing system and N content in the nutrient solution. *Acta Horticulturae* 361, 583–589.

Mastovska, K. (2013) Modern analysis of chemical contaminates in food. *Food Safety Magazine*, February/March 2013. Available at: http://www.foodsafetymagazine.com/magazine-archive1/februarymarch-2013/modern-analysis-of-chemical-contaminants-in-food/ (accessed 14 September 2020).

Mattheis, J.P. and Fellman, J.K. (1999) Preharvest factors influencing flavor of fresh fruit and vegetables. *Postharvest Biology and Technology* 15, 227–232.

Meilgaard, M., Civille, G.V. and Carr, B.T. (1999) *Sensory Evaluation Techniques*, 3rd edn. CRC Press, Boca Raton, Florida, pp. 195–208.

Meloan, C.E. and Pomeranz, Y. (1980) *Food Analysis Laboratory Experiments*, 2nd edn. AVI Publishing Company, Westport, Connecticut.

Mensinga, T.T., Speijers, G.J.A. and Meulenbelt, J. (2003) Health implications of exposure to environmental nitrogenous compounds. *Toxicology Reviews* 22(1), 41–51.

Morgan, L. (1996) Studies of the factors affecting the yield and quality of single truss tomatoes. PhD thesis, Massey University, Palmerston North, New Zealand.

Morgan, L. (2004) Tomato fruit flavour and quality evaluation. *The Tomato Magazine*, June 2004.

Morgan, L. (2008) Fruit quality and flavour. In: *Hydroponic Tomato Crop Production*. Suntec NZ Ltd, Tokomaru, New Zealand, pp. 146–159.

Morgan, L. (2012) *Hydroponic Salad Crop Production*. Suntec NZ Ltd, Tokomaru, New Zealand.

Narendra, V.G. and Hareesh, K.S. (2010) Quality inspection and grading of agricultural and food products by computer vision – a review. *International Journal of Computer Applications* 2(1), 43–65.

Niam, A.G. and Suhardiyanto, H. (2019) Root-zone cooling in tropical greenhouse: a review. *IOP Conference Series: Materials Science and Engineering* 557, 012044. https://doi.org/10.1088/1757-899X/557/1/012044

Nichols, M.A., Fisher, K.J., Morgan, L.S. and Simon, A. (1994) Osmotic stress, yield and quality of hydroponic tomatoes. *Acta Horticulturae* 361, 302–310.

OECD (Organisation for Economic Cooperation and Development) (2009) *Guidelines on Objective Tests to Determine Quality of Fruit and Vegetables, Dry and Dried Produce. OECD Fruit and Vegetable Scheme*. OECD, Paris. Available at: https://www.oecd.org/agriculture/fruit-vegetables/publications/Guidelines_on_Objective_Tests_2018.pdf (accessed 14 September 2020).

Olle, M. and Bender, I. (2009) Causes and control of calcium disorders in vegetables: a review. *Journal of Horticultural Science and Biotechnology* 84(6), 577–584.

Peterson, K.K., Willumsen, J. and Kaach, K. (1998) Composition and taste of tomato as affected by increased salinity and different salinity sources. *Journal of Horticultural Science and Biotechnology* 73(2), 205–215.

Porcel, R., Aroca, R. and Ruíz-Lozano, J.M. (2012) Salinity stress alleviation using arbuscular mycorrhizal fungi. A review. *Agronomy for Sustainable Development* 32(1), 181–200.

Qaryouti, M.M., Qawasmi, W., Hamdan, H. and Edwan, M. (2007) Influence of NaCl salinity stress on yield, plant water uptake and drainage water of tomato grown in soilless culture. *Acta Horticulturae* 747, 539–544.

Ramirez-Arias, J.A., Hernandez-Ibarra, U., Pineda-Pineda, J. and Fitz-Rodriguez, E. (2018) Horizontal and vertical hydroponic systems for strawberry production at high densities. *Acta Horticulturae* 1227, 331–337.

Ramirez-Gomez, H., Sandoval-Villa, M., Carrillo-Salazar, A. and Muratalla-Lua, A. (2012) Comparison of hydroponic systems in the strawberry production. *Acta Horticulturae* 947, 165–172.

Rosadi, R.A.B., Senge, M., Suhandy, D. and Tusi, A. (2014) The effect of EC levels of nutrient solution on the growth, yield and quality of tomatoes (*Solanum lycopersicum*) under the hydroponic system. *Journal of Agricultural Engineering and Biotechnology* 2(1), 7–12.

Rouphael, Y., Cardarelli, M., Rea, E., Battistelli, A. and Colla, G. (2006) Comparison of the subirrigation and drip-irrigation systems for greenhouse zucchini squash production using saline and non-saline nutrient solutions. *Agricultural Water Management* 82, 99–117.

Rubio, J.A., Garcia-Sanchez, F., Rubio, F. and Martinez, V. (2009) Yield, blossom end rot incidence, and fruit quality in pepper plants under moderate salinity are affected by K^+ and Ca^{2+} fertilisation. *Scientia Horticulturae* 119(2), 79–87.

Sakamoto, M. and Suzuki, T. (2015) Effect of root zone temperature on growth and quality of hydroponically grown red leaf lettuce (*Lactuca sativa* L. cv. Red Wave). *American Journal of Plant Sciences* 6(14), 2350–2360.

Sakamoto, M., Uenishi, M., Miyamoto, K. and Suzuki, T. (2016) Effect of root zone temperature on the growth and fruit quality of hydroponically grown strawberry. *Journal of Agricultural Science* 8(5), 122–131.

Santamaria, P. (2006) Nitrate in vegetables: toxicity, content, intake and EC regulation. *Journal of the Science of Food and Agriculture* 86(1), 10–17.

Savvas, D. and Lenz, F. (1994) Influence of salinity on the incidence of the physiological disorder 'internal fruit rot' in hydroponically grown eggplants. *Journal of Applied Botany* 68, 32–35.

Savvas, D., Gizas, G., Karras, G., Lydakis-Simantiris, N., Salahas, G., et al. (2007) Interactions between silicon and NaCl salinity in a soilless culture of roses in greenhouse. *European Journal of Horticultural Science* 72(2), 73–79.

Schmale, D.G. and Munkvold, G.P. (2009) Mycotoxins in crops: a threat to human and domestic animal health. *The Plant Health Instructor.* https://doi.org/10.1094/PHI-I-2009-0715-01

Schnitzler, W.H. and Gruda, M. (2002) Hydroponics and product quality. In: Savvas, D. and Passam, H. (eds) *Hydroponic Production of Vegetables*

and Ornamentals. Embryo Publications, Athens, pp. 373–411.

Sellitto, V.M., Golubkina, N.A., Pietrantonio, L., Cozzolino, E., Cuciniello, A., et al. (2019) Tomato yield, quality, mineral composition and antioxidants as affected by beneficial microorganisms under soil salinity induced by balanced nutrient solutions. *Agriculture* 9(5), 110. https://doi.org/10.3390/agriculture9050110

Skog, L.J. (1998) Chilling injury of horticultural crops. *Ontario Ministry of Agriculture, Food and Rural Affairs Fact Sheet*. Ontario Ministry of Agriculture, Food and Rural Affairs, Guelph, Ontario, Canada.

Sonneveld, C. and van der Burg, M.M. (1991) Sodium chloride salinity in fruit vegetable crops in soilless culture. *Netherlands Journal of Agricultural Science* 39, 115–122.

Stamatakis, A., Papadantonakis, N., Savvas, D. and Simantiris, N.L. (2003) Effects of silicon and salinity on fruit yield and quality of tomato grown hydroponically. *Acta Horticulturae* 609, 141–147.

Tandon, K.S., Baldwin, E.A. and Shewfelt, R.L. (2000) Aroma perception of individual volatile compounds in fresh tomatoes (*Lycopersicon esculentum* Mill.) as affected by the medium of evaluation. *Postharvest Biology and Technology* 20(3), 261–268.

United Fresh Produce Association Food Safety & Technology Council (2010) Microbial testing of fresh produce. Available at: https://www.united-fresh.org/content/uploads/2014/07/FST_MicroWhite-Paper.pdf (accessed 26 September 2020).

Valero, D. and Serrano, M. (2010) *Postharvest Biology and Technology for Preserving Fruit Quality*. CRC Press, Boca Raton, Florida, pp. 7–48.

Vallone, S., Lloyd, N.W., Ebeler, S.E. and Zakharov, F. (2012) Fruit volatile analysis using an electronic nose. *Journal of Visualized Experiments* (61), e3821.

Wang, S.Y. and Camp, M.J. (2000) Temperatures after bloom affect plant growth and fruit quality of strawberry. *Scientia Horticulturae* 85(3), 183–199.

Wu, M. and Kubota, C. (2008) Effects of high electrical conductivity of nutrient solution and its application timing on lycopene, chlorophyll and sugar concentrations of hydroponic tomatoes during ripening. *Scientia Horticulturae* 116(2), 122–129.

Yamaguchi, S. and Ninomiya, K. (2000) Umami and food palatability. *Journal of Nutrition* 130(4S Suppl.), 921S–926S.

Zhang, P., Senge, M. and Dai, Y. (2016) Effects of salinity stress on growth, yield, fruit quality and water use efficiency of tomato under hydroponic system. *Reviews in Agricultural Science* 4, 46–55.

14 Harvest and Postharvest Factors

14.1 Harvesting

Harvesting involves the gathering or removal of a mature crop, with minimum damage and losses, from where it has been grown and transporting it on either for direct consumption or into the postharvest handling chain for further storage and distribution. For many crops, harvesting at the correct stage of maturity has a direct bearing on the final quality and subsequent postharvest shelf-life of the product.

Most crops have a limited time frame for harvest at the optimum maturity level and will deteriorate if left unharvested. Crops harvested before reaching a suitable level of maturity may not ripen postharvest or develop a number of physiological disorders; others may not obtain a suitable quality for consumption, be low in sugars and volatile aromatic compounds, have unsuitable texture or colour and generally be unmarketable. Those harvested in an over-mature state risk a higher degree of harvest and postharvest handling damage such as bruising, compression injury and overly soft tissue more prone to decay. Over-mature crops also have a shorter postharvest life and may not be suitable for fresh market sales.

14.1.1 Harvest maturity

Determination of harvest maturity indices varies between different crops. Many fruiting crops such as strawberries, tomatoes, capsicums, grapes and melons may develop changes in skin colour and levels of soluble solids (measured as °Brix) in the flesh as they approach maturity, which can be used as harvest indicators. Others such as aubergine and cucumber may be harvested when they reach a certain size or weight as minimal colour changes occur at maturity. Determination of soluble solids is obtained by squeezing or puréeing fruit, filtering to obtain a clear extract and measuring this with a refractometer.

Visible degree of development is used to determine maturity levels in many crops such as cut flowers, these include the level of opening or bud tightness in roses. Colour levels and development in many crops such as tomato and capsicum are used to determine harvest timing as removal from the plant may occur anywhere from the mature green through to full red (or other coloration) depending on market destination. Mature green fruit may be harvested where a longer shelf-life and artificial ripening may be used before sale, green fruit are also firmer and less susceptible to handling bruising and other damage. Conversely, fruit such as tomatoes may be ripened to a certain degree of coloration such as 'breaker' or 'pink' to be marketed as 'vine ripened' and for more immediate local sales. Riper tomatoes and other fruit destined for fresh consumption are more prone to tissue damage during harvesting and postharvest handling operations and may require hand harvesting because of this.

14.1.2 Hand harvesting

High-quality, soft-textured greenhouse fruits and vegetables are most likely to be completely or partially hand harvested for fresh market sales. Hand harvesting has a major advantage over mechanization as workers, once adequately trained, can accurately determine produce quality and maturity, thus allowing selection of ripe

© L. Morgan 2021. *Hydroponics and Protected Cultivation* (L. Morgan)
DOI: 10.1079/9781789244830.0014

fruits and vegetables from those which are still immature. This is particularly important for crops which must be selectively harvested over a period of time without damaging remaining unripe fruits or plants. Careful hand harvesting also minimizes the risks of bruising and other tissue damage with soft fruits (Fig. 14.1) and leafy vegetables and allows grading to occur in the greenhouse for certain crops. Hand harvesting for some soft fruits such as strawberries and raspberries, which are particularly prone to damage, may also incorporate packing into market containers or punnets directly after removal from the plant, thus eliminating the need for further postharvest handling. Hand harvesting of many fruits and vegetables requires a minimum of capital investment compared with the mechanized forms of crop removal, however it is highly reliant on the supply of suitable numbers of workers during the harvest season.

Sanitation procedures, use of GAP and well-trained employees are essential where hand harvesting, grading and packaging of fresh produce are carried out. The success of hand harvesting is highly dependent on good management and training

of employees. Selection of fruits, vegetables and cut flowers at the correct stage of maturity with prevention of handling damage is essential; however, productivity, or rate of harvest, must also be maintained at an acceptable level. The correct method of removal of the harvested product from the plant is also a vital aspect of worker training both to protect the produce as well as prevent damage to the crop (Fig. 14.2).

Where hand harvesting is carried out a number of factors must be considered with regard to preventing produce damage. Many of these are centred around correct training and supervision of workers to ensure they develop the techniques and skills to handle fruits, vegetables and blooms correctly, thus preventing harvest damage and reducing food safety concerns (Figs 14.3 and 14.4). Bruising due to impacts, dropping into picking containers or bulk bins, over-compaction of harvest containers and fingernail or jewellery impacts on soft fruit surfaces are all common issues with hand harvesting. Incorrectly trained staff may also select fruit of the incorrect maturity, thus lowering shelf-life and increasing the requirement for further grading and sorting postharvest.

Fig. 14.1. Hand harvesting minimizes the risk of bruising and tissue damage in soft fruits such as aubergines.

Fig. 14.2. Hand harvesting and grading can assist with maintaining fruit quality.

14.1.3 Robotic harvesting of greenhouse crops

Use of robotic harvesters is a relatively new innovation, initially developed to harvest high-value fruit crops such as citrus, apples, melons and those grown under protected cultivation such as greenhouse tomatoes, capsicums, strawberries and cucumbers. The design and development of robots for agricultural applications raises issues not encountered in other industries where robotics is far more widespread: the robot has to operate in a highly unstructured environment in which no two scenes are the same; both the crop and the fruit are prone to mechanical damage and need to be handled with care; the robot has to operate under adverse climatic conditions such as high relative humidity and temperature as well as changing light conditions; and finally, to be cost-effective, the robot needs to meet high performance characteristics in terms of speed and success rate of the picking operation (van Henten *et al.*, 2002). The main reasons for adoption of robotic harvesting technology in protected cultivation have been the increasing cost of labour, difficulties in obtaining

Fig. 14.3. Harvest conveyor system.

Fig. 14.4. Packaged product.

and retaining a labour force for carrying out low-paid, repetitive and unskilled work, and increasing concern over fruit contamination and food safety when using human harvesters. The majority of the robotic harvesting systems in commercial use are based on vision processing systems, detecting changes in fruit colour which indicate harvest ripeness and differentiate between foliage, unripe and ripe fruit. Robotic greenhouse harvesters are fixed to autonomous vehicles which can access tight spaces and dense canopies without causing crop damage when moving down the aisles of a greenhouse. Some robotic harvesters for greenhouse use are mounted on heating pipe rails on the ground for guidance and support. These serve as a mobile platform for carrying power supplies, a pneumatic pump, electronic hardware for data acquisition and control, the camera vision systems and the manipulator arm with end-effector for cutting the fruit stem (van Henten *et al.*, 2002). Robotic harvesters are controlled by a central computer unit which keeps track of their position, orientation, speed and status and directs them in real time with global positioning systems, geographic information systems and continual sensing, allowing the robots to navigate their environment and carry out precision tasks (Gorvett, 2015). Once fruit of the correct colour has been identified by the computer vision systems for detection and three-dimensional (3D) imaging of the fruit, the manipulator or artificial hand grasps, cuts and removes the fruit from the plant and places it into a reception bin or on to a conveyor. Initially the first greenhouse robotic harvesters were developed for tomatoes, cucumbers and capsicums; however, other high-value crops are also being harvested with this new technology. The Agrobot SW 6010 strawberry harvester employs robotic manipulators with 60 robotic arms that are able to locate and identify ripe strawberries and move them to the packaging area via a conveyor system (Moran, 2016).

Apart from colour processing and recognition technology, some robotic harvesters have been developed which use other image-processing models to determine harvest maturity. These include size and geometric models to determine ripeness of fruit such as cucumbers which do not exhibit any major change in skin coloration to indicate harvest maturity and grow within a canopy of green leaves, making detection more difficult than with coloured fruit such as tomatoes. Detection of green fruit within the canopy is carried out via a double-camera system mounted into one wide-angle optical system. The detection of the fruit is achieved by using different filters on each of the cameras, one camera equipped with an 850 nm filter and the other with a 970 nm band. Whereas leaves show approximately the same reflectance at 850 and 970 nm, the reflectance of cucumber fruit at 850 nm is significantly higher than at 970 nm, making it possible to detect the cucumber fruit in the green environment by combining the images of both cameras (van Henten et al., 2002). To avoid issues with leaves, other plant parts and immature fruit obscuring mature fruit, images are taken from slightly different angles and perspectives so that a 3D scene reconstruction using standard triangulation techniques can be carried out; more than 95% of the cucumbers within a crop can be detected in this way (van Henten et al., 2002). Once a fruit is distinguished from the foliage, a geometric model of the cucumber volume can be used to determine the weight of the fruit with an estimated correlation of 97%, with harvest weight for cucumbers falling within the 300–600 g range.

The method of harvesting can have a significant effect on the quality of fruits and vegetables. While hand harvesting is often seen as the best way of preventing physical damage such as bruising, surface abrasions and cuts which can accelerate water and vitamin loss and increase occurrence of decay pathogens, this is not always the case. Modern developments in robotics and improvements to mechanical harvesting equipment can result in automated systems which handle produce with as much care as human workers, if not more in some cases.

14.2 Postharvest Handling, Grading and Storage

The aim of postharvest handling systems is to move produce from the grower through to the final consumer with minimal losses in quality and quantity and provide a uniform, year-round supply of fresh fruits and vegetables. The postharvest handling phase includes all stages of processing immediately following harvest and is characterized by various methods of pre-cooling, washing, cleaning, trimming, sorting, grading and packing.

With horticultural produce often being transported over large distances and passing through many handling systems before sale, attention to sanitation and food safety has become an increasing concern in the postharvest industry. Modern pack houses therefore incorporate a range of standards and procedures to ensure produce is of the highest compositional and safety quality for consumers. The use of strict guidelines for pack-house and field food-handling systems, GAP processes, correct storage, grading out of reject product and classification into maturity levels have all assisted with this.

Grading of fresh fruits, vegetables, herbs and cut flowers is an essential process in the postharvest handling chain and serves a number of purposes. Grading allows removal of produce which may be physically damaged or have pests or diseases already present that may carry over to infect healthy stored product. Grading also removes immature or over-mature fruits and vegetables which may require further ripening or different storage conditions to those at the optimum stage of maturity, these may also be destined for different markets. Sizing during the grading process assigns fruits and vegetables into different market classes, those with superior quality and optimum size receiving the premium grade classification and the highest returns from markets. At the consumer level, appropriate grading allows easy and reliable purchase of uniform fruits and vegetables without the need for inspection. Prices received from sale often reflect the grade or class of the product with most fruits, vegetables, blooms and other produce having well recognized and documented specifications for each grade; these may include not only weight, but also diameter, size, shape, colour, degree of defects, specific gravity, Brix (TSS or sugar levels), stem length (cut flowers) and stage of maturity.

14.2.1 Pack houses

Pack houses provide a facility for large volumes of produce to be handled rapidly, before deterioration can begin, under optimal temperatures with a high degree of hygiene and sanitation and comfortable conditions for workers. Many grading, sorting and packing facilities also incorporate cool stores, transportation bays and offices for commercial sales and administration by marketing and distribution companies.

Pack-house facilities and washing, grading/sorting lines should be arranged so that produce moves to a progressively cleaner area with each step in the process. This starts with the reception area where produce is delivered from the greenhouse, through to initial washing stages (or brushing for dry produce), rinsing, drying and on to grading lines, finishing at the final packaging area with the highest degree of sanitation and hygiene. Cross-contamination between these areas is avoided while graders, equipment, bins and other surfaces are designed to be regularly sanitized. Animals, insects, birds and rodents are controlled and excluded from pack-house areas to prevent contamination of produce and the spread of disease. Other hygiene measures include continual monitoring of the level of free chlorine in wash-tank water and enforcing worker hygiene procedures which may include the use of hair nets, gloves and face masks in some processing systems. Lighting systems, particularly those on grading tables and lines, should be sufficient to allow maximum visibility but prevent worker eye strain.

14.2.2 Washing, cleaning and sanitation

Many fruits and vegetables require washing during the postharvest handling process, this is largely to sanitize the surface of fruits such as tomatoes and with vegetable crops such as fresh-cut salad leaves. Washing may be carried out in large tanks through which the fruits float or via overhead high-volume sprays as the produce moves along the packing line. All water used during the washing and postharvest handling process must be of suitable potable quality and tested annually. A number of sanitizing agents may be added to wash water, with chlorine being the most widely used (sodium hypochlorite, calcium hypochlorite or liquid chlorine) at levels of 100–200 ppm, with a pH range of

6.8–7.2. The effectiveness of chlorinated water as a sanitizer is greatly affected by the pH of the solution: if the pH is too high (above 8.0) the chlorine acts slowly and a higher concentration is necessary to achieve a rapid kill of pathogens in the water; below 6.5 and the chlorine is too active and more corrosive of equipment (Sargent et al., 2000). The advantages of using chlorine to sanitize water during postharvest handling is that it is highly effective at killing a broad range of pathogens, is inexpensive and leaves little residue on produce surfaces. The main disadvantage is that if chlorine is added very regularly without a complete change of the wash water, high salt concentrations may occur which can damage sensitive produce. If dump or wash water is being recirculated it requires regular replacement on a daily basis or more frequently if organic material builds up rapidly. Other sanitation agents include hydrogen peroxide (3%), chlorine dioxide (1–5 ppm), peroxyacetic acid (80 ppm), acidified sodium chlorite, ozone at 0.5–2 ppm (requires a generator) and acetic acid (used for organic crops).

14.2.3 Size and shape grading

Produce sorted for size may be carried out before (pre-sizing) or after automated grading for colour or other variables depending on the commodity being handled. While highly undersized fruits are considered to be reject or cull grade, others which are closer to ideal weight may be classed into second or third grades for local sales or processing. Fruits and vegetables may be sized according to weight or dimensions such as diameter or length, or both. Spherical produce such as tomatoes is typically graded by size rather than by weight. In smaller postharvest handling systems this may be carried out manually by workers using size grade cards, rings or other methods of measurement.

14.2.4 Manual grading

Postharvest grading is the process of sorting produce into grades or categories based on quality and the recognized desirable characteristics of the commodity and variety being graded (Fig. 14.5). Sorting may be carried out manually by workers who assess fruit and vegetables for colour, size and grade and remove those with defects, or via automated computer-controlled systems. In smaller postharvest operations and for high-value or delicate crops, a static system may be used whereby produce is placed on a table for inspection by workers who grade each piece. A more common approach is the dynamic system whereby produce moves along a conveyor in front of trained workers who sort and remove defects, allowing only the highest-quality product to move through the line. Such systems rely on fully trained staff who accurately and efficiently remove all reject produce from the line rapidly with minimal errors. The number of human sorters required is dependent on the amount of product to be sorted per hour, amount of diverted product (culls and subgrades), the number of separations by colour, shape or defect, and the size of the product being sorted as smaller fruits and vegetables require more sorting decisions than larger

product (Thompson *et al.*, 2002). Sorting belts or grading lines require the produce to move in a single layer, slowly rotate and turn to provide as much surface visibility as possible for workers to carry out grading processes while not damaging fruit surfaces.

14.2.5 Colour sorting and grading

Colour sorting is carried out in produce which are harvested within a range of maturity such as tomato and capsicum where skin colour changes as fruit ripens and is therefore regarded as a quality index. While some fruit such as tomatoes may be automatically graded based on colour and size, other fruit with uniform coloration require other systems of classification. Colour sorting allows an assessment of maturity and may be carried out electronically; however, smaller operations may carry this process out by eye using standardized colour charts which exist for each different fruit, vegetable and/or cultivar being processed. Colour charts are typically used for manual grading of fruit such as tomatoes and

Fig. 14.5. Capsicum grading and packing.

peppers. Maturity indices may also be assessed during grading by using optical methods to measure skin colour which are based on the level of chlorophyll degradation or the development of colour pigments.

14.2.6 Automated colour and grading systems

Grading systems for fruit and vegetables are becoming increasing more automated, incorporating mechatronics, robotics, machine vision systems and near-infrared inspection systems which allow all sides of the produce to be inspected. Computer vision systems can simulate human vision to perceive the 3D features of spatial objects (Gao *et al.*, 2014). These grading lines use captured fruit images as the basis of determining shape and size. Shape may be described with six feature parameters: radius index, continuity index, curvature index, symmetry of radius index, symmetry of continuity index and symmetry of curvature index (Gao *et al.*, 2014). Electronic sizers capture several different images of each piece of produce and calculate volume based on these data. Weight sizers weigh each piece as it moves over a load cell, these data are sent to a central controller unit, where a computer program sends a signal for the fruit to be dropped on to a cross-conveyor system or collection bin. Large-scale pack houses often have a number of lanes operating simultaneously achieving high flow-through rates during the automatic sizing and grading process. Electronic sizers can be reprogrammed easily to allow for changes in product size distribution and the handling of different fruits and vegetables (Thompson *et al.*, 2002). An additional advantage of electronic size grading lines is that records are obtained of the weight, grade and number of fruits obtained for each size grade, allowing for accurate accounting of packed produce.

Colour imaging processing has become an important feature of many fruit graders and different models exist based on the fruit being graded. Fruit bruising and other surface defects may also be assessed via electronic graders using colour imaging. Defective or bruised regions of the fruit surface have a different spectral reflectivity from healthy tissue and this can be used to detect and remove fruit from the grading line. Grading using machine vision is achieved in six steps: image acquisition, ground colour classification, defect segmentation, calyx and stem recognition, defects characterization and, finally, fruit classification into quality classes (Leemans *et al.*, 2002).

14.2.7 Grading other produce – cut flowers

A number of crops require specific postharvest handling and grading procedures and cannot be processed in the same way as many fruits and vegetables. Cut flower blooms are typically hand harvested and may be graded in the greenhouse or transported to a central pack house for sorting and application of postharvest treatments to prolong shelf-life. Many cut flower blooms are sorted into size grades based on bloom size, degree of maturity or opening, and stem length/diameter. Blooms may be rejected if they contain pest or disease damage, split stems, physiological or colour disorders. During postharvest handling or immediately after harvest, leaves are removed from the lower stem portion in most species; flowers may be recut to length after bunching and placed in a hydrating solution before packaging for market (Fig. 14.6).

Fig. 14.6. Cut flowers in postharvest storage.

Many cut flowers are pre-cooled before grading and packaging while being maintained in buckets or containers of sanitized water. During the postharvest sorting, grading, preconditioning treatment, packaging and transportation phases, some species of cut flowers such as snapdragons and gladioli must be maintained in an upright position to prevent the tips from curving.

14.3   Fresh-Cut Salad Processing

While many hydroponic salad crops are sold as 'whole heads' either with roots intact or cut from the base, processing into salad mixes onsite has become increasingly more common for hydroponic producers. Where fresh produce is to undergo further processing, specific guidelines must be followed to ensure food safety, shelf-life and quality of the final product. Fresh salad mixes usually fall under local regulations for the preparation, packaging and holding of food as they can pose a significant food safety risk.

Steps for fresh-cut salad processing start with harvesting at the correct stage of maturity during the coolest time of day when tissues are most turgid, this is typically early in the morning (Fig. 14.7). Lettuce foliage contains higher levels of water at this time which is known to prolong shelf-life and assist in the prevention of postharvest wilting. Plants should be cut with a sharp, stainless steel knife or scissors; sharp cutting instruments minimize tissue damage and should be disinfected regularly. During harvesting procedures, full food safety measures should be taken; these include use of disposable gloves, regular disinfection of harvesting bins/containers (or use of disposable bin liners), no smoking or food/drink consumption during the harvesting process or near the crop, and immediate removal of crop debris after harvesting. After harvesting produce should be transported immediately to the salad processing area as the speed at which lettuce and other greens are cooled, washed and packaged is a factor which determines shelf-life.

Fig. 14.7. Lettuce and salad greens should be harvested as early in the day as possible while tissues are turgid.

After harvest, washing and packaging should follow guidelines for food safety handling of fresh, uncooked produce as foodborne bacteria such as *Escherichia coli* and *Salmonella* are a risk with fresh-cut salad products. This includes use of materials which require disinfecting on a daily basis including stainless steel benches, wash and rinse facilities, concrete floors with built-in drains, head coverings and use of disposable gloves for all workers, and exclusion of pests and insects. Sodium hypochlorite solution may be used to disinfect all surfaces including floors. Once transported to the processing area, lettuce heads, baby salad leaf, cut herbs and other greens should be sorted and sliced as required. The finer produce such as head lettuce is sliced, the shorter will be the shelf-life due to an increased area of cut surface. Fine, sharp knives are used for this process which are regularly disinfected, carried out on a stainless steel or other non-porous work surface.

After sorting and preparation, fresh-cut lettuce and baby leaf are placed into a first wash tank containing a disinfection agent at the correct pH and temperature (Fig. 14.8). This first process reduces microbial contamination by up to 99% and is typically maintained at 1–2°C with 150–200 ppm of chlorine and a pH of 6–7. Contact time required is 2–3 min. Cut lettuce releases latex sap which reduces the disinfection potential of the wash water over time via organic loading, requiring regular water changes and monitoring of chlorine levels with use of chlorine test strips. Temperature of the wash water must be maintained at 1–2°C as this is the first and main method of cooling the cut product. Addition of large volumes of warm cut product or baby salad leaves will increase the temperature of the wash water rapidly, ice may be added to maintain temperature and wash tanks should be as large as possible to slow the rate of temperature increase. After the first wash tank, the cut produce is removed and placed in a second (rinse) tank also maintained at 1–2°C. This tank contains water of a potable standard and citric acid may be added to this tank at a

Fig. 14.8. Salad processing line.

rate of 0.1% with a contact time of 1 min. Some salad processes choose to carry out a triple wash process with a third rinse tank.

After removal from the rinse process, cut product is typically dried using either small 'salad spinners' or automatic centrifuges. Excessive drying results in moisture loss from the cut product and a more rapid rate of postharvest wilting, while under-drying can result in an increased rate of rots and moulds. Cut leaves at this stage are extremely fragile and susceptible to crushing and bruising – packing needs to be carried out carefully to avoid damaging cells. After drying, the salad product is packed into plastic packages or bulk boxes as rapidly as possible to prevent rewarming of the foliage. Bulk packages often consist of cardboard cartons with plastic liners into which salad mixes are placed based on customer requirements. Smaller plastic packs for consumers may use 'modified atmosphere' (MA) package materials. These packs are commonly used for salad mixes and are manufactured from selective films which help maintain the produce at a desired lower O_2 concentration and higher CO_2 concentration than exist in air. Lower O_2 levels prolong shelf-life and prevent oxidation which causes browning, as well as reducing the rate of respiration and retarding microbial growth. MA bags should be sealed to form a 'pillow' which cushions the cut product against crushing damage during transit and storage. All packaging should be marked with the date and batch number.

Once packaged, salad products must be immediately stored and transported at low temperatures (1–3°C) to maintain shelf-life, preferably in the dark (packed into cartons). Maintenance of the cool chain is essential for cut lettuce products to prevent a loss of shelf-life and microbial growth. It is also important that the temperature is stable; any freezing damage (below −0.2°C) will result in tissue breakdown, water soaking, browning, wilting and loss of product quality despite the use of MA packaging. Warming/cooling phases during the storage/transport/marketing process also cause tissue breakdown and loss of shelf-life. Once MA packaging is opened by the consumer, the product needs to be held under refrigeration and consumed within 24 h.

14.4   Shelf-Life Evaluation

Shelf-life testing may be carried out on samples of certain lines, grades of fruit or packaged products processed through the pack house to determine how long these will store under standardized conditions, the development of any bruising or other mechanical injury which may show after packing, the development of postharvest rots, loss in coloration, and wilting or shrivelling. Samples of uniform size, shape, colour, stage of maturity and grade are selected and placed into a temperature- and humidity-controlled room or chamber and observed daily for changes in postharvest quality. This allows pack-house operations to determine if any handling damage such as tissue bruising, which can take several days to develop, is occurring during the postharvest phase as well as the overall shelf-life of each line of fruit being packed.

14.5   Packaging

For both manual and automatic sorting and grading, the line ends at the packaging area where produce of different classifications is placed into cartons, retail packages or bulk bins for storage and transportation to markets or processors (Fig. 14.9). Reject grades which are suitable for processing are often simply placed into large bulk containers for shipment. Higher grades for export and premium fruits, vegetables or blooms may be individually wrapped and placed into compartmented trays which are loaded into cartons; this immobilizes the product, reducing the chance of impact injuries during transportation (Figs 14.10 and 14.11). Packaging may be carried out manually, particularly where a certain presentation or additional checking may be required on premium grades, even where grading and sorting were carried out automatically, or mechanical packaging systems may be in

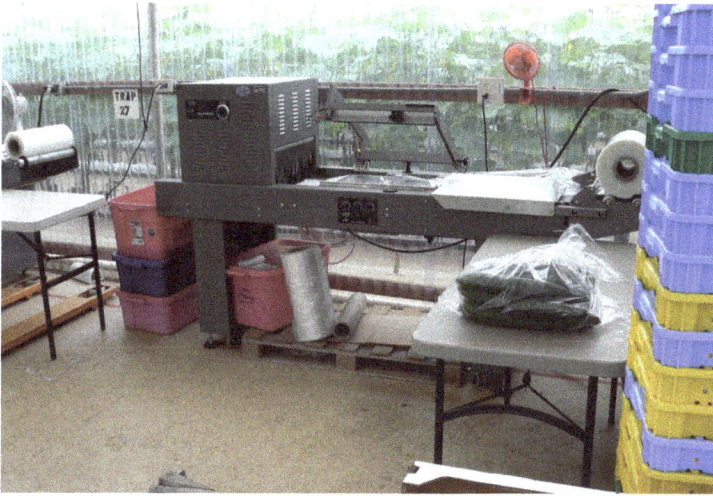

Fig. 14.9. Cucumber shrink wrapping.

Fig. 14.10. Clamshell packing of lettuce.

Fig. 14.11. Packing of high-quality tomato fruit.

place, particularly where large volumes of produce are being handled.

14.6 Postharvest Cooling

An important procedure during the postharvest handling phase of many fruits and vegetables is rapid cooling or initial removal of field heat. Cooling, in conjunction with refrigeration during subsequent handling operations, provides a 'cold chain' from pack house to final market or supermarket to maximize postharvest life and control diseases and pests (Sargent *et al.*, 2000). Cooling to the ideal temperature assists in maintaining quality and prolonging storage life by reducing the rate of respiration, ripening and softening, retarding many rot pathogens, and if carried out at the correct humidity, by lowering the loss of moisture from the harvested produce. Removal of field heat may occur immediately after harvest before produce is sorted or graded. This allows produce entering the pack-house facility to be stored under ideal temperature conditions while awaiting sorting, grading and packing. Field heat may also be removed during the initial water dump or wash phases with use of temperature-controlled water. Finally, cooling may occur for the first subsequent time after the produce has been put through the grading line and packaged into cartons. Many fruits benefit from immediate cooling postharvest, even when rewarming will occur during subsequent handling: deterioration is proportional to the total exposure time to warm temperature, not the pattern of cooling and warming (Kader, 1992). Commercial cooling for perishable crops has been defined as the rapid removal of at least 7/8 of the field heat from the crop by a compatible cooling method, which is known as the '7/8 cooling time', within a fairly short amount of time (Sargent *et al.*, 2000). The rate of removal of field heat is also an important aspect of cooling and the efficiency of cooling is dependent on time, temperature and contact. A number of cooling methods may be used during the postharvest handling phase; these include icing, hydro-cooling, room cooling, vacuum cooling and forced-air cooling. Postharvest or pre-cooling is in addition to the cool storage and high humidity conditions maintained after grading and packing.

14.7 Postharvest Handling Damage

Damage caused during the postharvest handling, grading and sorting processes can include compaction, cuts, punctures, abrasions, impact and bruising injuries. Soft fruits and vegetables such as strawberries and tomatoes are more prone to handling damage than hardier commodities. Impact damage and bruising typically occur where fruit drop on to hard surfaces, with the degree of bruising relative to the height of the fall. Impact bruising affects the flesh, resulting in brown spots under the skin that may not be visible from the surface (Kader, 1992). Tomatoes are sensitive to a physiological disorder known as internal bruising which only becomes apparent after the fruit reaches the full-red ripeness stage. Internal bruising occurs when a tomato fruit receives an impact above the locule during harvesting or postharvest handling, which disrupts the normal ripening in the locular gel. Tomato fruit at the breaker stage of maturity, dropped once from a 10 cm height on to an unpadded surface, developed internal bruising in 73% of fruit, indicating that packing lines should be designed to prevent such impact drops during the handling process (Sargent *et al.*, 2000). Mechanical injuries also speed up the rate of ethylene production and accelerate fruit ripening and senescence; they may also result in an increased occurrence of rot pathogens and in rapid water loss from the fruit.

Mechanical injury during the grading and sorting process can be minimized by designing packing lines and pack-house procedures that prevent excessive drops and rolls at transfer points on the packing lines, particularly over unpadded surfaces. Other causes of mechanical injury on handling lines include protrusions at impact surfaces, directional changes in product flow, unmatched speeds of adjoining components, excessive transfer plate angles, excessive rotational velocities of brushes and rollers and excessive line speed due to undersized capacity (Sargent *et al.*, 2000). During packing processes, compaction of produce inside cartons due to overfilling or mechanical injury from movement in underfilled containers are other causes of produce damage. Postharvest decay of many fruits and vegetables may originate in the handling and grading process where produce receives mechanical injury which permits entry of rot pathogens. During sorting and grading an important procedure is the removal of damaged produce and also those which are already exhibiting rots or other pathogens to prevent further contamination during storage.

14.8 GAP – Good Agricultural Practices in Postharvest Handling

GAP are a series of guidelines used worldwide to ensure the safety of fresh produce for human consumption. These guidelines were developed in response to the increasing number of foodborne disease outbreaks that have occurred from contaminated fresh produce. Since many greenhouse-grown fruits and vegetables are consumed raw without further washing or cooking, GAP principles should be applied to all postharvest operations to reduce the risk of foodborne disease contamination. Postharvest GAP recommendations are largely directed at the use of suitable water and sanitation of equipment and suitable worker hygiene. GAP procedures state that during the postharvest handling process, all water used in washing, cooling, processing and icing is of potable quality; this must be established through regular testing. Furthermore, washing water should use an acceptable level of active sanitizer compounds. All surfaces in the postharvest handling area which come into contact with fresh produce must also be sanitized on a regular basis, including harvesting bins, work surfaces, grader lines, storage areas, sorting bins and transportation vehicles. Since many foodborne disease pathogens can be spread by workers, strict guidelines for hand washing, use of gloves, hair coverings and protective clothing, and separate areas for eating and drinking are also part of the GAP procedure. Control and monitoring of pack-house pests including birds and rodents that may contaminate produce is another aspect of GAP guidelines. GAP protocols and compliance are recorded through accurate documentation which ensures that each batch of produce can be traced back to the source of any illness which may occur. By tracing back contaminated produce to specific products, distributors, pack houses or greenhouses, recalls of contaminated produce can be more specific and accurate.

14.9 Postharvest Storage

Postharvest storage aims to preserve the desirable characteristics and nutritional quality of produce for as long as is required before consumption. Along with this, food safety during storage is a major consideration as is consumer perception of appearance, taste and, increasingly these days, convenience and nutrition.

14.9.1 Postharvest physiology during storage

Horticultural produce such as fresh fruits, vegetables and flowers consist of living tissue which continues to respire postharvest. The objective of postharvest storage is to reduce losses in quantity and quality of horticultural crops between harvest and consumption by delaying senescence and water loss, preventing decay and storage disorders while eventually obtaining the correct degree of ripeness for consumption. Respiration is the process by which stored organic materials (carbohydrates, proteins and fats) are broken down into simple end products with the release of energy (Kader, 1992). During this process O_2 is consumed and CO_2 is produced along with the loss of stored food reserves. The rate of respiration during postharvest storage is dependent on a number of factors including the plant organ, maturity level, species, temperature, postharvest treatments, gaseous environment maintained around the product and for some commodities, the presence of ethylene. The rate of respiration determines the speed of senescence, quality aspects such as sweetness and firmness and dry weight of the stored product, thus reducing the rate of respiration is the main method of prolonging storage life.

Ethylene is a plant hormone that regulates aspects of plant growth and development such as ripening, senescence and abscission and plays a major role in postharvest storage and physiology. Ethylene in postharvest storage may be produced by the plant material itself or originate from exogenous sources in the environment; however, exposure speeds up the rate of ripening and reduces shelf-life in many products. Control over ethylene exposure and production through the use of cool temperatures and CA storage, as well as ethylene inhibitors, are important aspects of postharvest storage. Compositional changes in horticultural produce during postharvest storage are varied and while some such as increases in sugar content during fruit ripening are acceptable, others reduce shelf-life and result in produce being discarded. Loss of green coloration in vegetables is a common postharvest issue, while development of other pigments postharvest include tissue browning which may occur in a wide range of crops. As ripening progresses, the breakdown of pectins and polysaccharides results in softening and an increased likelihood of bruising and impact damage during storage and transport.

14.9.2 Storage systems

Storage of high-quality fresh greenhouse produce includes many considerations, some of which are specific to the crop. These include factors such as:

1. Postharvest quality control. Only high-quality produce, free from disease, decay, rots, pest damage, mechanical damage and physiological disorders, and of the correct maturity, should be stored. In some crops, maturity testing is required to determine optimum storability postharvest.
2. The correct postharvest treatments should be applied before storage. This may include sanitation, removal of field heat, washing, drying, trimming, packaging, heat treatments or other options for prolonging storage life and quality.
3. The correct environment for storage of each crop should be provided. This includes temperature, humidity and ventilation, which are monitored and controlled consistently throughout the storage process. Temperature control is vital and storage rooms should not be overloaded or containers stacked too closely. Ventilation and air movement, control and prevention of ethylene sources and a high degree of hygiene are required where food crops are to be stored.
4. Pests including rodents, birds and insects should be excluded from the storage area as these will contaminate the crop and cause postharvest losses.

5. Containers for storage must be clean, hygienic, strong enough to support the weight of the crop and facilitate stacking inside the storage facility.
6. Where different crops are to be stored in the same facility, care should be taken to avoid storing ethylene-sensitive products with those that produce ethylene.
7. Stored produce should be regularly inspected, and any diseased or damaged produce be removed immediately. Produce should also be examined for damage caused by excessive moisture loss, temperature injury or other storage disorders.

14.9.3 Refrigeration and cool storage

Refrigerated storage is one of the main methods for extending the postharvest life and maintaining product quality for many fresh horticultural commodities; however, temperature and humidity must be evenly controlled within the correct range for the produce being stored. Lowering temperature, often at the same time as increasing humidity level, slows the rate of respiration of tissue, reduces water loss and oxidation of cut surfaces and retains turgidity for a greater length of time, assisting in the preservation of colours, flavour and nutritional levels. Refrigerated storage facilities are typically well insulated and have equipment capable of not only maintaining the correct temperature but also extracting the heat generated by crops with a high rate of respiration. Cold storage is carried out for a wide range of fresh produce from fruit, herbs and vegetables to flowers and ornamentals. Fresh produce is typically pre-cooled immediately after harvesting to remove field heat and obtain optimum temperature levels as quickly as possible; this may be carried out via a number of different methods depending on the product being handled. Once pre-cooled, product is transported to the cool storage facility and maintained at the optimum storage temperature which varies according to the commodity or mix of produce being stored and level of maturity. Optimum storage temperatures are maintained as low as possible, but still above the level which is known to cause chilling injury for that particular crop. Along with temperature, relative humidity is run at levels which reduce water loss, but at the same time do not promote development of decay pathogens. Most leafy vegetables, brassica crops and berries grown in greenhouses are stored under conditions of 0–2°C and 90–98% relative humidity. Many tropical fruits and fruit vegetables are stored at 7–10°C and 85–95% relative humidity and melons at 13–18°C and 85–95% relative humidity. Under the correct cold storage conditions, shelf-life may be extended significantly compared with ambient conditions.

Low-temperature storage is a commonly used treatment to control enzymatic activity, oxidation and reduce the metabolic rate and subsequent tissue browning of a number of fresh-cut products such as fruit and vegetables that are cut or sliced postharvest and packaged. These include salad packs, shredded lettuce and other green vegetables.

14.9.4 Controlled and modified atmosphere storage

Controlled and modified atmosphere storage refers to adjustment of the gaseous composition directly around the produce to reduce the rate of tissue respiration, delay senescence and ripening, reduce sensitivity to ethylene and chilling injury and help alleviate certain physiological disorders and pathogens. Both controlled and modified atmosphere storage involve reducing the levels of O_2 and elevating CO_2 concentrations, the only difference between CA storage and MA storage being the degree of control over these aspects.

CA storage involves maintaining atmospheric O_2 concentration below 5% and CO_2 concentration above 1% (air being 78% N_2, 21% O_2 and 0.03% CO_2) and is used in addition to the maintenance of optimum temperatures and humidity for the commodity being stored. These levels of low O_2 and high CO_2 can be obtained in a number of ways; the simplest involves filling a sealed room with produce and allowing the O_2 level to fall

naturally with respiration. When the required level of O_2 has been reached, it is monitored and maintained by introduction of fresh air from outside so that it does not fall below a set level that may cause tissue damage. CO_2 level also increases in the sealed room with respiration until it reaches the required level, at this point excess CO_2 is removed via scrubbers. This method of obtaining and controlling the gaseous atmosphere inside a storage room is termed 'product generated' as the gas levels are created by product respiration; however, it can take some hours for this to develop and thus reduce maximum storage life. Other CA storage facilities use a higher degree of technology to create and monitor the gaseous conditions surrounding the produce. The most common method is to inject N_2 gas until the O_2 has reached the required low level, then bring in outside air as required to maintain the correct values during storage. The atmosphere in CA stores is then constantly analysed for CO_2 and O_2 levels using an infra-red gas analyser to measure CO_2 and a paramagnetic analyser for O_2 (Thompson, 2016). Gas levels are monitored and controlled via computer along with temperature and humidity levels throughout the postharvest storage process. Very accurate control over O_2 levels is vital so that tissue damage and disorders do not occur.

The biochemical effects of CA storage include a reduced rate of respiration induced by low O_2 and an inhibition of ACC synthase (key regulatory site of ethylene biosynthesis) by elevated CO_2. Sensitivity to ethylene is reduced at O_2 levels below 8% and/or CO_2 levels above 1%. The correct level of CA conditions also slows chlorophyll loss (yellowing), biosynthesis of carotenoids (yellow and orange colours) and anthocyanins (red and blue colours), and biosynthesis and oxidation of phenolic compounds (browning). Controlled atmospheres slow down the activity of cell wall-degrading enzymes involved in softening and enzymes involved in lignification and may influence flavour by reducing loss of acidity, starch to sugar conversions, sugar interconversions and biosynthesis of flavour volatiles as well as improving retention of ascorbic acid and other vitamins (Kader, 1986).

Apart from biochemical changes, CA storage conditions can have an effect on postharvest pathogens and in some cases may inhibit the development of fungal and bacterial decay. Elevated CO_2 levels (10–15%) have been shown to significantly inhibit the development of *Botrytis* rot on strawberries and other perishables, while low O_2 (below 1%) can control insects in some fresh produce (Kader 1986).

While controlled and modified atmosphere storage has significant benefits for maintenance of storage life and quality for a wide variety of produce, if not carried out correctly detrimental effects may occur. These include the development of brown spotting on lettuce and increased chilling injury. Other fruits may fail to ripen correctly, tomatoes may develop irregular ripening if exposed to O_2 levels below 2%. Off flavours and odours may also develop at low O_2 levels (anaerobic respiration) or at very high CO_2 levels.

14.9.5 Modified atmosphere packaging

MA storage describes the formation of a modified gaseous mixture around the produce inside packaging constructed from specifically permeable films that allow a reduced O_2 and increased CO_2 atmosphere to develop. This modified atmosphere may be permitted to develop naturally once the produce is sealed inside the packaging and respiration depletes O_2 and increases CO_2, or packages may be flushed with N_2 gas to remove excess O_2 and speed up the development of the modified atmosphere. Reducing O_2 around the produce, particularly that which has been pre-cut before packaging, such as shredded lettuce prone to postharvest browning, reduces the rate of respiration and oxidation of the cut surfaces. Studies have found that the correct modified atmosphere inside packages of shredded iceberg lettuce at 5°C inhibited browning over a 10-day period (Heimdal et al., 1995) and prolonged visual quality after storage for 2 weeks at 2.8°C compared with unpackaged samples which scored significantly lower (Garcia and Barrett, 2002). MA packaging has also proven to be efficient in controlling microbial build-up during

storage (King *et al.*, 1991). Once the modified atmosphere has developed inside the packaging the gaseous mix consists of largely N_2 and CO_2 and very little O_2. With MA packaging it is important to avoid damaging low levels of O_2 or high levels of CO_2 which lead to anaerobic respiration, resulting in the development of off flavours and odours and increasing susceptibility to decay (Garcia and Barrett, 2002).

14.10 Postharvest Disorders

14.10.1 Temperature injury

Injury due to temperatures which are either too high or too low for optimal storage can result in a range of symptoms and an overall reduction in postharvest storage life. Many fruit vegetables such as tomato, capsicum and aubergine are susceptible to chilling injury if exposed to temperatures below 5°C (i.e. under refrigeration) with damage occurring either in the postharvest storage phase or at the consumer level. Incorrect temperature management during postharvest storage can result from poorly run and monitored cold stores, by lack of maintenance of the cold chain during transit to and from cool stores, and after purchase by the consumer. Some produce such as tropical fruits are highly sensitive to chilling injury which may occur at temperatures above freezing, while many others with an optimal storage temperature of 0°C can suffer damage if tissue freezing occurs when conditions drop below this level. Chilling damage is dependent on time, temperature and tissue: the longer the period of time of exposure and the lower the temperature, the greater the damage will be as damage is cumulative in nature. Other factors such as level of maturity and tissue ripeness affect sensitivity to chilling injury. Common symptoms of chilling or freezing injury include surface lesions, water soaking, internal discoloration, failure to ripen, accelerated senescence, increased decay, compositional changes, waterlogging and glassy areas in the flesh.

Heat injury may be caused by exposure to direct sunlight or by excessively high temperatures during storage. Symptoms of heat injury often include bleaching, surface burning or scalding, uneven ripening, excessive softening and desiccation (Kader, 1992).

14.10.2 Ethylene injury

A number of fruits produce small quantities of ethylene; however, many fruits, cut flowers and some vegetables are highly responsive to the presence of ethylene in the storage environment and may be damaged by exposure to concentrations as low as 1 ppm. Ethylene contamination may originate from a number of sources in the postharvest process. These include being stored in close proximity to ripening fruit such as apples or bananas that naturally produce ethylene in mixed produce facilities or at the retail level, from the exhaust of combustion engines (i.e. forklifts) or as a by-product of gas heating systems. The effects of ethylene on stored produce can be difficult to differentiate from other storage disorders or the natural senescence process, but include premature chlorophyll degradation, yellowing of green tissues, calyx abscission, russet spot on lettuce and other vegetables, acceleration of fruit softening, petal drop in cut flowers and a rapid rate of fruit ripening (El-Ramady et al., 2015).

Ethylene inhibitors may be used postharvest to inhibit ethylene production or action during ripening and storage of certain fruits and ornamentals such as cut flowers. 1-Methylcyclopropene (1-MCP), aminoethoxyvinylglycine (AVG), silver nitrate, silver thiosulfate, cycloheximide and benzothiadiazole are some of the chemical compounds used for this purpose.

14.10.3 Other postharvest storage disorders

Other postharvest storage disorders are often commodity related and include those which originate from preharvest nutritional imbalances; for example, blossom end rot of tomato fruit and tipburn of inner lettuce leaves and other sensitive greens which results from a tissue calcium deficiency. Calcium deficiencies may also induce premature softening, reduce textural quality and increase the rate of senescence of fruits and vegetables. The development of off flavours and off aromas can be caused by incorrect gaseous environments in CA and MA storage such as low O_2 or high CO_2 resulting in anaerobic respiration. Disorders may originate from other incorrect storage conditions, these include the greening of Belgian endive when exposed to light, upward curvature of asparagus spears during storage and seed germination inside fruits such as tomatoes and peppers induced by incorrect temperature control. Postharvest losses in nutritional quality, particularly vitamin C (ascorbic acid), can be substantial and increase with physical damage, extended storage, high temperatures, low relative humidity and chilling injury (Kader, 1992).

14.10.4 Storage decay

Most postharvest diseases of fresh produce in storage are caused by fungi and bacteria, with the most important fungal genera being Penicillium, Aspergillus, Geotrichum, Botrytis, Fusarium, Alternaria, Colletotrichum, Dothiorella, Lasiodiplodia, Phomopsis, Phytophthora, Pythium, Rhizopus, Mucor, Sclerotium and Rhizoctonia; common bacterial soft rots are caused by various species such as Erwinia, Pseudomonas, Bacillus, Lactobacillus and Xanthomonas (Coates and Johnson, 1997).

Postharvest diseases of fruit and vegetables may be classified based on when infection was initiated. Latent infections usually occur prior to harvest; however, the disease remains dormant until a later stage during storage when the host tissue becomes suitable for infection to proceed. Ripening and softening of produce tissue during storage are often the trigger for reactivation of a latent infection, the most common being Botrytis cinerea (grey mould) and anthracnose of various fruits. The second type of postharvest disease occurrence results from injury such as surface wounds, harvesting cuts, bruising or insect damage which permit infection to take place. Diseases resulting from physical damage include those caused by *Penicillium* spp. (blue and green moulds), *Rhizopus stolonifer* and bacteria such as *Erwinia carotovora* (Coates and Johnson, 1997).

The main methods of control over postharvest decay organisms include application of preharvest fungicides to reduce pathogen inoculum levels which develop later in storage, sanitation and hygiene throughout the harvesting, postharvest and storage process including surface sterilant dips and sprays for some commodities, use of heat treatments, irradiation, prevention of injury, use of the correct storage conditions including temperature and humidity control, and in some instances controlled atmosphere or modified atmospheres which may inhibit some pathogens. One of the most important methods of pathogen control, apart from temperature, is maintaining the correct relative humidity. For most commodities, the correct humidity is required to prevent excessive moisture loss; however, high humidity promotes the development of many pathogens, so a trade-off develops between maintaining turgidity and controlling disease occurrence.

14.11 Food Safety and Hygiene

Many food safety outbreaks in recent years have been traced back to workers contaminating produce during the harvesting or postharvest handling operations. Staff may contaminate fresh fruit and vegetables by touching these and transferring human pathogens. All harvesting and pack-house staff should be trained in worker hygiene practices such as hand washing or use of gloves and wearing of hair nets, protective clothing, footwear and face masks where appropriate and those who are ill not coming into contact with produce. Hand hygiene is of particular importance for a number of crops such as fresh berries where foodborne outbreaks linked to poor personal hand hygiene have been reported in strawberries and highbush blueberries (Rodriguez *et al.*, 2011). GAP recommendations state that workers should wash hands before beginning harvesting, but also after using restrooms, eating, drinking, if moving between

harvesting sites, after using agrochemicals or being in contact with plant debris, growing media or other surfaces which may contain microorganisms. This requires not only thorough worker training and education before harvesting begins, but continual retraining, signage and reminders throughout the harvesting period. Suitable facilities for regular hand washing are needed or hand sanitation chemicals and/or disposable gloves provided in the greenhouse where harvesting, particularly for sensitive crops such as berries.

During the harvesting process, only clean and sanitized knives and other equipment should be used to prevent pathogen transfer and harvested product should be placed directly into clean containers. Plastic harvest bins are recommended as these can be more thoroughly sanitized than wooden surfaces. Any water which is used to wash harvest bins, contact surfaces or produce should be clean and treated with sanitation compounds to prevent contamination with microbial agents. All transport containers and produce contact surfaces such as conveyor belts, dump tanks, packing and grading lines need to be cleaned and regularly sanitized with approved cleaning chemicals; this is often a sodium hypochlorite solution of at least 200 ppm, however other compounds may also be used for this purpose. Organic matter such as plant debris on floors and produce dump areas should be removed before sterilization agents are used. Cool store facilities also require attention to sanitation and hygiene as some pathogens such as *Listeria* can multiply at refrigeration and storage temperatures and contaminate produce through condensation dripping from walls and ceilings.

14.12 Ready-to-Eat, Minimally Processed Produce

Much of the bacterial contamination and issues related to food illness originate from ready-to-eat produce, sprouts and fruits and vegetables which are consumed raw with no or little cooking. The ready-to-eat market has grown considerably over the last few decades as the demand for pre-prepared salads and vegetables, fruit and other produce has increased rapidly. The damage to the outer, protective layers of fruits and vegetables that occurs in the processing steps of peeling, slicing, shredding and juicing allows bacteria to enter tissues and proliferate where pH values and storage temperatures allow. Steps to prevent bacterial contamination and growth during processing typically involve the use of sterilization agents and other compounds in the wash water. The effectiveness of these steps for microbial control are, however, reliant on a number of factors such as the concentration of chlorine (available free chlorine levels), temperature and pH (optimal levels are pH 6–7.5) as well as the amount of organic material present, type of produce and microorganisms which may be present. While chlorination and other similar compounds, when applied correctly, control most bacterial pathogens, it has limited effectiveness against protozoan oocysts and viruses and some bacterial spore resistance also exists (Parish *et al.*, 2003).

14.13   Certification and Food Safety Systems

Food safety systems are designed to promote and regulate food production and processing practices to reduce risk of food-related illness or contamination of food. These include GAP (Good Agricultural Practices), GMP (Good Manufacturing Practices) and HACCP (Hazard Analysis and Critical Control Points). Many larger markets for produce including supermarket chains and exporters/importers are increasingly insisting suppliers utilize these programmes and regulations with associated documentation to ensure food safety. In particular, a HACCP programme is recommended for all suppliers of lightly processed produce (Kader, 1992) which is most prone to food safety outbreaks.

HACCP, which often incorporates GAP and GMP into its structure, is designed to identify potential food safety hazards, which may be physical, chemical or microbial, at each step in the production and processing system for a particular product or crop. Once identification of each potential food safety hazard has occurred, steps are put into place to establish critical limits, monitoring procedures and correction activities should a food safety hazard be found to occur. Record-keeping systems are used to document the HACCP programme and procedures are established to ensure monitoring and assessment are ongoing. All areas of production are evaluated for HACCP analysis, including growing and harvesting operations, pack-house operations, packing material and storage areas and all steps in the distribution to final market. While HACCP is useful in detecting and controlling potential food safety hazards (critical control points) from fresh and processed products, it is only effective with hazards which are preventable and controllable.

GAP incorporate codes, standards and regulations relating to food safety as a major aspect of this programme; however, other production factors such as economic viability, environmental sustainability and social acceptability as well as food quality are also included in GAP programmes to varying degrees. GAP was initially developed by the food industry and producer organizations but has now been accepted by governments and other non-governmental organizations as a widely recognized and effective programme. It has been recognized that adoption of GAP helps improve the safety and quality of food crops and helps reduce the risk of non-compliance particularly regarding pesticide residues, maximum levels of mycotoxins as well as other chemical, microbiological and physical contamination hazards. Some of the disadvantages of implementing GAP programmes are the high level of record keeping, documentation and certification which can increase production costs, particularly for smaller growers and processors, and the requirement for GAP certification may exclude some producers from export market opportunities.

GAP guidelines for crop producers focus on four components of growing and processing:

soil, water, human handling and surfaces. GAP guidelines for water use incorporate the requirement for water used in all crop processes to be free of microbial contaminants and regular water testing to provide documentation that the water is of suitable quality. Human handling of fresh produce is also an important aspect of GAP regulations which include guidelines for hand washing and general hygiene, providing suitable hand washing and toilet facilities for all workers, and a number of other steps to prevent worker contamination of produce. GAP guidelines for surfaces which may come into contact with crops at any time during the production, postharvest handling and processing and distribution chains include keeping facilities clean, sanitation of equipment, bins and rinse water to reduce bacterial species, control of animal contamination sources such as pets, wildlife, birds, insects and rodents, as well as correct temperature control and management postharvest. Specific food safety plans are also part of GAP regulations which are customized to each operation implementing GAP procedures. These identify specific risks for each stage of the operation, including monitoring and documentation of each stage of production and processing stages, and have a requirement to be regularly revised and updated.

GMP guidelines and regulations apply to facilities which are processing and packaging raw produce to be sold on to wholesalers and retailers for distribution to consumers. This includes operations which prepare 'ready-to-eat' products such as salad mixes, sliced or shredded fruits and vegetables such as lettuce.

14.13.1 Documentation and recall programmes

Many fruit and vegetable pack houses and processors now use computer trace and track programmes, bar codes and documentation that allow any contaminated batch of produce to be traced back to its source of production and packing. This allows a recall of any produce lines if a food safety issue is detected and traceability to the source of contamination. Recall and trace back programmes are not mandatory in all countries and regions, however they are becoming an increasingly more common aspect of food production and distribution. Produce line recalls may be carried out by produce production and packaging companies or by regulatory authorities such as the FDA in response to a suspected food safety issue or outbreak. The FDA has defined three recall classifications and FDA actions. Class I is an emergency situation involving the removal of products from the market that could lead to an immediate or long-term life-threatening situation and involve a direct cause–effect relationship (e.g. *Clostridium botulinum* in the product). Class II is a priority situation in which the consequences may be immediate or long-term and may be potentially life-threatening or hazardous to health (e.g. *Salmonella*) and Class III is a routine situation in which life-threatening consequences are remote or non-existent (Gorny and Zagory, 2016).

14.14   Postharvest Developments

The more recent incorporation of robotics and image processing and sensing technology into harvest mechanization has seen greenhouse crops harvested selectively with no or minimal requirement for workers. While these are still very recent and emerging technologies, they are likely to lead the way with large-scale harvesting of an increasing range of high-value crops. New innovations in the postharvest handling industry are centred around improved standardization of grades on an international scale to take account of the increasing volumes of exported fresh produce and the large distances much of this often travels to the final consumer. Other innovations are also aimed at increasing the sustainability of postharvest technologies through a more rational use of energy and refrigeration, the recycling of water, the application of non-destructive technologies for quality evaluation, the reduction of chemicals as well as by decreasing the amount of standard packaging and the introduction of environmentally friendly or recyclable materials (Tonutti, 2013). Improved methods of ensuring food safety and increased uptake of innovative guidance and reporting systems such GAP are another rapidly expanding aspect of the postharvest handling chain which will take an increasingly more important role in fresh produce processing and distribution.

Current and ongoing trends in postharvest storage for many fresh commodities focus on the continued reduction of many chemical treatments, an increased concern with food security and safety, and prevention of losses during the storage process.

Continued research into CA and MA storage conditions for a wider range of fresh fruits and vegetables is also a current trend, with the object of determining the optimal gaseous, temperature and humidity levels not only based on factors such as species, but also on variables such as crop maturity, nutrition and cultivar. Along with the control and reduction in sensitivity to ethylene postharvest, advances in storage technology are seeing many crops being stored with a high compositional quality maintained for considerably longer than they have in the past. An increased understanding of the physiology and biochemical changes that occur in postharvest storage is also seeing innovative new advances in packaging materials and methods. Many of these are based on 'active' and 'intelligent' packaging technology. The use of innovative, new packaging films incorporating active compounds that increase shelf-life, improve sensory quality and assist with food safety is becoming more common, particularly for fresh-cut produce. These may include the use of antimicrobial compounds incorporated into the packaging itself which assist in the prevention of growth of both spoilage organisms and human pathogens. Use of ethylene absorption compounds in plastic films is another innovative feature of some postharvest storage packaging for many products which are sensitive to ethylene.

Other innovations are linked to the changing trends in the way many greenhouse-produced fresh fruits and vegetables are marketed with an increased emphasis on ready-to-eat, pre-cut, pre-prepared produce including fresh salad packs, pre-peeled and sliced fruits and vegetables. These products pose new challenges to postharvest storage and packaging, with prevention of tissue browning and maintenance of nutritional and flavour quality in fresh-cut products creating the need for increasingly sophisticated technology to maintain consumer acceptability. Maintenance of nutritional quality during postharvest storage is also an emerging trend as producers growing fruits and vegetables with increased antioxidant activity, higher vitamin and mineral contents and improved flavour do not want to see these degrade during postharvest storage.

References

Coates, L.M. and Johnson, G.I. (1997) Postharvest diseases of fruit and vegetables. In: Brown, J. and Ogle, H. (eds) *Plant Pathogens and Plant Diseases*. Rockvale Publications, Cambridge, Massachusetts, pp. 533–547.

El-Ramady, H.R., Domokos-Szabolcsy, E., Abdalla, N.A., Taha, H.S. and Fari, M. (2015) Postharvest management of fruits and vegetables storage. In: Lichtfouse, E. (ed.) *Sustainable Agriculture Reviews*. Sustainable Agriculture Reviews, Vol. 15. Springer, Cham, Switzerland, pp. 65–152.

Gao, H., Cai, J. and Liu, X. (2014) Automatic grading of the post-harvest fruit: a review. In: Li, D. and Zhao, C. (eds) *Computer and Computing Technologies in Agriculture III. CCTA 2009. IFIP Advances in Information and Communication Technology*, Vol. 317. Springer, Berlin/Heidelberg, Germany, pp. 141–146.

Garcia, E. and Barrett, D.M. (2002) Preservative treatments for fresh-cut fruits and vegetables. In: Lamikanra, O. (ed.) *Fresh-Cut Fruits and Vegetables: Science, Technology, and Market*. CRC Press, Boca Raton, Florida, pp. 267–304.

Gorny, J.R. and Zagory, D. (2016) Food safety. In: Gross, K.C., Wang, C.Y. and Saltveit, M. (eds) The commercial storage of fruits, vegetables, and florist and nursery stock. *Agriculture Handbook No. 66*. US Department of Agriculture, Agricultural Research Service, Beltsville, Maryland, pp. 149–165. Available at: https://www.ars.usda.gov/ARSUserFiles/oc/np/CommercialStorage/CommercialStorage.pdf (accessed 26 September 2020).

Gorvett, Z. (2015) World's first pepper picking robot heralds new era. *Horizon – The EU Research and*

Innovation Magazine, 28 May 2015. Available at: https://horizon-magazine.eu/article/world-s-first-pepper-picking-robot-heralds-new-era.html (accessed 14 September 2020).

Heimdal, H., Kuhn, B.F., Poll, L. and Larsen, L.M. (1995) Biochemical changes and sensory quality of shredded and MA-packaged iceberg lettuce. *Journal of Food Science* 60(6), 1265–1268.

Kader, A.A. (1986) Biochemical and physiological basis for effects of controlled and modified atmospheres on fruits and vegetables. *Food Technology* 405, 99–100, 102–104.

Kader, A.A. (1992) *Postharvest technology of horticultural crops,* 2nd edn. *Publication 3311*. University of California, Division of Agriculture and Natural Resources, Davis, California.

King, A.D. Jr, Magnuson, J.A., Torok, R. and Goodman, N. (1991) Microbial flora and storage quality of partially processed lettuce. *Journal of Food Science* 56(2), 459–461.

Leemans, V., Magein, H. and Destain, M.F. (2002) On-line fruit grading according to their external quality using machine vision. *Biosystems Engineering* 83(4), 397–404.

Moran, N. (2016) Harvest more profits. *Produce Grower Magazine*, April 2016. Available at: https://www.producegrower.com/article/harvest-more-profits-web/ (accessed 14 September 2020).

Parish, M.E., Beuchat, L.R., Suslow, T.V., Harris, L.J., Garrett, E.H., *et al.* (2003) Methods to reduce/eliminate pathogens from fresh and fresh-cut produce. *Comprehensive Reviews in Food Science and Food Safety* 2(Suppl.), 161–173.

Rodriguez, M.A., Valero, A., Posada-Izquierdo, G.D., Carrasco, E. and Zurera, G. (2011) Evaluation of food handler practices and microbiological status of ready to eat foods in long term care facilities in the Andalusia region of Spain. *Journal of Food Production* 74(9), 1504–1512.

Sargent, S.A., Ritenour, M.A., Brecht, J.K. and Bartz, J.A. (2000) Handling, cooling and sanitation techniques for maintaining postharvest quality. In: *Vegetable production handbook. Document HS719*. University of Florida Extension, Institute of Food and Agricultural Science, Gainesville, Florida, pp. 97–109. Available at: https://ufdcimages.uflib.ufl.edu/IR/00/00/16/76/00001/CV11500.pdf (accessed 14 September 2020).

Thompson, A.K. (2016) Controlled atmosphere storage. Fruit and Vegetable Storage: Hypobaric, Hyperbaric and Controlled Atmosphere. *SpringerBriefs in Food, Health and Nutrition*. Springer, Cham, Switzerland, pp. 21–36.

Thompson, J.F., Mitchan, E.J. and Mitchell, F.G. (2002) Preparation of fresh market crops. In: Kader, A.A. (ed.) *Postharvest technology of horticultural crops*, 3rd edn. *Publication* 3529. University of California, Division of Agriculture and Natural Resources, Davis, California, pp. 67–80.

Tonutti, P. (2013) Innovations in storage technology and postharvest science. *Acta Horticulturae* 1012, 323–330.

Van Henten, E.J., Hemming, J., van Tuijl, B.A.J., Kornet, J.G., Meuleman, J., *et al.* (2002) An autonomous robot for harvesting cucumbers in greenhouses. *Autonomous Robots* 13, 241–258.

Index

Page numbers in **bold** type refer to figures and tables.

abiotic disorders
  abiotic factors, crop damage effects 184–188
  caused by cultural practices 188
  identification (diagnosis) 184, 188–189
  postharvest injuries and disorders 284–285
  see also physiological disorders
abscission, premature (fruitlets) 188, 207, 212
acidity/pH, produce testing 256–257
aeration 69–70, 160–161, 165
  see also oxygenation
aerobic processing, organic matter 103–104
  see also composting
aeroponics 9, 71–72, 112, 132
  in plant factory systems 233, 237
affective consumer panel testing 258
agrochemicals
  contamination of water supplies 137
  disposal of containers 58
  phytotoxicity 187–188
air conditioning units 32, 238
air filled porosity 87–88
air movement
  boundary layer disturbance 35, 238
  effects on transplant development 128
  fan assisted 31, 35–36, 179
  see also ventilation
air pollution control 7, 238
algae 44, 51, 55, 85
alkalinity, water supplies 136–137, 140, 154, 155
ammonia/ammonium
  levels in anaerobic digestates 103
  nitrification by bacteria 103–104
  physiological effects on plants 144, 152

ammonium molybdate 155
ammonium nitrate 152, 153
ammonium phosphate 153
anaerobic processing 103–104
ancient civilizations, horticulture 1, 4–5
animals
  exclusion from food production areas 48–49,
    223, 231
  fodder for, hydroponic production systems
    74–75
  as sources of organic fertilizers 107–108
Antarctica, indoor crop production 230–231
anthocyanin content 241, 252, 253
anti drip (anti fog) additives, plastic films 17
antioxidants 241, 249, 253, 261
aphids 17, **172**, 172–173, **173**, 184
aquaponics 72–73, 103
  considered as organic 9, 74, 101–102, 108
  system design 25, 73–74, 108
arbuscular mycorrhizal fungi (AMF) 249
aroma, chemicals and detection 259–260
artificial lighting
  control in greenhouse production 6, 33–34
  equipment technologies 34, 72, 239–241
  needs in high latitude regions 13, 32, 199
  provided for seedling development 127–128
  as sole light source in indoor systems 27,
    34, 232
  supplementary, energy efficiency 34, 40–41
Asia
  expansion of protected cultivation 3, 11
  green leafy vegetable crops 69, 212, 236
  plant factories development 230

automation
    brushing of seedlings 128
    failure alarms and backup systems 39, 65
    hydroponic irrigation systems 39, 126
    monitoring sensors and equipment adjustment
        38, 39, 63–64, 159
    in plant factories 235, 237, 241–242
    produce colour analysis and grading 254, 271, 275
    propagation operations 118, **122**, 124, 125, 130
    solution culture systems 75
    see also computer based control; robotic systems

Bacillus spp., beneficial effects 96, 97
    B. thuringiensis (Bt, biocontrol agent) 113, 176, 178
back pain prevention (workers) 59
bacterial diseases 183
    bacterial wilt of tomato (*Ralstonia solanacearum*)
        130, 183, 199
    postharvest spoilage pathogens 217, 231–232, 285
banker plants 178
bare root transplants 122, 220, 225
bark, composted 82, 89
batch feeding, irrigation nutrients 94, **94**
bato bucket (Dutch) system 82, 198
Beauveria bassiana (biopesticide) 178
bees, role in pollination 202–203, 222, **222**
beneficial microbes 65, 96–98
    disease antagonists 111, 113, 162, 170
    inoculated biocontrol agents 170, 182
    role in slow sand filtration 164–165
beneficial mineral elements 143, 147–149
bioactive compounds (phytochemicals) 241, 260–261
bioassays
    for growth media 90, **90**, **91**
    water quality 142
biochar products 84
biodegradable materials 3, 123–124
biodigester systems 103–104, 114
biofilms 51, 109–110, 111
biofiltration *see* slow sand filtration
biofortification (selenium) 149
biofuels 44–45
biological control agents
    entomopathogenic nematodes 176
    insect parasites and predators 171–172, 173,
        176, 177
    microbial 170, 176, 178, 179
    predatory mites 174, 175, 176, 177–178
biomass 31, 44–45
    heating, for biochar products 84
biosecurity **49**, 52–54
blossom end rot (BER) 189–190, 207
bolting 191, **191**, 216–217
boron
    fertilizer sources (boric acid, borax) 155
    levels in water supplies 141

plant nutritional requirements 146–147
    toxicity symptoms from high levels 187
botanical pesticides 113, 172
Botrytis cinerea (grey mould) 178–179, 223
    disposal of diseased material 50–51, 179
    flower blight of roses 226
    stored produce damage 284, 285
Brix testing (total soluble solids) 247–248, 254–256,
        255, 256
browning, stored produce 282, 283, 284
bruising
    optical assessment 254, 275
    risks and avoidance 269, **269**, 280
brushing (mechanical conditioning) 128
bulk density, substrates 86–87

calcium
    deficiency effects in storage 285
    fertilizers 153, 154
    levels in water supplies 138, 140, 141
    plant nutritional requirements 145, 207, 248–249
    uptake and transport problems 145, 189–191,
        190, 225, 249–250
calcium nitrate 89, 149, 152–153
calcium superphosphate 154
calibration, pH meters 159
capillary watering systems 93
capsicum production
    crop nutrition 207
    harvesting and yields 205, 208
    pests, diseases and disorders 172, 189, 192,
        207–208
    plant training and support 205, 206, **206**
    pollination and fruit set 206
    propagation 205
    varieties and production systems **204**, 204–205
carbon dioxide (CO_2) enrichment 6, 38–39, 101, 199,
        209–210
    in controlled atmosphere storage 283–284
    efficiency and control in plant factories 238–239
carbon filters, for water supplies 163–164
caterpillars 175–176
cation exchange capacity (CEC) 77, 83, 89
CELSS (controlled ecological life support systems) 6, 229
certification processes 112, 287
chelation of trace elements 146, 154–155, 156
    organic chelating agents 156
chemical injury (phytotoxic chemicals) 187–188
chicory production 196, **197**
chilling injury 185, 211, 252–253, 284
China
    expansion of plastic clad protected cropping
        3, 11
    geothermal heating of greenhouses 42
    traditional greenhouses (passive solar) 1, 14
chloramines (water treatment chemicals) 138–139

chlorine
  chloride levels in water, effects on crops 141, 147
  chlorination of water for sanitation 48, 138,
        142, 165, 273
  negative impacts on beneficial microbes 98
  toxicity symptoms from chlorination 147,
        165, 187
chlorosis, causes
  ammonium toxicity 152
  mineral deficiency 146, 147, 189
  pest infestation/virus disease 175, 183
  response to waterlogging 85
city/municipal water supplies 138–139, 162
cladding materials
  for low and high tunnels 20
  mesh screens and nets 18–19
  plastic and glass, properties 16–18, 30
  technological development trends and impacts
        2–3, 18
cleaning
  between crops, NFT systems 64
  drip irrigation equipment 92
  of surfaces in greenhouse facilities 50, 51
cleft (wedge) grafting 131
climate control, greenhouse production 6, 230
cloche covers 6, 19, **20**, 27
clones, propagation of 132, 133
closed plant production systems 1, 4, 6, 13–14
  climate control 32, 230
  collection and recirculation of nutrient solution
        78, 78–79, **79**
  fully enclosed (CPPS) 4, 229, 231–232, 238
  importance of water quality 136
  role in environmental pollution prevention 26,
        161, 231
  see also plant factories
coatings, seed 97, 119–120
coconut fibre based media (coir, coco) **78**, 80–81,
        84–85, 224
cold frames 20–21
colour sorting, grading and analysis 254, 271,
        274–275
compliance programmes 49
compositional quality see quality assessment
composting
  of green waste crops 56
  pest and disease risks and suppression 51,
        106, 107
  of tree bark, for hydroponic substrates 82, 89
  uses of compost 58, 97, 106–107
  see also vermiculture
compression testing 262
computer based control 6, 12, 30
  gas composition in CA storage 283
  monitoring data, records 39, 75
  in plant factories 237, 238
  plant physiology modelling 95

consumer interests
  perceptions of plant factory crops 232, 243
  produce quality components affecting
        choice 246, 247, 254, 257
  in sustainability and environmental issues 101, 115
contact dermatitis 59
container capacity 81, 88
container (potting) mixes 8, 80, 82, 105
contaminants
  food produce 262–263, 281
  water supply 137, 138–139, 141–142
controlled atmosphere (CA) storage 283–284
controlled environment agriculture (CEA) 6, 27
cooling methods and equipment 32, **33**, 277, 278,
        279–280
Cooper, Dr. Allen 6, 61
copper
  crop damage, sensitivity 187
  fertilizer sources 155
  levels in water supplies 141
  plant nutritional requirements 147
corn salad (Valerianella locusta) 148
cracking, fruit skin 192
crooking (cucumber defect) **192**, 212
crop protection
  disease control and prevention 178–183
  from disorders, by abiotic factor management
        184–192
  from pests, control measures 170–177
  sanitation and hygiene measures 47, 49–51, 184
cucumbers
  benefits of grafting 130, 131
  bioassay subjects for substrate evaluation 90, **91**
  crop nutrition 210
  cultivars in hydroponic production 208, **208**
  environmental requirements 63, 209–210
  harvesting and yields 210–211, 271–272
  pests, diseases and physiological disorders 191,
        192, 211–212
  plant training and support 210, **211**
  propagation and plant densities 208–209
cucurbitacins (toxins) 263
cultural practices
  crop damage risks 188
  produce quality improvement 248–250
culture methods, microbial 262
cut flowers
  harvesting and grading 226, **275**, 275–276
  types cultivated 224
  see also rose production
cuttings 132, 220

damping off, seedlings 85, 126, 182
decay, stored produce 285–286
deep water culture, DWC (float systems) 66–71,
        67, **69**, 126

deficiency symptoms 66, **70**, **144**, 144–147, 160
deficit irrigation 96, 201, 248
degradation, growing substrates 88
descaling, irrigation systems 51, 52, 166
descriptive sensory analysis 258–259
dew point 37
DFT (deep flow technique) systems 66, **67**, 67–68, **68**, 69–71
difference testing (sensory evaluation) 258
diffuse light (radiation turbidity) 17
direct covers *see* floating mulches
direct seeding of crops 121
disbudding, roses 225
diseases
    management and control
        allowable organic control options 112–113
        induced systemic resistance 111
        outbreak minimization in recirculating systems 65, 71, 170
        prevention measures, seedling disease 126
        resistant crop varieties 180, 183, 198, 223
        suppression in mature compost 106, 107
        transmission, and spread prevention 50–51, 52, 162, 166
    types
        common diseases of hydroponic crops 170, 178–183
        diagnosis 188–189
        postharvest spoilage pathogens 217, 231–232, 285–286
        of roots, impacts on nutrient uptake 160
disinfection
    of discharged wastewater 55–57
    hand sanitizers 54
    of irrigation water, methods 52, 142, 162–166
    of soil, alternatives for organic systems 101
    of surfaces and equipment 50, 277
disposal of waste *see* waste management
DLI (daily light integral) 33, 34
    requirements of specific crops 199, 214, **215**, 222, 251
DO (dissolved oxygen) *see* oxygenation
downy mildew 179, 180, 226
drip irrigation 91–92, **92**
    nutrient dosing methods 94, **94**, **95**
    organic production systems 102, 108–109, **109**
    in staggered level planting systems 221
    system design and layout 92–93, **93**, 209, **209**
dry matter percentage, produce 256
DTPA (chelating agent) 154, 156
Dutch bucket system 82, 198

ebb and flow delivery systems 93, 126–127, 236
Eden Project, UK 4, **4**, **5**
EDTA/EDDHA (chelating agents) 154, 156
effective pore space, substrates 87

electrical conductivity (EC) 89
    impacts on fruit compositional quality 249
    management, effects on crops 80, 128–129, 158, 201, 207
    measurement 39, 90, **156**, 156–157
    recommended levels for water supply 140, **157**, 157–158
    unreliability for organic system nutrient monitoring 109, 110, 156
ELISA test kits 262–263
emissions
    from biomass energy, carbon neutrality 44
    carbon dioxide (atmospheric) 3, 15
    from incineration of plastics 58
    regulation and standards 26, 56
energy use
    efficiency improvement needs 3–4, 39–40, 242–243
    greenhouse power load estimates 15
    input costs and consumption 30, 40, 232, 239
    reduction and conservation methods 15, 30, 40–41
    sources of energy 15, 40, 41–45
environmental concerns
    aquifer depletion 137
    excessive fertilizer use 161
    plastic waste 3, 58
    soil/water pollution from intensive agriculture 26, 56, 231
    use of fossil fuels 3, 4, 15, 41
epinasty 85, 188
Escherichia coli (foodborne pathogen) 48, 52, 262, 277
essential nutrients, plant growth 143–147
ethylene
    accelerated biosynthesis in water stress 186
    buildup in oversaturated conditions 85
    from incomplete fuel combustion, crop damage 39, 188
    postharvest exposure and effects 281–282, 285
evaporative cooling 32, **33**
expanded clay 79, 83

fan (forced) ventilation 35, **36**
fertilizers
    energy costs of production 4
    grade (compositional) 155–156
    nutrient formulation for stock solutions 143, 149–152, **151**, 236
    organic, sources 8–9, 74, 101, 107–108
    pollution risks of intensive use 26, 231
    soluble types commonly used 143, 152–155
    as supplements to correct deficiencies 111, 156
    use minimization in closed systems 161, 198–199, 237
field heat, produce 280, 282

filtration
    biofilters in organic systems 109, 110
    for nematode pest removal 177
    prefiltration before disinfection treatments
        162, 166
    in recirculating hydroponic systems 91, 92
    water treatment 52, 56, 163–165
firmness, produce testing 258–259, 261–262
flavour quality 256, 257–258, 259–260
float systems *see* deep water culture
floating mulches
    frost and pest protection uses 23, 26
    materials 20, 23
flood and drain delivery systems 93, 126–127, 236
fluorescent lamps 34, 239
fodder systems, hydroponic 74–75
fogponic systems 71
foliar fertilizers 105, 129, 156
food safety
    compliance programmes 49, 142, 263, 281,
        287–288
    foodborne pathogens, human health risks
        47–48, 52, 231, 262
    fresh cut products, hygiene measures 276–277
    guidelines for use of reclaimed water 139
    hygiene in postharvest handling 273, 281, 286
    management and design for, in greenhouses 24,
        48, 48–49
food security, global and local 230
foot baths 50, 53, 54
formulation, nutrient stock solutions 149–152, **151**
fossil fuels
    burning for carbon dioxide enrichment 38–39
    current reliance on 15, 41
    depletion and costs 3, 4, 41, 45
    factors favouring use 41
    incomplete combustion, crop damage 39, 188
fresh produce
    benefits of hand harvesting 268–269, **269**, **270**
    fresh cut products, quality 257, 276–278, 289
    perishability and waste 232, 268
frost damage, protective measures
    aluminized screens 19
    floating row covers (frost cloth) 23
fruit quality
    benefits of higher salinity/EC levels 248–249
    potassium effects on 144–145, 150, 203,
        222–223, 250
    shape deformity disorders 183, 191–192, **192**,
        207–208
    size control by truss pruning (tomatoes) 203
    skin damage, causes 174, **174**, 185, 192
    soft texture problems 223
    sugars and ripeness, testing for 254–256, **255**, **256**
fruit set 201, 206, 252
fungus gnats (*Bradysia* spp.) 51, 176, 177
Fusarium wilt 183

GAP (Good Agricultural Practices) compliance 49,
        262, 281, 287–288
gardening, home/hobby 9, 233
GC/MS analysis 260, 263
genetic improvement (selection) 246, 247–248,
        253–254
genetically modified organisms 133, 235
geothermal energy 41–42
Gericke, Frederick 5, 25
germination
    chambers, and temperature control 125
    problems and solutions 120, 125–126, 214
glasshouses
    heat loss and heating costs 30
    history of development 1–2
    light transmission benefits of glass 16, 18
    use and design in Northern Europe 3, 11, 16
grading, produce quality 272
    colour and size/shape classes 273, 274–275
    manual, operating systems 273–274
    risks of mechanical injury during 280
    testing and evaluation methods 254–262
    at time of harvest (strawberries) 224, 269, **270**
grafting 121, 130–131, 188
    tomato plants 130, 131, 199
gravels (growing medium) 73, 74, 80, 83–84
gravity fed irrigation systems 93–94
green waste, uses 56, 84, 104, 108
greenhouse production
    benefits of soilless systems 7–8
    control of growing environment 6, 13–14, 30,
        35, 39
    design considerations 12–18, **22**, 23–24, 40
    global extent 3, 11, 196
    major pests 170–177
    operations management 47–54, 58–59
    organic systems 100–101, 109, **110**
    waste disposal issues 54–58
growing medium (substrate)
    allowed in organic systems 101, **101**, 129
    chemical properties 88–90
    early peat based mixes 8
    evaluation, testing methods 89–90, **90**, **91**
    form and containment 79–80
    mineral materials 80, 82–84
    organic materials 80–82, 83, 84, 106–107
    physical properties 77, 86–88, 104
    requirements for crop production 77, 85–86
    reuse for second crops 26, 27, 80, 209
    wetting agents 125, 129, 166
growth balance, vegetative and generative 96,
        201–202, 205, 210
gutter connected structures
    commercial greenhouse designs 3, 13,
        13, **22**, 24
    energy efficiency 40
    gutter height 11, 13, **14**

HACCP (Hazard Analysis and Critical Control
        Points) 49, 262, 287
hand harvesting 268–269, **269**, **270**
hanging gutter systems 79–80, 198
'hard' water 52, 136, 138
hardening off, seedling transplants 21, 96, 122,
        128, 185
harvest maturity indications 208, 254, 268
harvesting
        frequency, for fruiting crops 210, 223
        immature and vine ripened fruits compared
                258, 268
        methods, manual and mechanical 268–272, **269**
        optimum time of day for 192, 219, 224,
                226, 276
        plant factory produce 235–236
health and safety, occupational 58–59, 82
heat bank materials 32
heat exchangers 42, 166
heat exhaustion, workers 59
heat injury, plant damage 184–185, 252, 284–285
heat treatment
        of nutrient solutions, for disease control 165–166
        solid organic waste 57–58
        for wastewater disinfection 55
heating
        costs and efficiency 40
        for grafts and cuttings 131, 132
        hot beds, traditional and modern methods 23
        methods and sources 30–32, **31**
        regulations for organic systems 100–101
        requirements in cold climates 12–13, 14
heavy metals
        accumulation risk in hydroponics 139, 152
        food contamination regulations 263
herb crops
        physiological disorders 190, 191
        solution culture systems 62
        types and market demands 212–213, 218
herbicide damage 187
hermetic storage, seeds 121
HID (high intensity discharge) lamps 34, 40, 127
high tunnels (poly tunnels) 19–20, **21**
Hoagland, D. R. 5
honeydew contamination 171, 172
horizontal airflow (HAF) fans 35, **36**
hot beds (hot boxes) 23
hot water pipe heating 31, **31**, 41–42
human activity
        biosecurity measures and risks 53–54, 177
        crop and produce handling 48
        health problems, greenhouse workers 59
humic acid 103, 104, 108
humidity
        control in closed and semi closed systems 14,
                36–38, **37**
        fogging/misting equipment 18, 32, 38

influence on choice of growth medium 85–86
levels for cold storage of produce 283
maintenance in aeroponics 71
needs of grafts and plantlets 131, 133–134
plant physiological effects 37, 214
reduction for protection from disease 179, 180,
        183, 286
hybrid seed, F1 and F2 118–119, 198
hybrid systems, organic/hydroponic 113
hydroculture *see* solution culture systems
hydrogen peroxide 142, 165, 187
hydroponic systems
        benefits and challenges 6, 7
        crops grown commercially 196
        history 4–6
        organic status, debate on possibility 8–9,
                101–102, 112
        resource requirements and siting 23–24
        scale of operations 7, 63, 74–75, 196
        water quality and supplies 136–139, 142
        see also outdoor hydroponic systems; solution
                culture systems; substrate based
                hydroponic systems
hygiene 47
        facilities and equipment for staff 54
        for greenhouse crop protection 50, 53, 176, 177
        in handling of fresh produce 48, 273, 281, 286
        plant factory systems 234
        requirements in propagation 126, 132

image processing technologies 242, 254,
        271–272, 275
immunoassays 262
indeterminate growth habit 200, **201**, 205, 210
induced systemic disease resistance 111
infrared/near infrared radiation
        blocking films, for tropical cladding 16
        capture, solar energy PV systems 43
        used in automated optical produce grading 275
injectors, nutrient dosing 94, **95**
inoculant products, microbial 65, 97, 102
        mycorrhizal fungi 249
        use of vermicast 104, 105
insect pests 170–174, 175–176, 184
insulation, in protected cultivation systems 12–13,
        40, 238
integrated pest management (IPM) 113, 171,
        177–178
internal bruising, tomato 280
iron
        chelates, for hydroponic fertilizers 154–155,
                162–163
        deficiency chlorosis, identification of cause 189
        as micronutrient, requirements 146
        in water supply, problems caused by 137, 141
iron sulfate 154

irrigation
    blockages, and prevention measures 91–92, 137
    design for outdoor systems 26
    equipment sanitation programmes 51
    methods and control in hydroponic systems 91–96
    problems caused by irregular supply 186, 192
    seedling transplant production 126–127
    solar power use for pumps 42
    supply requirements, hydroponic systems 142
    water sources and treatment 51–52, 94, 137, 138, 139

Japan, plant factories 129, 229–230, 232, 233, 242

Kew Gardens, London 1, **2**
kieserite 154

labelling
    fresh cut produce 49, 278
    information on stored seeds 120, 121
    organic certification 112
    water samples for analysis 139
labour
    manual operations, and mechanization 125,
        130, 269, 270–271
    occupational health and safety 58–59, 273
    productivity and efficiency 24
    requirements in plant factories 233, 235, 242
    see also staff training
landfill waste disposal 57, 58
layering, tomato plants 200, **201**
leaf miners (Liriomyza spp.) 176
LECA (light expanded clay aggregates) 79, 83
LED lighting technology
    advantages of use in plant factories 239–240,
        240, 241
    energy efficiency 34, 40–41, 199, 229, 239
    installation costs 241
    spectra beneficial to seedlings 127–128
lettuce production
    bolting, and its prevention 191, **191**, 216–217
    crop nutrition **151**, 214, 216, 236
    environmental requirements 214, **215**, 216, **216**
    harvesting and handling 213, 214, 217, **276**
    pests, diseases and physiological disorders **190**,
        190 191, 216
    plant factory productivity 233, 240–241
    propagation 214
    solution culture systems 62, **68**, 71, 75, 213
    types grown, market preferences 212, 213–214, **215**
light
    energy harvesting by solar panels 42–43
    gradients in tiered vertical systems 72, 240
    growth responses to LED wavelengths 239–240, 241
    levels and colours, effects on seedlings 127–128

    low and high levels, effects on crops 32–33,
        159–160, 185–186, 214, 250–252
    optimization, by design features 33
    transmission, glass and plastics 16, 18
    see also artificial lighting; DLI
lime, for pH adjustment 88, 89
Listeria (foodborne pathogen) 48, 262, 286
'living greens' products 62, 213, 236
low cost methods
    basic plastic clad structures 11, 16, 19–20
    composting, benefits and problems 106–107
    outdoor hydroponic systems 26–27
    propagation, use of paper pots 124
    water disinfection 52, 56–57
low tunnels (row covers) 3, 19, **20**
lycopene content 250, 253, 261

magnesium
    levels in water supplies 138, 140, 141
    plant nutritional requirements 145–146, 207
magnesium nitrate 154
magnesium sulfate (Epsom salts) 154
management
    commercial greenhouse operations 24, 47–54, 58–59
    horticultural waste materials 54–55, 57–58
    moisture and nutrients in substrates 80, 84–86
    organic systems 108–110, 113–115
    solution culture systems 61, 65–66, 69–71
    water quality 51–52, 54–57, 136
manganese
    fertilizer sources 155
    levels in water supplies 141
    plant nutritional requirements 146
marketing
    large and small scale systems 7
    maintenance of produce cool chain 278,
        279–280, 284
    onsite sales, biosecurity risks 54
    premium price produce 11–12, 114, 232, 241, 272
    produce with cosmetic defects 57
    quality requirements 27, 212, 246–247, 258,
        287–288
    year round/out of season production 208,
        219–220, 224, 231
medicinal crops, production systems 72, 232, 234–235
Mediterranean region 3, 11, 40
membrane filtration 163
methane (biogas), used for heating 40, 44, 103
Mexico, organic production 100
microbial populations
    in hydroponic substrates 96–98, 102, 107, 109
    mineralization of organic matter 102–104, 111
    produce contamination 262
    role in biofiltration 164–165
    supplemented by inoculation 65, 97, 102, 104
    in vermicast 104

microbubble generation 70
microgreens
    harvesting 219
    NFT system cultivation 62–63, 218, **218**
    sowing rates and yields 218–219, **219**
    types, uses and characteristics **217**, 217–218
micronutrients *see* trace elements
micropropagation *see* tissue culture
mildew diseases 148, **179**, 179–180, 226
mineral wool *see* stone wool substrates
minerals, in water supplies 140–141
mites
    impacts and control problems 223, 226
    pest species **174**, 174–175, **175**, 212
    predatory, as biocontrol agents 174, 175
modified atmosphere (MA) packaging 121, 278, 284
moisture control
    hydroponic substrates 80, 84–86, 94–96
    stored seeds 121
molecular farming (vaccine production) 234–235
molybdenum, plant nutritional requirements 147, 155
monopotassium phosphate 154, 158
moveable systems
    floats, in deep water culture 68
    mobile plant factory units **231**, 232, 234
    NFT channels 64
    robotic harvesters **270**, **271**, 271–272
    seedling transplant benches/racks 126, 235
    use of conveyors, and automation 235, **235**
multilevel systems
    hydroponic fodder production 74–75
    suitable crop characteristics 234
    system design 72, 235
municipal solid waste, processing and use 106, 113
mycotoxins, testing for 262–263

NASA, research programmes 6, 27, 229
National Organic Program, NOP (USA) 100, 112
natural (passive) ventilation 35
    high tunnels 20, **21**, 38
needle seeders 125
nematodes
    as biocontrol agents 176
    as pests 176–177
Netherlands
    environmental regulations 26, 54, 101
    geothermal heating systems 42
    organic production 100
    production area under glass 1
    Venlo greenhouse design 11, 16
nets (and net structures)
    bird netting (strawberries) 223
    insect proof mesh 18
    materials, colours and properties 19
    shading and wind/hail protection 18–19
NFT (nutrient film technique)
    construction components 61, **62**, 63–65, **64**
    crops 61–63, **63**
    disease control methods 162
    early development 6, 61
    nutrient and oxygenation management 102
    outdoor structures **7**, 25, **25**, 27, 64
    used in aquaponic systems 73–74
niche market crops 9, 196
nitrates
    compared with ammonium as N supply 151–152
    content in produce, reduction of 237, 264
    fertilizers, hydroponic formulations 144, 153, 154
    in leafy greens, health risks 263–264
    levels in water supplies 140–141
    from organic matter mineralization 102–104
    removal from wastewater, methods 56–57, 73
    sources for plant nutrition in aquaponics 72
nitric acid, for pH control 144, 155, 158
nitrification bacteria 103–104, 110, 111
nitrogen
    accumulation in leafy vegetables 264
    deficiency symptoms 144, 160
    nitrate and ammonium sources 151–152, 153
    plant growth requirements 143–144, 207
non ionic surfactants 166, 182
nursery plants
    factors influencing establishment 130, 188
    grafting, at seedling stage 121
    hardening off 21, 96, 122, 128, 185
    health screening before planting out 50, 52
    planting out, compared with direct seeding 121
    plug plants as planting stock 26, 122–123, 220
    see also transplant production
nutrient solutions
    chilling, in tropical conditions 71, 75, 181, 214, 253
    concentrate products for nutrient amendment 102, 111, 150
    early development of standard solutions 5
    flow rate effects 64, 74, 164
    formulations of stock solutions 149–152, **151**
    heating, for root zone warming 32, 71
    management, organic hydroponics 110, 111
    natural extract sources for organic systems 74, 101, 107–108
    nutrient level analysis and adjustment 65–66, 70–71, 94, 143, 150
    sterilization in recirculating systems 51, 71
nutrient uptake *see* plant nutrition
nutritional quality
    contributory factors, analysis for 260
    improvement aims 247, 289
    plant factory produce 241, 243

occupational health and safety 58–59, 273
off flavours, produce 257, 258, 284, 285
OMRI (Organic Materials Review Institute) 102, 111–112

open hydroponic systems 6, 77–78, 136, 142
open pollinated seed 119, 198
organic material
    hydroponic substrates 80–82, 83, 84, 89
    management in slow sand filtration 165
    mineralization, in aquaponics 73, 108
    nutrient sources for organic production 101,
        107–108, 114
    processing (mineralization) methods 102–105,
        106–107
    waste disposal 53, 57–58, 106
organic production
    certification of hydroponic systems 8–9, 74,
        101–102, 112
    global extent and regulations 100–101
    management challenges 108–110, 113–115
    nutrient input sources 101, 102–108
    pest and disease control 111, 112–113
    of seedling transplants 129–130
    solution culture systems 74, 102, 108, 109–110
ornamental crops 7, 9
    propagation 119, 132
    sensitivity to ethylene contamination 188
    using plant factory type installations 234
outdoor hydroponic systems 6, **7**, 25–26
    commercial production **25**, 26–27
    excess rainfall protection 64
    protection by windbreaks 24–25
oversaturation 84, 85
    causing seed germination failure 125–126
    and susceptibility to pathogen infection 181, 186
    tolerance, variation between crops 85–86, **86**
oxalic acid toxicity, rhubarb 263
oxygenation
    biological oxygen demand (BOD) in systems 74,
        102, 164–165
    within containers, substrate based hydroponics 88
    dissolved oxygen (DO), saturation levels 61, 69
    effects on plant nutrient uptake 160–161
    methods for increasing, in solution culture 66,
        69–70
    reduction for controlled atmosphere storage 283–284
    requirements of longer term NFT crops 63, 64, 209
    suppression of *Pythium* infection 181
oxytropism, roots 85
ozone, water disinfection treatment 56, 163

pack houses
    documentation and traceability 288
    facilities arrangement and hygiene 272–273, 281
    treatment of waste washing water 55–56
packaging
    designs for soft fruit and vegetables 224,
        278–279, **279**
    fresh cut produce 49, 278
    innovative 'intelligent' technologies 289

    modified atmosphere (MA) 121, 278, 284
    for moisture loss protection 211, **212**, 217
    seeds (for propagation) 121
pak choi production 113, **213**, 236
paper chain pots 124
PAR (photosynthetically active radiation)
    measurement and monitoring 33
    in solar energy PV systems 43
    transmission through greenhouse cladding 16
particle size distribution, substrates 87
passive solar greenhouses 1, 14–15, 43
PCR testing 262
peat based media
    chemical properties 88, 89
    demand and sustainability 8, 81–82
    physical properties 82
pelleting, seeds 119, **120**, 214
penetrometer testing 261–262
people, movement of *see* human activity
peppers *see* capsicum production
perlite 81, 82, 84, 224
pesticides
    botanical and microbial 113, 172
    crop damage from incorrect application 187
    development of pest resistance to 171, 173, 175
    regulations to avoid food contamination 48, 263
    waste disposal, safety issues 58
pests
    common pests in greenhouse crops 170–177
    control by UV blocking greenhouse clad-
        dings 16–17, 184
    exclusion methods (mesh/covers) 18, 23, 171,
        173, **174**
    organic control options 112–113
    produce contamination risks 48–49
    transmission, and spread prevention 50, 51
PFAL systems *see* plant factories
pH
    adjustment methods 110, 140, 155, 158, 159
    buffering 106, 152
    levels in digested organic materials 103, 104
    measurement 90, **158**, 158–159, 256
    range in hydroponic substrates 88–89
    related to nutrient uptake by plants 88, 146
phase change materials (PCMs) 14–15
phosphoric acid, for pH control 155, 158
phosphorus
    fertilizers supplying phosphate 153, 154
    phosphate levels in water systems 55, 56, 141
    plant nutritional requirements 145, 160, 207
photovoltaic (PV) cells 42–43
physiological disorders
    bolting tendencies 191, **191**, 216–217
    calcium related 144, 145, 189–191, **190**
    caused by ammonium nitrate 152
    fruit quality defects 191–192, **192**, 212, 280
    reduced by beneficial mineral elements 143

Phytophthora spp. (pathogens) 182, 223
    spread and control measures 165, 166
phytotrons 229
plant based organic inputs 108
plant factories
    advantages and criticisms 230–233
    costs and returns 232, 233
    crop nutrition and water supplies 236–237
    environmental control 238–239, 242
    history and future development 229–230,
        242–243
    lighting 239–241, **240**
    produce, types and quality 234–235
    robotics and automation 233, 235, 241–242
    scales of operation 233–234
    for seedling transplant production 129, 234
    vertical and moveable systems **231**, 232, 234,
        235, **235**
    worker and visitor hygiene practices 54, 238
plant growth regulators, chemical 128, 188
plant health *see* crop protection
plant nutrition
    beneficial elements 143, 147–149
    early research 5
    mineral availability, from organic materials 102,
        104, 107–108
    monitored by tissue analysis 161, **161**, 189
    nutrient uptake, influencing factors 88, 146,
        159–161, 203
    requirements, essential elements 143–147
    risks of imbalance in solution culture 65–66, **66**,
        70, 152
    role in disease protection 179
    seedlings 128–129, 199, 205
plastics
    cladding, for low cost greenhouse structures 11,
        16, 196
    lifespan and disposal 3, 17, 30, 58
    light transmission properties 16–17
    liners for wetland water treatment systems 56–57
    rigid structured materials 16, 17–18
    shrink wrapping of produce 211, **212**, **279**
    surface tension, modification by surfactants 17
plate seeders 125
plug (cell) transplants 26, 121–122, 123, 130, 220
pollination
    assistance, mechanical methods 203, 221, 222
    prevention, for parthenocarpic cultivars 208
    use of portable beehives 202–203, 206, 222, **222**
polyethylene
    degradation by UV, and inhibitors 2–3, 17, 26
    high density (HDPE) 19, 64
    inflatable ducts for warm air heating 31
    low density (LDPE) film 16, 17
    perforated, used in row covers 20, 23
    single and double layers, insulation value 40
    spectrum selective additives 16–17

polymer seed coating 119–120
polypropylene materials
    uses 19, 23, 24, 26
    waste disposal and recycling 58
polystyrene
    disposal 58
    in float culture systems 66, 68, 126
    transplant production cell trays 123
polyurethane foam 79, 84
pond systems 69, 71
porosity, substrates 87–88
postharvest processing 272, 288–289
    grading 272, 273–276
    guidelines and standards 272, 281, 286
    handling damage 224, 280
    importance of rapid chilling 217, 224, 279–280
    losses and wastage in transport 232
    ready to eat produce 286–287, 289
    washing of fresh produce 48, 55–56, 236, 273,
        277–278
    see also storage
potassium
    effects on fruit quality 144–145, 203, 222–223, 250
    hydroponic fertilizer formulations 144, 150
    importance of winter supply levels 159–160, 207
    leakage from roots at low oxygen levels 161
    low potassium vegetables for dialysis patients 237
    requirements and deficiency symptoms **144**,
        144–145
    uptake competition with calcium 225, 250
potassium nitrate (saltpetre) 153
potassium sulfate 153–154
pots
    air–water gradients, effects on root growth 88
    biodegradable 123–124
    cell (plug) trays, size and materials 122–123
    root system containment, NFT systems 62, **62**,
        124, 236–237
    stacked/staggered, for strawberries 220–221, **221**
    types of container for loose substrates 80,
        124–125
powdery mildew **179**, 179–180, 226
    beneficial effects of silicon 148
pre-conditioning, seedling transplants 128
pressure compensating drippers 91
priming, seeds 120
propagation
    annual renewal rates, long term crops 225
    from seed 118–121, **119**, 125–126
    tissue culture 132–134
    vegetative (cuttings) 131–132, 220
protected cropping
    current trends and innovations 3–4
    energy input costs and efficiency 30, 233
    history of development 1–3, **2**, 229–230
    objectives 1
    structures, variety of 11–12

protective clothing 48, 53, 54, 238, 281
pruning
    effects on fruit quality 192
    equipment disinfection 50, 54, 184
    plant framework establishment (roses) 225
    removal of damaged/diseased plant material
        50–51
    side shoot and leaf removal 200, **200**, 206, 221
    for vegetative/generative growth steering
        201–202, 205, 210
Pseudomonas spp.
    growth promoting properties 96, 97
    plant pathogens 183
    as root disease biocontrol agents 182
pumice 82
pumps
    in aeroponic systems 71
    capacity and head height, NFT systems 65
    capacity, drip irrigation systems 92
    power supply issues and solutions 66, 70, 75
    solar powered 42
Pythium (pathogen)
    damping off in seedlings 126
    root rot 180–182, **181, 182**
    spread and control measures 162, 164, 165, 166

quality assessment
    challenges, for compost products 106
    compositional standards 246, 247, 256
    development of testing methods 246–247
    quantitative tests and grading 254–257,
        260–262
    sensory evaluation 247, 257–259
quality of produce
    attributes, internal and external 246, 247
    influencing factors in crop production
        benefits of UV light treatment 241
        effects of beneficial microbes 97, 249
        improvement by deficit irrigation 96, 248
        light level requirements 17, 250–252, **251**
    maintenance during storage 281–284
    marketing and regulation issues
        cleanliness 26, 27, 231
        consumer acceptance of defects 57
        food safety certification programmes 49,
            262, 263, 281, 287–288
        perceptions of plant factory produce
            232, 243
    see also fruit quality
quarantine, planting material 50, 52–53

raft systems
    in aquaponics 73, 108
    for seedlings 67, 70, 126
rain covers 18, 25–26

rainwater, collection and quality 137–138
ready to eat produce 286–287, 289
recalls, contaminated produce 231, 281, 288
recirculating systems
    closed substrate based systems 77, **78**,
        78–79, **79**
    disease risks and prevention 51, 65, 162, 170
    intermittent and continuous flow 61, 62–63
    nutrient imbalance risks 65–66, **66**, 70
reclaimed water 55–57, 139
recycling
    economic viability 58
    green waste, for manufactured substrates 84
    of substrates after use 80
    of waste produce, to livestock or food
        products 57
    of water, treatments needed 55, 56, 77, 162, 237
refractometers 254–255, **255**, 268
refrigeration (cool storage) 282–283, 284
renewable energy sources 3, 41
    biomass/biofuels 40, 44–45, 103
    geothermal 41–42
    solar 42–43
    wind power 44
repetitive strain injury, RSI (workers) 59
reservoirs (solution culture)
    aeroponic systems 71
    automated and manual adjustments 75
    size recommended for NFT systems **64**, 64–65
    solution oxygenation in DWC/DFT systems 70
resource use efficiency 3–4, 15–16, 196, 231
reverse osmosis (RO) 79, 138, 141, 142, 237
rhizobacteria 96, 97, 103, 182
rice hulls 83, 84
robotic systems
    extraction of transplant plugs from trays 130
    harvesting 270–272
    plant factory operations 233, 242
rockwool 80, 164
root crops, direct seeding 121
root systems
    biofilm activity, in organic systems 111
    chilling, in tropical climates 75, 214, 222, 253
    damage during transplanting 122
    development in plug cells 123
    mineral uptake 143, 160
    misting, in aeroponics 71
    oxytropism 85
    pest damage 176, 177
    root pathogens 180–183, 223
    rooting of stem cuttings 132
    volume, challenges in NFT systems 63
root zone conditions
    abiotic stress factors 63, 186–187, 210
    rhizosphere microbial community 111, 170,
        181–182
rootstocks, for grafted plants 131, 199

rose production
        crop nutrition 225–226
        cultivation and planting material 224–225
        environmental requirements 225
        harvesting 226
        pests, diseases and disorders 180, 226
        pruning and plant management 225
row covers
        floating (floating mulches) 20, 23, 26
        low tunnel structures 19, **20**
runners, strawberry 131–132, **132**, 220

salad greens
        direct seeded, high density production 121,
                217–218
        fresh cut salad mixes, processing 217,
                276–278, **277**
        mesclun/baby leaf products 62–63, **63**, 69
        types and varieties 212, **213**
        see also lettuce production
salinity
        causing physiological drought in plants 186, 248
        of composted material 106
        effects on yield and fruit quality 248–249
        estimated by EC measurement 89, 156
        of irrigation water sources 52, 79, 136
Salmonella (foodborne pathogen) 48, 262, 277, 288
sand
        growing medium 8, 80, 83–84
        water/nutrient solution filtration 52, 56, 97,
                164–165
sanitation
        pack house procedures 272–273, **274**, 286
        routine procedures for crop protection 49–51, 179
        standards and guidelines 47, **49**, 281
        sterile culture for micropropagation 133
        wastewater treatment methods 55–57
sawdust, as hydroponic substrate 82
scoria 82, **83**
screen structures and materials 18–19
Second World War, interest in hydroponics 5–6, 25
sedimentation, for mineral removal 56
seedlings
        diseases of 126
        grafting 121, 131, 199
        nutrition 128–129, 199, 205, 219
        used in bioassays 90, **90**, **91**, 142
        see also microgreens; transplant production
seeds
        advantages of use for propagation 118
        hybrid and open pollinated types 118–119, 198
        organically certified 129
        pelleting, coating and priming 119–120, **120**
        sanitary measures before sprouting 48
        sowing (seeding methods) 121, **122**, 125
        storage life 120–121

suitability for microgreens production 218–219
        treatment with microbial inoculants 97
selenium 149
semi closed greenhouse systems 14, 79
semi determinate growth (tomatoes) 200, **202**
sensors
        for greenhouse humidity control 38
        nutrient solution analysis 159, 237
        placement and maintenance 39
        plant factory microclimate control 238
        plant growth and physiology 39, 238, 242
sensory evaluation 247, 257–259
shade houses 6, 18, **19**
shading
        effects on nutrient uptake 160
        excessive, effects on plant growth 185, 251
        for light and temperature control 32, 34
        materials and equipment 34–35
shelf life
        effects of handling damage 211
        evaluation, testing processes 278
        'living greens' (sold with intact roots) 62, 213, 236
        physiological changes in storage 281–282, 283
        of plant factory produce 231–232
        reduced by abiotic factors in crop growth 184, 185
        of stored seeds 120–121
signage, hygiene and biosecurity **49**, 53, 286
silicon, effects on plant growth 148–149, 225–226, 248
site selection, greenhouses 15, 23–24, 40
size grading, electronic 275
slabs, growth substrates **78**, 79, 81, **81**, 84
slope, solution culture growing channels 63, 64
slow sand filtration (of water) 52, 56, 97, 164–165, 182
snowfall, problems for greenhouses 13
sodium
        accumulation in recirculating systems 79, 136
        crop tolerance 141
        in hydroponic substrates 89, 106
        levels in water supplies 52, 136, 138, 141
sodium hypochlorite (disinfectant) 55, 165, 187, 277
sodium molybdate 155
soft rot, bacterial (Erwinia carotovora) 183
soilless cultivation
        categories and terminology 4, 6
        historical development 4–6
        open and closed systems 6, 77–79, 136
soilborne diseases 7, 26, 101
solar energy
        for air heating, heat bank materials 32, 43
        excess, screening and harvesting 43
        passive, in greenhouse design 14–15
        solar panels (PV systems) 42–43
solar thermal collectors 43
solarization (sterilization method) 80, 83, 101
solution culture systems
        advantages and problems 61, 66, 213
        benefits of flow turbulence 181

deep water/flow systems (DWC/DFT) 66–71,
    67, **68**
definition 6
intensive production methods 71–72, 236–237
nutrient film (NFT) systems 61–66, **62**, 74
see also aquaponics
Spain
    annual organic waste from horticulture 57
    organic greenhouse production 100
    plastic greenhouse design (Parral type) 11
    roof mounted solar panels, effects of use 42
speedling float system 68–69, 126
splice grafting 131
splitting, fruit 192
sprouts
    food safety issues 48, 142
    wheat/barley, for animal fodder 74–75
staff training
    hygiene practices 54, 286
    in laboratory testing methods 263
    manual harvesting and sorting 268–269, 274
    pest and disease diagnosis 52
stock solutions, hydroponic 149–152, **151**, 236
stone wool substrates 80, 85, 121, 124
storage
    postharvest plant physiology 281–282
    produce disorders and decay 284–286
    seeds (for propagation) 120–121
    storage conditions and care systems
        282–284, 289
strawberry production
    crop nutrition 222–223, 249
    environmental requirements 221, 222, 223, 252
    flowering and pollination 221–222, **222**
    harvest and postharvest handling 223–224,
        224, 269, **270**
    pests, diseases and disorders 190, 223
    planting density, staggered trough systems
        220–221, **221**
    propagation 131–132, **132**, 220
    soilless outdoor bag culture 26
    water supply management 85, **86**
    year round production systems 219–220
stress (in plants)
    and disease susceptibility 180
    interactions of abiotic factors 184, 185, 189
    tolerance, beneficial effects of silicon 148
structural stability, substrates 82, 88
substrate based hydroponic systems
    definition 6
    microbial activity and populations 96–98
    moisture (watering) management 84–86, 94–96
    nutrient delivery methods 91–94
    open and closed systems 77–79, 198–199
    organic systems, management 108–110
    planting density and layout 220–221, **221**
    technology development trends 7–8

types of substrate 77, 79–84, 198
    see also growing medium
sulfates
    allowable in organic systems 111
    fertilizers, commonly used 153–154, 155
sulfur
    fungicides, crop damage risks 187
    levels in water supplies 141
    plant nutritional requirements 146
sunscald 185, 186, 207, 252
surface water 137
surfactants 166, 182
sustainability issues
    improvement by use of recycled products
        84, 113
    new technologies in greenhouse design 15–16
    plant factory systems 230, 232
    postharvest handling technologies 288–289
    public demand for sustainable production
        101, 115
    reliance of horticulture on peat 8
    see also environmental concerns

taste qualities 256, 257
TDS (total dissolved solids) meters 157
temperate zone climates
    energy use and efficiency 40
    greenhouse design requirements 12–13
    seasonal pests, diseases and disorders 175,
        180, 192
temperature
    causes of crop damage/injury 184–185, 252–253
    day/night differential (DIF) control 128, 191,
        201–202
    effects on nutrient uptake by plants 159, 160
    passive control methods 14–15
    postharvest produce requirements 278, 284–285
    related to saturation vapour pressure 37
    requirements for pollination and fruit set 206
temporary coatings, greenhouse cladding 16, 35
texture attributes, testing 259, 261–262
thermal dormancy, seed (lettuce) 214
thermal screens 15, **22**, 30
    as insulation, energy cost savings 40, 43
    used for shading 34, 43
thrips 17, **173**, 173–174, **174**, 184
tiered vertical systems *see* multilevel systems
time of day
    for crop harvesting 192, 219, 224, 226, 276
    effect on nutrient uptake 160
tipburn **190**
    caused by high temperatures 185
    control in plant factories 238
    predisposition in high humidity 214
    related to calcium deficiency 144, 145, 190–191
tissue analysis, plants 161, **161**, 189

tissue culture
    laboratory biosecurity and hygiene 54, 133, 134
    techniques and methods 133–134
    uses and benefits 132–133
titanium 149
titratable acidity (produce) 256–257
tomato production
    benefits and process of grafting 130–131, 199
    crop nutrition 150, **151**, 160, 203, 249
    environmental requirements 199
    growth steering 201–202
    pests and diseases 174, 183, 203–204
    physiological disorders 189–190, 191, 192
    plant training and support 199–200, **200,
        201, 202**
    pollination and fruit development 202–203
    propagation 199
    varieties and production systems **12**, 196–199,
        197, 198, 204
tomato russet mite *(Aculops lycopersicae)* **174**,
    174–175
tomato spotted wilt virus 51, 173, 184
total porosity 87
total soluble solids (Brix testing) 247–248, 254–256,
    255, 256
total waste (complete) aquaponics 74, 108
toxic materials
    agrochemical container waste 58
    in bark and sawdust 82
    excessive levels of mineral ions 136, 186–187
    naturally occurring toxins in plants 263
    phytotoxic chemicals 187–188
    unprocessed organic material 102
    water treatment chloramines 138–139
trace elements
    chelation 146, 154–155, 156
    requirements as micronutrients 146–147, 150
    separate stock solution, for adjustment 150
    soil based and hydroponic crops compared 148
    toxicity of high levels 136
training systems (plant support)
    capsicums 205, 206, **206**
    cucumbers 210, **211**
    tomatoes 199–200, **200, 201, 202**
transplant production
    bench/rack systems 126, **127**
    environmental conditions, control 127–128
    growing medium 8, 82, 123, 124–125
    nutrient solution delivery methods 93, 126–127
    organic production 129–130
    pre conditioning (height control) 128
    seedling delivery systems 121–125, **123, 124**
    solution culture methods 66–67, 68–69, **69**
    use of plant factories 129, 234
    see also nursery plants
transplant shock 122, 188
trees

as biomass energy source 44
species for windbreaks 24
Trialeurodes vaporariorum (greenhouse whitefly)
    170–172, **171**
Trichoderma spp., beneficial effects 96, 97, 179, 182
tropical/subtropical zones
    crop diseases 183
    greenhouse cooling methods 32, **33**
    near infrared blocking cladding additives 16
    production of cool season crops 75, 214, 222, 253
    protected cropping structures **12**, 18–19
    tropical crop sensitivity to chilling 185
tunnel structures
    cost and use by small farmers 2
    design, low and high types 19–20, **20, 21**
    energy efficiency 40
turbidity, radiation (diffuse light) 17
twin skin (inflatable) plastic cladding 12, 15, 30, 251
two spotted mite *(Tetranychus urticae)* 174, 175, 212

UC standard growing mixes 8
ultraviolet (UV) radiation
    blocking materials, effect on pest numbers
        16–17
    disinfection of water supplies 162–163
    effects on phytochemical levels in produce 241
unheated cropping structures 11, 12, 16
United States (USA)
    certification of organic hydroponics 8, 74,
        101–102, 112
    energy sources for protected cropping 41
    hydroponics research and development 5–6, 230
    organic production, market value 100
urban agriculture
    land scarcity solutions 4, 230, **231**, 232
    use of soilless systems 6, 7
    see also plant factories
urea, as fertilizer 153, 156
USDA (US Department of Agriculture) 100, 112, 256

'V cordon' training systems 206, 210
vacuum seeding 125
vapour pressure deficit (VPD) 37–38
vegetative growth control 96, 144, 201–202
vegetative propagation 131–132, **132**, 220
Venlo greenhouse design 11, 16, **22**
ventilation
    needs in vertical systems 72
    plant growth and health benefits 35, 190
    requirements of stored produce 282
    types in greenhouse design 20, **22**, 35–36, **36**
vermiculite 82, 83
vermiculture (vermicomposting)
    solid vermicast product, uses 104–105, 129
    vermiliquer, nutrient composition 105–106

vertical air movement fans **22**, **37**
vertical production systems
    efficient use of restricted space 4, 27, **73**, 220, 229
    light penetration and levels, effects 72, 240,
        251–252
    suitable crops for 64, 72, 234
    system design 221, **221**, **235**, 235–236
Verticillium wilt 182–183
vibration wand seeders 125
Victorian glasshouse construction 1, **2**
virus diseases 170, 183–184, 204
    transmission by insect pest vectors 171, 172,
        173, 184
visitors
    members of public, hygiene practices 54
    onsite registration records 53
vitamin C content 249, 253, 260, 285
volatiles testing (aroma) 259–260

wasabi 9, 196, **197**
washing
    of hands, staff personal hygiene 48, 50, 54, 286
    postharvest produce 48, 55–56, 236, 273, 277–278
waste heat utilization systems 31, 40
waste management 47, 103, 115
    liquid wastes 53, 54–57
    solid wastes 53, 55, 57–58, 84, 113
wastewater
    effluent composition, nutrient levels 55
    from fish farming, used in aquaponics 73–74
    runoff disposal options 54–55
    treatment and recycling 55–57, 162
water analysis, tests and reporting 136, 137, 139–141
water holding capacity, substrates 84–85, 87, 88
water quality
    environmental protection regulations 26, 54–55, 56
    groundwater nutrient loading, effects 55
    hardness, variation and effects 52, 136, 138
    importance in hydroponic systems 136, 237
    pathogen control treatments 52, 161–166
    problems and treatments, irrigation water
        51–52, 137, 138–139, 142, 186–187
    testing for contamination 48, 53, 141–142

waterlogging, plant symptoms 85, 186
wavelength (of light), effects on plant growth
    239–240, 241
weed control 51, 171, 187
well water 136–137, 147, 162
western flower thrips (*Frankliniella occidentalis*) 51,
    173, 178
wetland systems, wastewater treatment 56–57
wetting agents 125, 129, 166
whitefly 17, 170–172, **171**, 184
wilting, causes of
    fungal wilt diseases 180, **181**, 182–183
    over and under watering 85, 186
    transplant shock 122
wind generated energy 44
windbreaks 24–25
Woodward, John 5
workers *see* labour

yellow sticky traps (pest monitoring) 171, 173, **173**, 176
yields
    benefits of technological development 204, 208
    effect of light levels 33
    improved by grafting 130–131
    maximization, optimal light wavelengths
        239–240, 241
    outdoor hydroponic and field systems
        compared 27
    in plant factory systems 230, 233
    positive impacts of beneficial microbes
        96, 97, 113

zero gravity plant production 27, 229
zinc
    fertilizer sources 155
    levels in water supplies 141
    plant nutritional requirements 147
zoospores
    effect of surfactants 166, 182
    killed by chlorination 165
    role in pathogen spread, hydroponic
        systems 170, 180–181

www.ingramcontent.com/pod-product-compliance
Lightning Source LLC
Chambersburg PA
CBHW040137200326
41458CB00025B/6297